PUBLICATIONS OF THE NEWTON INSTITUTE

T0282364

Lectures on Solar and Planetary Dynamos

Publications of the Newton Institute

Edited by P. Goddard
Deputy Director, Isaac Newton Institute for Mathematical Sciences

The Isaac Newton Institute for Mathematical Sciences of the University of Cambridge exists to stimulate research in all branches of the mathematical sciences, including pure mathematics, statistics, applied mathematics, theoretical physics, theoretical computer science, mathematical biology and economics. The four six-month long research programmes it runs each year bring together leading mathematical scientists from all over the world to exchange ideas through seminars, teaching and informal interaction.

Associated with the programmes are two types of publication. The first contains lecture courses, aimed at making the latest developments accessible to a wider audience and providing an entry to the area. The second contains proceedings of workshops and conferences focusing on the most topical aspects of the subjects.

LECTURES ON SOLAR AND PLANETARY DYNAMOS

edited by
M. R. E. Proctor
University of Cambridge
and
A. D. Gilbert
University of Exeter

CAMBRIDGE UNIVERSITY PRESS
Cambridge, New York, Melbourne, Madrid, Cape Town, Singapore, São Paulo

Cambridge University Press
The Edinburgh Building, Cambridge CB2 8RU, UK

Published in the United States of America by Cambridge University Press, New York

www.cambridge.org
Information on this title: www.cambridge.org/9780521461429

© Cambridge University Press 1994

First published 1994

A catalogue record for this publication is available from the British Library

ISBN 978-0-521-46142-9 hardback
ISBN 978-0-521-46704-9 paperback

Transferred to digital printing 2008

Contents

Preface

The mass of mathematical truth is obvious and imposing; its practical applications, the bridges and steam-engines and dynamos, obtrude themselves on the dullest imagination.

— G.H.Hardy

This volume contains the texts of the invited lectures presented at the NATO Advanced Study Institute 'Theory of Solar and Planetary Dynamos' held at the Isaac Newton Institute for Mathematical Sciences in Cambridge from September 20 to October 2 1992. Its companion volume 'Solar and Planetary Dynamos', containing the texts of the contributed papers, has recently been published in the same series as the present one, and contains a full list of participants and their addresses. It is a measure of the recent growth of the subject that one volume has proved insufficient to contain all the material presented at the meeting: indeed, dynamo theory now acts as an interface between such diverse areas of mathematical interest as bifurcation theory, Hamiltonian mechanics, turbulence theory, large-scale computational fluid dynamics and asymptotic methods, as well as providing a forum for the interchange of ideas between astrophysicists, geophysicists and those concerned with the industrial applications of magnetohydrodynamics.

The topics of the lectures cover almost all the principal parts of the subject. Authors were asked to give reviews of a pedagogical nature. Earlier chapters cover relatively fundamental aspects of the subject; later chapters treat more specialised topics. Although each chapter is self-contained, there are cross-references to other lectures where appropriate; in addition. the Editors have striven to maintain uniformity of notation and style, in the hope that the resulting complete text will find favour as a unified work of reference, rather than as a disparate set of reviews. Each chapter has its own bibliography, but there is a general index.

The Editors wish to record their grateful thanks to the staff of the Isaac Newton Institute, who coped gracefully with many bureaucratic vicissitudes, and especially to the Director, Peter Goddard, for making the whole event possible as part of the programme at the Institute. They are also indebted to

Alastair Rucklidge and Paul Matthews, the coeditors of the previous volume, for giving generously of their expertise in the preparation of the final version of the manuscript.

M.R.E. Proctor
A.D. Gilbert

Cambridge, Exeter, March 1994

Scientific Organising Committee

A. Brandenburg	(Denmark)	NORDITA, Copenhagen
S. Childress	(USA)	Courant Institute, New York University
H.K. Moffatt	(UK)	DAMTP, University of Cambridge
M.R.E. Proctor (Director)	(UK)	DAMTP, University of Cambridge

Local Organising Committee
A.D. Gilbert
P.C. Matthews
M.R.E. Proctor (Chairman)
A.M. Rucklidge

Introduction

Magnetic fields are observed to exist wherever there is matter in the visible universe; they exist on planetary, stellar and galactic length-scales, indeed wherever there is a sufficiently large mass of rotating conducting fluid. Dynamo theory is that branch of fluid mechanics that seeks to explain both the origin of these magnetic fields, and the manner of their variation in space and time. The subject has exerted a profound challenge, and great advances have been made over the last few decades. Nevertheless, acute problems remain in relation to both planetary and stellar magnetism. The NATO Advanced Study Institute on Stellar and Planetary Dynamos, and the six-month Dynamo Theory Programme of the Isaac Newton Institute of which it formed part, set out both to review the present state of knowledge in this broad field, and to define the critical problems that now demand attention.

The problem of the origin of the Earth's magnetic field has challenged the imagination of great scientists of past centuries. Edmund Halley showed extraordinary prescience three hundred years ago when, in considering the possible causes of the secular variation of the geomagnetic field, he wrote:

" ... the external parts of the globe may well be reckoned as the shell, and the internal as a nucleus or inner globe included within ours, with a fluid medium between ... only this outer Sphere having its turbinating motion some small matter either swifter or slower than the inner Ball."

This view of the inner structure of the Earth was not confirmed till Jeffreys' discovery in 1926 of the liquid outer core and Bullen's discovery in 1946 of the solid inner core. The liquid outer core is the seat of geomagnetic dynamo action, and attention is increasingly focussed on the role of the solid inner core whose slow growth provides an important source of energy for (compositional) convective motion in the outer core.

This convective motion in the Earth (and in the liquid cores of similar planets) is strongly influenced by Coriolis forces associated with the Earth's rotation. There are two important consequences: first, the convection is helical in character, i.e., a buoyant parcel of fluid tends to follow a helical path, rather than to rise vertically, an effect that is antisymmetric about the equatorial plane; secondly, since rising or falling blobs of fluid have a tendency to conserve angular momentum, a state of 'differential rotation' is established in which (it may be expected) inner regions of the liquid core rotate slightly more rapidly than outer regions.

These two ingredients of core motion — helicity and differential rotation — are now recognized as the basic ingredients of dynamo action, i.e., of the spontaneous conversion of the kinetic energy of core motion to magnetic energy. The magnetic field results from currents that are driven by electromagnetic induction in the liquid region. This field exists not only in the core, but extends (as a potential field) to the external space, and is indeed the magnetic field that is detected by compass or magnetometer here on the Earth's surface. Observations of the time variation of the surface field thus give us indirect information, as Halley correctly guessed, concerning "those inner depths of the earth, to the knowledge of which we have so few means of access" (Maxwell 1873, Treatise on Electricity and Magnetism).

It may be natural to suppose that, if dynamo action is to occur, then it should saturate when the magnetic energy is of the same order of magnitude as the kinetic energy of the liquid motions responsible for the process. Remarkably, this is far from true: the magnetic energy stored in the Earth's field is several orders of magnitude greater than that of the core motions (relative to the Earth's rotating frame). These core motions merely play a catalytic role in converting the gravitational (or possibly rotational) energy of the Earth to magnetic form.

While the key problem for the geodynamo is concerned with the dynamics of a rapidly rotating fluid system, and the catalytic conversion of energy alluded to above, the key problem for the *solar* dynamo is the very nature of the dynamo process itself. Here again we are dealing with a rotating convecting system, the convection zone of the Sun extending from its visible surface down about one fifth of the radial distance to the center. However the convective velocities are very much greater than in the case of the Earth (of the order of kilometres per second, as compared with millimetres per second!) so that the physics of the inductive process, although still governed by the classical equations of magnetohydrodynamics, is now quite different. Induction is now so dominant that ordinary molecular resistive effects become relatively extremely weak; it is tempting to neglect them altogether, but this is a temptation that must itself be resisted! Without resistive effects, the topology of a magnetic field is perfectly conserved for all time, and magnetic field evolution as actually observed on the solar surface is impossible. Just as in the problem of turbulence, in which viscous effects, although extremely weak, remain of central importance, so here with resistive effects which remain effective on smaller and smaller length-scales as the resistivity parameter is allowed to tend to zero. On these very small scales, the topological (or 'frozen-field') contraints are broken, and the magnetic field can re-assemble itself to provide a smoothed out dynamo action on large scales.

Analysis of this type of behaviour is what has come to be known as 'fast dynamo theory' — *fast* because it occurs on the time-scale characteristic of turn-over of convective eddies, and independent of resistivity no matter how small this may be. The problems that arise are extremely delicate, both analytically and numerically; but they are important because the whole corpus of knowledge concerning stellar (and galactic) dynamos rests ultimately upon a correct understanding of the fast dynamo type of process.

There are other difficulties that beset even the solar dynamo. The first relates to the spatial location of the dynamo process: if the field is generated throughout the convection zone, then the 'magnetic buoyancy' mechanism, first identified by E.N. Parker in 1955, causes it to erupt to the surface much more rapidly than is consistent with observations. This has led to the hypothesis that the field is in fact generated, again by a combination of the effects of helicity and differential rotation, in a relatively thin layer near the bottom of the convection zone where the effects of magnetic buoyancy can be slowed down if not entirely eliminated. But decreasing the volume in which dynamo action occurs must also decrease its global effectiveness: can such a 'layer' dynamo really explain the solar magnetic field as observed at the surface?

The second difficulty relates to the problem of turbulent diffusivity, on which the operation of all solar dynamo models depend. If a magnetic field grows, then it suppresses the very turbulence on which it feeds, via in part the mechanism of turbulent diffusivity. Can such a process be self-sustaining?

These and many related problems are discussed in the lectures contained in this volume, and in a companion volume of contributed papers. As far as the Earth is concerned we have now moved a long way from Schuster's (1912) statement that "the difficulties which stand in the way of basing terrestrial magnetism on electric currents inside the earth are insurmountable"; indeed, the dynamo explanation for the Earth's field is now almost universally accepted by geophysicists. The situation for solar magnetism is very different; however appealing the dynamo mechanism may be, there are still profound difficulties in the way of a complete understanding of the process. It is hoped that these two volumes will give a good picture of the present state of affairs in a subject which may be expected to continue to challenge and stimulate for some time yet!

H.K. Moffatt

CHAPTER 1

Fundamentals of Dynamo Theory

PAUL H. ROBERTS

Department of Mathematics
University of California
Los Angeles, CA 90024, USA

1.1 INTRODUCTION

1.1.1 Cosmic Dynamos

Many stars, planets and galaxies possess magnetic fields whose origins are not easily explained. Even the 'solid' planets cannot be sufficiently ferromagnetic to account for their magnetism, for the bulk of their interiors are above the Curie temperature at which permanent magnetism disappears; obviously the stars and galaxies cannot be ferromagnetic at all. Nor are the magnetic fields transient phenomena that just happen to be present today. Palaeomagnetism, the study of magnetic fields 'fossilized' in rocks at the time of their formation in the remote geological past, shows that Earth's magnetic field has existed at much its present strength for the past 3×10^9 years, at least. But, unless they contain sources of electric current, conducting bodies of spatial dimension, $\mathcal{L}$, can retain their magnetic fields only for times of the order of the *electromagnetic diffusion time* $\tau_\eta = \mathcal{L}^2/\eta$, where η is the magnetic diffusivity of their constituents; $\eta = 1/\mu_0\sigma$, where σ is their electrical conductivity and μ_0 is the permeability of free space. (SI units are used throughout.) Being proportional to $\mathcal{L}^2$, this time may be very considerable, but is as nothing compared with the ages of the bodies concerned. For example, Earth contains a highly conducting region, its core, of radius about $\mathcal{L} = 3.48 \times 10^6$ m, and its conductivity is about 4×10^5 S/m. This gives $\eta \approx 2$ m^2/s and $\tau_\eta \approx 200,000$ years. Similarly, it is thought that the magnetic fields of other planets cannot be fossil relics of their birth. A mechanism is required to maintain them.

The necessity for a source of magnetic field is less evident in the case of stars because of their much greater dimensions, $\mathcal{L}$. For example, if we use a magnetic diffusivity based on the probable electrical conductivity of the solar plasma, we find that τ_η is of order 10^{12} years. This estimate is, however, too simplistic: the solar magnetic field is created in the solar convection zone,

1

M.R.E. Proctor & A.D. Gilbert (eds.)
Lectures on Solar and Planetary Dynamos, 1–58
©1994 Cambridge University Press.

which is highly turbulent. In such an environment, the magnetic field varies on a length scale, l^{turb}, comparable with that of the motion and vastly less than the depth $(2 \times 10^8$ m) of the convection zone. The molecular η is reduced almost to irrelevance, and is replaced by η^{turb}, which is of order $l^{turb} v^{turb}$, where v^{turb} is a typical turbulent velocity. The resulting $\tau_\eta^{turb} = \mathcal{L}^2/\eta^{turb}$ turns out to be only of the order of a decade. And it certainly seems to be true that the Sun is able to purge its surface magnetic field every decade or so, and to then create a new magnetic field of opposite polarity that lasts for a similar duration. Many stars of solar type exhibit similar cycles, suggesting again that their effective magnetic decay times are measured in decades rather than trillions of years. It is only for the relatively quiet magnetic A-stars that a 'fossil-field' origin is at all tenable, and even this has been called into question (see Krause & Meinel 1988). Moreover, to accept a fossil explanation of their origin is only to bury the question at a deeper level since one must wonder how the proto-stellar material from which the star was formed acquired the magnetism that we detect in the star today.

The search for an origin of the magnetism of planets and solar-type stars has led to an explanation that depends on the existence in these bodies of highly conducting material. It invokes a mechanism that postpones, perhaps indefinitely, the free decay of magnetic field. It supposes that motions in conducting regions induce from the prevailing magnetic field an electromotive force (emf) that creates currents that generate the inducing magnetic field itself. At first sight, this proposal, first made by Sir Joseph Larmor (1919), seems suspiciously like hoisting oneself up by his own shoelaces, but there is really no conflict with the demands of energy conservation: the energy lost to heat through the electrical resistance of the conductor is made good by the rate, $\mathbf{V} \cdot (\mathbf{J} \times \mathbf{B})$ (per unit volume), at which the fluid motion generates electromagnetic energy, i.e., by its rate of working against the Lorentz force, $\mathbf{J} \times \mathbf{B}$. (Here $\mathbf{B}$ is the magnetic field, $\mathbf{J}$ is the electric current density and $\mathbf{V}$ is the velocity of the conductor.) Thus the device is not a perpetual motion machine! Analysis of this mechanism has engendered one of the most exciting branches of magnetohydrodynamics (MHD), one that has both electrodynamical and hydrodynamical facets. In this chapter we shall concentrate on the electrodynamic aspects. We attempt to answer the question, 'Can a body of conducting fluid indefinitely maintain a magnetic field, entirely through the emfs created by a *specified* fluid motion in its interior?' The use of the word 'entirely' is crucial. It signifies that the body, V, of conductor owes its magnetism solely to the electric currents flowing within it, that these currents are generated by the emfs of the motions in V alone, unsupplemented by potential differences created thermoelectrically or electrochemically (i.e., 'batteries'), and that there are no external sources of magnetic field in the exterior, $\hat{V}$, of V, not even (in the mathematical sense) 'sources at infinity' (see section 1.3.1 below). When we say 'specified fluid motion', we mean that we shall regard **V** as given; in the parlance of the subject, we shall study in this chapter only *kinematic dynamo theory*. MHD dynamo theory, which will be discussed later in this book, attempts to determine both **B** *and* **V** jointly by solving the full MHD equations for some given power source, such as thermal buoyancy. This involves solving coupled nonlinear vector partial differential equations and is a considerably more difficult undertaking than kinematic theory, which involves the solution of a linear vector partial differential equation. The full MHD problem is sometimes called 'the fully self-consistent dynamo problem'.

1.1.2 A Simple Dynamo

It may be objected that the answer to the question we have just posed was answered by Werner Siemens more than a century ago. Man-made dynamos in power stations manufacture their own magnetic fields and widely distribute the associated electric currents to domestic and industrial users. The simplest such machine, the homopolar dynamo, is illustrated in figure 1.1. Here D is a solid conducting disk rotating with angular velocity Ω about its axis of symmetry, A'A, parallel to which we at first (figure 1.1a) suppose there is a uniform magnetic field $\mathbf{B}_0$. Because of the velocity, $\mathbf{V} = \Omega \times \mathbf{x}$, of the conductor, an emf $\mathbf{V} \times \mathbf{B}_0$ exists in the disk which would drive a current if there were a closed circuit in which a current could flow. (Here $\mathbf{x}$ is the position measured from an origin on A'A.) Instead, positive charges (P) build up in D and on its rim, R, and negative charges (N) are set up on the axle, A'A. In this way an electric field **E** is created that cancels out the emf, as of course is necessary if no current flows.

Now let a conducting wire, W, join A'A to R, electrical contact being maintained through sliding contacts (or 'brushes') at S_1 and S_2 (figure 1.1b). The electric charges that in figure 1.1(a) remained on the rim and axle can now be 'drawn off' as an electric current, I, along W; the circuit has been completed, and the electromotive force has become motive! The current I can be directed to perform a useful task such as powering the electric light bulb shown in figure 1.1(b), the concomitant expenditure of electrical energy being made good by the energy source that maintains the rotation of D. It is this conversion of mechanical energy to electrical energy that defines what is meant by the word 'dynamo'.

Suppose next that, instead of taking the shortest route between S_1 and S_2, the wire W is wound in the plane of the disk round, and just outside R, making almost one complete loop around R before being led to the two brushes (figure 1.1c). The direction of the winding chosen is such that the magnetic field **B** produced by the current I in this loop re-inforces the applied magnetic field $\mathbf{B}_0$. In the plane of D, both are perpendicular to the disk, in

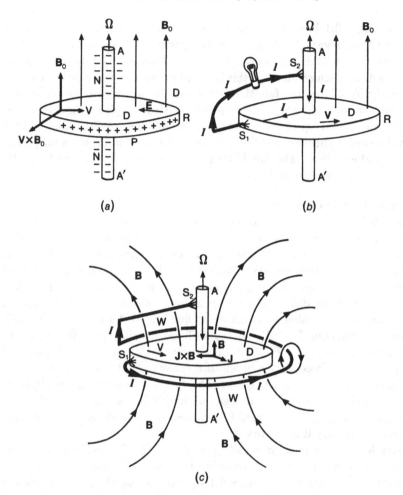

Figure 1.1 The homopolar dynamo: (a) charge distribution, (b) current flow and (c) magnetic field; for details see text.

the same sense: A' → A. Unlike $\mathbf{B}_0$, the magnetic field $\mathbf{B}$ is not uniform across D, so that the emf it creates does not depend on $\mathbf{x}$ in the same way that $\mathbf{V} \times \mathbf{B}_0$ does. Nevertheless, it has the same outward direction in D: from A'A towards R. It may, if Ω is large enough, produce a significant current in W. Indeed, it is plausible and true that a value, Ω_c, of Ω exists at which $\mathbf{B}_0$ may be removed, I being then driven by the emf $\mathbf{V} \times \mathbf{B}$ alone. The device can then dispense with the externally applied 'excitation', $\mathbf{B}_0$, except as a 'seed magnetic field' to start the induction process off. Thereafter it sustains the magnetic field, $\mathbf{B}$, and the current I for as long as Ω_c is maintained. Such a machine, which provides its own source of 'excitation', is called a *self-excited dynamo*. It is with these that this book is concerned. This chapter deals only with the kinematic aspects of self-excited dynamos, but it is worth noting that (ignoring other sources of energy loss, such as friction in the axle bearings) the homopolar dynamo can, when $\Omega = \Omega_c$, maintain a magnetic field of infinitesimal strength, and does not require a power source to do so. When power is available to drive the motion of the disk, the final steady state of the system is decided by the strength of that power source, the steady state magnetic field and currents being such that the electrical power dissipated through the electrical resistance of the circuit matches the power input. This balance is characteristic of all working dynamos.

Particularly important for the success of the homopolar self-excited dynamo is its structural asymmetry. It is easily seen that, if W is led around R in the opposite sense from that shown in figure 1.1(c), *or* if the direction of Ω is reversed, then **B** opposes $\mathbf{B_0}$ and self-excitation is impossible no matter how large Ω is. Indeed, the device then destroys, rather than creates, magnetic field! Like all commercial machines, the homopolar dynamo is deliberately constructed to be multiply-connected and highly asymmetric. This stands in bleak contrast to the nearly spherical structure of stars and planets. If the homopolar dynamo were embedded in a spherical mass of stagnant material of the same electrical conductivity, its multiple-connectedness and asymmetry would be lost, the induced currents created by the motion of the disk would be short-circuited, and dynamo action would cease. The asymmetry of man-made dynamos is crucial for their success, but naturally-occurring bodies such as the planets, stars and galaxies do not have asymmetrical structures. If they are to act as dynamos, the asymmetry of their internal motions must, in some far from obvious way, compensate for their lack of structural asymmetry. It is now apparent that the question posed above was not precisely what we intended. It is more appropriate to ask, 'Can a symmetric body of homogeneous conductor, such as a planetary core, a stellar convection zone, or a galactic disk, maintain a magnetic field indefinitely, entirely through specified fluid motions taking place in its interior?' This *homogeneous dynamo* problem is far more relevant to the natural sciences, and is far harder to answer than our previous question.

1.1.3 Examples

There are by now very many applications of homogeneous dynamo theory. We may mention particularly:

A. Geomagnetism. It is known from the study of palaeomagnetism that the geomagnetic field is essentially as old as the Earth. Dynamo theory is the only tenable explanation of its origin. The MHD of the Earth's core and the geodynamo have been reviewed in Jacobs (1987), volume 2 of a recent series of books devoted to geomagnetism. For more recent reviews, see Braginsky (1991), Roberts (1988) and Soward (1991).

B. *Planetary magnetism.* Several other planets in the solar system possess self-generated magnetic fields, presumably also created by dynamos operating in conducting regions within them. They show however bewildering differences in character (see Russell 1987). Venus, the sister planet of the Earth, is apparently non-magnetic but there are indications that Mercury may be sustaining a magnetic field at present, despite its small size. It has been established that the Moon had an operating core dynamo early in its history. The 'magnetic tilt angle' of a planet is the angle between its rotation vector and the axis of the dipole component of its magnetic field. This shows wide and perplexing differences, from less than 1° for Saturn, through about 11.3° for the Earth, to about 60° for Uranus! (See also Ruzmaikin & Starchenko 1991.) Except perhaps in the case of Mercury, it is generally agreed that dynamo theory should be able to explain the existence and character of the magnetic planets, and to account for why some planets, and most if not all satellites, do not have dynamos within them. Despite the spacecraft exploration of the past 30 years, observations are still too sparse to provide a very complete picture of planetary magnetism, and most of today's 'theories' amount to little more than speculations.

C. *Solar magnetism.* It was long conjectured, from the appearance of coronal streamers seen during total eclipses, that the Sun might be magnetic, but this was not established for certain until 1908, when Hale announced the results of his measurements of the Zeeman splitting of spectral lines, which established that sunspots are the seats of magnetic fields of as much as 0.3 T. He recognized that the 11 year sunspot cycle indicated a 22 year periodicity in the magnetic field of the Sun. Since that time, the solar surface has been intensively observed and many theoretical investigations of both the global magnetism of the Sun and its detailed features have been made. It became accepted (see above) that the diffusivity of the magnetic field is so greatly enhanced by the turbulence taking place in the solar convection zone that the molecular diffusivity was essentially irrelevant. Its place was taken by a turbulent diffusivity, η^{turb} (see section 1.5 below), and the corresponding magnetic diffusion time, τ_η^{turb}, was taken to be comparable with the 22 years of the solar cycle, i.e., the Sun creates and destroys its field every 22 years by some kind of solar dynamo. At first this dynamo was thought to be operating throughout the convection zone of the Sun, and Parker (1957) provided a persuasive explanation of the solar cycle (see section 1.4.3 below, and also Steenbeck & Krause 1969). This explanation was kinematic, and required fluid motions that resembled neither those deduced from later integrations of the full MHD dynamo problem nor those inferred from helioseismology (in which the observed surface motions of the Sun are analysed to extract information about its internal structure from the sound waves that pass through it). As a result, the picture of the solar dynamo today is quite different: it shows a dynamo acting principally near the bottom of the convection zone and the top of the radiative core beneath it; e.g., see Cox, Livingstone & Matthews (1991). Despite this dramatic change of model, it is seldom contested that the solar field is created by dynamo action.

D. Stellar magnetism. Other main sequence stars, besides the Sun, possess magnetic fields, and dynamo theories to explain them have been developed which take into account their differing states of rotation and the systematic variation in stellar structure along the main sequence. (See, for example, Tuominen, Moss & Rüdiger 1991, Weiss, 1989, Cattaneo, Hughes & Weiss 1991.) The peculiar (Ap) stars raise special problems, and it is here that the explanation in terms of fossil magnetic fields stands at least a chance of being tenable, although dynamo mechanisms have again been postulated (Krause & Meinel 1988). Pulsars are neutron stars that have surface dipole fields of strength typically between 10^8 and 3×10^9 T. It is conceivable that these are fossil fields conserved during the core collapse that produces the stars themselves, but this view encounters difficulties. First, it is hard to account for fields even as large as 10^7 T as a fossil field remnant. Second, the progenitor stars, and the neutron stars themselves when they form, are violently convective, and are likely to destroy magnetic flux rapidly unless dynamo action occurs (see section 1.5). It is not easy to understand why only some 3% of white dwarfs have surface magnetic fields stronger than 10^2 T. Explanations, that invoke dynamo action in the progenitors of the dwarfs, have been proffered.

E. Interstellar medium; galaxies. It has been known for more than 40 years that magnetic fields pervade interstellar space, probably brought there with the material ejected from the supernova explosions of earlier stars. Fermi proposed in 1949 that a general 'toroidal' magnetic field existed in the plane of the Galaxy. The angular momentum carried by an interstellar cloud intensifies as the cloud collapses gravitationally during star formation, and this proves a very severe obstacle to star formation. It has been suggested that the interstellar magnetic field is a seed, amplified by a transient dynamo that exists during the collapse of the cloud. The magnetic field so created transfers the angular momentum of the cloud outwards, so allowing the process of star formation to proceed to completion. See for example Bisnovatyi-Kogan, Ruzmaikin & Sunyaev (1973). It was at first thought that the lifetime of interstellar magnetic fields would be comparable with that of the Galaxy, but Parker (1971) pointed out that, because the turbulence of the interstellar medium drastically reduced the effective τ_η, the lifetime of the galactic field would be much shorter than that of the Galaxy, and that some contemporary process was needed to explain it, in fact a galactic dynamo. It is now known through radio observations that very many galaxies possess magnetic fields (e.g., see Beck, Kronberg & Wielebinski 1990), and this has engendered a considerable body of theory leading to numerous galactic dynamo models; see for example Elstner, Meinel & Beck (1992) and Ruzmaikin, Sokoloff & Shukurov (1988a, b).

F. Accretion disks. An accretion disk surrounding a black hole needs, like the nascent star, a mechanism that ejects angular momentum, from the inflowing material, and for this purpose a magnetic field has been invoked; the possibility that this is dynamo maintained appears to be gaining ground; see for example Vishniac & Diamond (1993)

G. Liquid metals. There have been several attempts to demonstrate fluid dynamos in the laboratory; see for example Gailitis *et al.* (1989). There is also the possibility (Bevir 1973) that dynamo action might occur spontaneously in the liquid sodium circulating within a fast reactor, much to the detriment of its performance!

The basic electrodynamics of dynamo theory is expounded in section 1.2 of this chapter. Section 1.3 sets up the mathematics of the kinematic dynamo problem, and establishes some necessary conditions for dynamo action to occur. That these conditions are not sufficient is demonstrated by a proof of Cowling's theorem, which shows that self-excited dynamos having purely axisymmetric magnetic fields do not exist. Section 1.4 focusses on dynamos that *do* work. Numerical and asymptotic approaches are described, and the so-called $\alpha\omega$-dynamo is introduced. Mean field theory is developed in section 1.5, the mean referring to averages over an ensemble of turbulently moving conductors. The so-called α^2-dynamo is also defined in that section.

1.2 BASIC ELECTROMAGNETIC THEORY

1.2.1 Non-relativistic Electromagnetism
Pre-Maxwell equations and boundary conditions. In this book we shall be concerned with conductors moving at speeds, V, small compared with the speed of light in vacuo, c_0. This means that we can, and should, adopt the approximate electromagnetic theory that applies when the rate of change of the electromagnetic field is small compared with $c_0/\mathcal{L}$, namely the theory in use before Maxwell introduced his famous displacement current. The pointwise form of the electromagnetic equations were then

$$\nabla \times \mathbf{H} = \mathbf{J}, \qquad \nabla \times \mathbf{E} = -\partial_t \mathbf{B}, \qquad (1.2.1,2)$$
$$\nabla \cdot \mathbf{B} = 0, \qquad \nabla \cdot \mathbf{D} = \vartheta. \qquad (1.2.3,4)$$

Here the electric current density, $\mathbf{J}$, and the free charge density, ϑ, are the *sources* that create the magnetic field, $\mathbf{B}$, the electric field, $\mathbf{E}$, the magnetizing force, $\mathbf{H}$, and the electric displacement, $\mathbf{D}$; ∂_t denotes differentiation with respect to the time, t. All fields in $(2.1-4)^1$ are measured in SI units. Equations (2.1) and (2.2) are respectively Ampère's law and Faraday's law.

The sources in (2.1, 2) must satisfy charge conservation, and in pre-Maxwell theory this requires that

$$\nabla \cdot \mathbf{J} = 0, \tag{1.2.5}$$

an obvious consequence of (2.1). Similarly (2.2) shows that, if (2.3) holds for any t, it holds for all t.

Under the Galilean transformation

$$\mathbf{x}' = \mathbf{x} - \mathbf{U}t, \qquad t' = t \tag{1.2.6, 7}$$

(where $\mathbf{U}$ is constant) the fields and sources of pre-Maxwell theory transform as

$$\mathbf{B}' = \mathbf{B}, \qquad \mathbf{E}' = \mathbf{E} + \mathbf{U} \times \mathbf{B}, \tag{1.2.8, 9}$$

$$\mathbf{H}' = \mathbf{H}, \qquad \mathbf{D}' = \mathbf{D} + \mathbf{U} \times \mathbf{H}/c_0^2, \tag{1.2.10, 11}$$

$$\mathbf{J}' = \mathbf{J}, \qquad \vartheta' = \vartheta + \nabla \cdot (\mathbf{U} \times \mathbf{H})/c_0^2, \tag{1.2.12, 13}$$

where $c_0 = (\mu_0 \epsilon_0)^{-1/2}$, and ϵ_0 is the permittivity of free space. According to (2.6, 7), $\partial_t' = \partial_t - \mathbf{U} \cdot \nabla$, from which it is readily verified that the primed variables also obey (2.1–5).

We shall be particularly interested in systems in which, on some surface or surfaces S, the properties of the material change abruptly. We exclude cases in which S separates the same material in different physical states and through which the material passes, changing state as it does so, i.e., a shock discontinuity. Dynamos involving shocks have not been much studied. In fact, the principles of dynamo action are so well illustrated by ignoring the compressibility of the medium that most often the working fluid is supposed to be incompressible; see section 1.3.1 below. From the same integral laws that lead to (2.1–5), it may be shown that

$$[\mathbf{n} \times \mathbf{H}] = \mathbf{C}, \qquad [\mathbf{n} \times (\mathbf{E} + \mathbf{U} \times \mathbf{B})] = 0, \quad \text{on S}, \tag{1.2.14, 15}$$

$$[\mathbf{n} \cdot \mathbf{B}] = 0, \qquad\qquad\qquad [\mathbf{n} \cdot \mathbf{D}] = \Sigma, \quad \text{on S}, \tag{1.2.16, 17}$$

$$[\mathbf{n} \cdot \mathbf{J}] = -\nabla_s \cdot \mathbf{C}, \quad \text{on S}, \tag{1.2.18}$$

where $\mathbf{C}$ and Σ are the surface current and free surface charge on S; $\nabla_s\cdot$ is the two-dimensional surface divergence on S; $[Q] = Q^{(1)} - Q^{(2)}$ is the difference between the limiting values of any quantity Q as the point, P, of S concerned in (2.14–18) is approached from sides 1 and 2; the unit vector, $\mathbf{n}$, to S at P is directed *out* of side 2, i.e., *into* side 1; $\mathbf{U}$ is now the velocity of P. By

---

[1] Equation $(x.y.z)$ is referred to as $(y.z)$ in chapter x.

(2.16), only the component of **U** along **n** is involved in (2.15), but the origin (2.9) of (2.15) is clearer when (2.15) is written as shown. It may be seen from (2.11, 17) that the surface charge, Σ', in the frame moving with P is

$$\Sigma' = \Sigma + \mathbf{n} \cdot [\![\mathbf{U} \times \mathbf{H}]\!]/c_0^2. \tag{1.2.19}$$

Equations (2.13, 19) may also be written (see (2.14)) as

$$\vartheta' = \vartheta - \mathbf{U} \cdot \mathbf{J}/c_0^2, \qquad \Sigma' = \Sigma - \mathbf{U} \cdot \mathbf{C}/c_0^2. \tag{1.2.20, 21}$$

Conservation of charge is as fundamental as conservation of mass, and it may seem anomalous that, according to (2.20, 21), ϑ and Σ are frame-dependent (unless **J** happens to be zero). This paradox, which is also faced by the full Maxwell theory, was resolved early in the development of electromagnetic theory by Lorentz and Minkowski. Restricting attention to the case in which S is stationary in the primed frame, it is clear enough that the *net* charge,

$$Q \equiv \int_V \vartheta \, dV + \oint_S \Sigma \, dS, \tag{1.2.22}$$

of an isolated body, V, surrounded by an insulator $\hat{V}$ (e.g., free space) is frame-independent since, according to (2.3, 9) and the divergence theorem,

$$Q' \equiv \int_V \vartheta' \, dV + \oint_S \Sigma' \, dS = Q + c_0^{-2} \oint_S \widehat{\mathbf{H}} \times d\mathbf{S}, \tag{1.2.23}$$

where the surface integral involves **H** on the $\hat{V}$ side of V, i.e., side 1 in the notation of (2.14–18). Since the body is isolated, $\widehat{\mathbf{H}}$ vanishes at infinity so that, by an application of the divergence theorem, the surface integral in (2.23) is the integral over $\hat{V}$ of $\widehat{\mathbf{J}}$, which is zero. Thus $Q = Q'$.

To understand the difference between ϑ and ϑ', we should ask how charge density could be measured at a point P within V. A collector inserted to withdraw charge from a small domain, v, of volume v surrounding P would gather the charge ϑv, equal to $\vartheta' v$, but *only* if $\mathbf{J} = 0$ at P. Otherwise the current flowing across the surface, s, of v would be continually drawn off by the collector and the measurement would be corrupted. (This would in any case be true, no matter how small v, if a point source of current existed at P, but we shall exclude that exceptional case.) To overcome the difficulty of **J** not being zero at P, we could isolate v by surrounding it with a thin film of insulator which would prevent currents crossing into v from the outside. That would, of course, alter the current and charge distribution everywhere, and the charge $q^\dagger = \vartheta^\dagger v$ reaching the collector might differ substantially from the previous ϑv and $\vartheta' v$. Nevertheless, the $q^\dagger$ so measured would be unambiguous since it differs from $q'^\dagger$ by $-\mathbf{U} \cdot \mathbf{I}/c_0^{-2}$ where

$$\mathbf{I} = \oint_s d\mathbf{S} \times \mathbf{H}^\dagger = \int_v \mathbf{J}^\dagger dV = \oint_s \mathbf{x}(\mathbf{J}^\dagger \cdot d\mathbf{S}) = 0. \tag{1.2.24}$$

The steps in (2.24) use (2.5, 14, 18) and the divergence theorem, together with the fact that $\mathbf{J}^\dagger = \mathbf{0}$ in the insulating film; as previously, the dagger indicates the altered state, after the film has been added. For simplicity, we have also supposed that the material is finitely conducting, so that (see below) $\mathbf{C}^\dagger = \mathbf{0}$ on s.

In short, the frame dependence of ϑ and Σ does not conflict with the demands of charge conservation.

1.2.2 Closure and Consequences

Constitutive equations. The pre-Maxwell equations must be supplemented by relations that define the physical constitution of the medium in which they are applied. We shall suppose the medium is isotropic and homogeneous so that, at a point P within it, and in a reference frame moving with the velocity $\mathbf{U} = \mathbf{V}(\mathbf{x}_P)$ of P,

$$\mathbf{H}' = \frac{\mathbf{B}'}{\mu}, \qquad \mathbf{D}' = \epsilon \mathbf{E}', \qquad \mathbf{J}' = \sigma \mathbf{E}', \qquad (1.2.25, 26, 27)$$

where the permeability μ, the permittivity ϵ and the electrical conductivity σ would be replaced by tensors if the material were non-isotropic and by functions of position if it were non-homogeneous. For simplicity, and because it is approximately true in many applications, we shall suppose that $\mu = \mu_0$, the permeability of free space. Equations (2.25–27) apply only at P but, when translated back to the laboratory frame with the help of (2.8–12), they assume forms that apply for all P, namely

$$\mathbf{H} = \frac{\mathbf{B}}{\mu}, \qquad \mathbf{D} = \epsilon \mathbf{E} + (\epsilon - \epsilon_0)\mathbf{V} \times \mathbf{B}, \qquad \mathbf{J} = \sigma(\mathbf{E} + \mathbf{V} \times \mathbf{B}). \quad (1.2.28, 29, 30)$$

We shall henceforward, with the help of (2.28), eliminate $\mathbf{H}$ from the discussion; we shall find (2.29) of little relevance, but (2.30), which is the generalisation of Ohm's law for a moving conductor, is centrally important. It is often useful to rewrite it in one of the equivalent forms

$$\mathbf{E} = -\mathbf{V} \times \mathbf{B} + \frac{\mathbf{J}}{\sigma}, \qquad \mathbf{E} = -\mathbf{V} \times \mathbf{B} + \eta \nabla \times \mathbf{B}. \qquad (1.2.31, 32)$$

According to (2.29, 32),

$$\vartheta = -\epsilon_0 \nabla \cdot (\mathbf{V} \times \mathbf{B}). \qquad (1.2.33)$$

It should be noted that, unlike the case of a stationary conductor treated in many textbooks on electromagnetism, where it is shown that a nonzero ϑ would disappear in a time of order ϵ_0/σ, the free charge in a moving conductor attains the generally nonzero value (2.33) instantaneously according to pre-Maxwell theory.

Induction equation. The term $\mathbf{V} \times \mathbf{B}$ appearing in (2.29–32) is known as the electromotive force, or emf for short. In contrast to EHD (electrohydrodynamics) where a generally large $\mathbf{E}$ is externally applied, $\mathbf{E}$ owes its existence in MHD to the emf, so that $\mathcal{E} = O(\mathcal{V}\mathcal{B})$, where $\mathcal{E}$, $\mathcal{V}$ and $\mathcal{B}$ are typical magnitudes of $\mathbf{E}$, $\mathbf{V}$ and $\mathbf{B}$. We are concerned with situations in which $\mathcal{V} \ll c_0$, so that $\mathcal{E} \ll c_0\mathcal{B}$, in contrast to electromagnetic radiation for which $\mathcal{E} = O(c_0\mathcal{B})$. It is for this reason that we are able to use the pre-Maxwell equations rather than the full Maxwell equations, why we can use non-relativistic equations of motion, why we shall consider light to propagate instantaneously, and why we are concerned with Galilean rather than Lorentz transformations. It is also the reason why $+\partial_t \vartheta$ could be omitted from the left-hand side of (2.5), why the advection, $\vartheta\mathbf{V}$, of free charge did not contribute to $\mathbf{J}$ in (2.30), and why the momentum, $\mathbf{D} \times \mathbf{B}$, of the electromagnetic field could be ignored. It is in fact possible to derive (2.1–5) and the entire subsequent development of pre-Maxwell theory by assuming that $\mathcal{E} = O(\mathcal{V}\mathcal{B})$ and, after expanding all terms in powers of $\mathcal{V}/c_0$, substituting into the full Maxwell theory and subsequently eliminating all effects of relative order $(\mathcal{V}/c_0)^2$ or smaller.

According to (2.2, 32),

$$\partial_t\mathbf{B} = \nabla \times (\mathbf{V} \times \mathbf{B}) + \eta\nabla^2\mathbf{B}. \tag{1.2.34}$$

The fact that this electromagnetic *induction equation* involves $\mathbf{B}$ alone casts the magnetic field into the central role in MHD, to such an extent that 'the field' is usually used and understood to mean 'the magnetic field'. Once (2.3, 34) have been solved for given $\mathbf{V}$, subject to appropriate boundary conditions, $\mathbf{E}$ can, if desired, be extracted from (2.32), $\mathbf{D}$ from (2.29), and ϑ from (2.33). This is, however, a matter of secondary importance.

If $\mathbf{B}$ were discontinuous across any surface S, that discontinuity would, for nonzero η, be instantaneously obliterated, according to (2.34). Then $\mathbf{C} = \mathbf{0}$, and (2.14, 18) could be replaced by

$$\text{if} \quad \eta \neq 0, \quad [\![\mathbf{B}]\!] = \mathbf{0}, \quad [\![\mathbf{n} \cdot \mathbf{J}]\!] = 0, \quad \text{on} \quad \text{S}, \tag{1.2.35, 36}$$

the first of which implies the second according to (2.1, 28). By (2.36), $[\![\mathbf{n} \cdot \nabla \times \mathbf{B}]\!] = 0$ on S, but (2.15, 32, 35) show that $[\![\mathbf{n} \times \nabla \times \mathbf{B}]\!]$ is generally discontinuous on S, unless both η and $\mathbf{V}$ are continuous.

Magnetic Reynolds number. When $\mathbf{V} = \mathbf{0}$, (2.34) reduces to a vector diffusion equation and, provided no applied sources exist to maintain $\mathbf{B}$, it will disappear in a time of the order of the *electromagnetic diffusion time*, $\tau_\eta = \mathcal{L}^2/\eta = \mu_0\sigma\mathcal{L}^2$, where $\mathcal{L}$ is its characteristic length scale (see section 1.1). Also relevant is the *advection time*, $\tau_V = \mathcal{L}/\mathcal{V}$, and the ratio, τ_η/τ_V, which is the *magnetic Reynolds number*,

$$R = \mathcal{V}\mathcal{L}/\eta = \mu_0\sigma\mathcal{V}\mathcal{L}. \tag{1.2.37}$$

Two non-dimensional forms of (2.34) are prevalent, differing by the scaling of t. If $t \to \tau_\eta t$, we have

$$\partial_t \mathbf{B} = R\nabla \times (\mathbf{V} \times \mathbf{B}) + \nabla^2 \mathbf{B}, \qquad (1.2.38)$$

but if $t \to \tau_V t$, we have instead

$$\partial_t \mathbf{B} = \nabla \times (\mathbf{V} \times \mathbf{B}) + R^{-1}\nabla^2 \mathbf{B}. \qquad (1.2.39)$$

In both cases $\mathbf{x} \to \mathcal{L}\mathbf{x}$, $\mathbf{V} \to \mathcal{V}\mathbf{V}$ and R measures the ratio of the terms on the right-hand sides. The induction equation, however written, is translationally and rotationally invariant.

Energy equation. Of special significance is the energy equation that follows from (2.1, 2, 28):

$$\partial_t U_M + \nabla \cdot \mathbf{P} = -\mathbf{E} \cdot \mathbf{J}, \qquad (1.2.40)$$

where

$$U_M = B^2/2\mu_0, \qquad \mathbf{P} = \mathbf{E} \times \mathbf{B}/\mu_0 \qquad (1.2.41, 42)$$

are the electromagnetic energy density and the electromagnetic energy flux, sometimes called 'Poynting's vector'; the electric energy density is negligible compared with the magnetic energy density which therefore alone provides U_M. By (2.31) we may rewrite (2.40) as

$$\partial_t U_M + \nabla \cdot \mathbf{P} = -\mathbf{V} \cdot \mathbf{L} - \epsilon_\eta, \qquad (1.2.43)$$

where

$$\mathbf{L} = \mathbf{J} \times \mathbf{B}, \qquad \epsilon_\eta = J^2/\sigma \qquad (1.2.44, 45)$$

are the Lorentz force and the ohmic dissipation, sometimes also called 'Joule heating'; ϵ_η represents the irreversible dissipation of electromagnetic energy into heat through the electrical resistance of the medium. The term $\mathbf{V} \cdot \mathbf{L}$ on the right-hand side of (2.43) is the rate at which the Lorentz force does work in accelerating the medium. If $\mathbf{V} \cdot \mathbf{L} < 0$, it represents the creation of electromagnetic energy from the kinetic energy of the motion $\mathbf{V}$. In a working dynamo, the integral of $-\mathbf{V} \cdot \mathbf{L}$ over the system is positive and large enough to make good the ohmic losses, i.e., the integral of ϵ_η over the system. Since $-\mathbf{V} \cdot \mathbf{L} = O(\mathcal{V}B^2/\mu_0^2)$ and $\epsilon_\eta = O(B^2/\mu_0^2\sigma)$ by (2.1, 28), this can happen only if R is sufficiently large (see section 1.3.3).

Electromagnetic stress. According to (2.1, 28, 44), $\mathbf{L}$ is the divergence of the magnetic stress tensor, m_{ij}:

$$L_i = \nabla_j m_{ji}, \qquad m_{ij} = \frac{B_i B_j}{\mu_0} - \frac{B^2}{2\mu_0}\delta_{ij}. \qquad (1.2.46, 47)$$

The stress exerted over the surface element $\mathbf{dS}$ by the first term on the right-hand side of (2.47) is $\mathbf{B}(\mathbf{B} \cdot \mathbf{dS})/\mu_0$ and therefore represents a tension in a magnetic flux tube of B^2/μ_0 per unit area of tube. The second term on the right-hand side of (2.47) is an isotropic *magnetic pressure* of $B^2/2\mu_0$; for a field of 1 T, this is about 4 atm. The sum of the kinetic pressure, p, and the magnetic pressure is the *total pressure* $P = p + B^2/2\mu_0$. The total nonviscous stress in the fluid is $-P\delta_{ij} + B_i B_j/\mu_0$. It may be noted that the electric stresses are $O(\mathcal{V}/c_0)^2$ times smaller than the magnetic stresses and are therefore negligible. For the same reason, the body force, $\vartheta \mathbf{E}$, exerted by the electric field on the free charges does not appear in (2.44). It may be remarked that the magnetic stresses comprising m_{ij} can be effective in transporting (angular) momentum from one (part of a) body to another, even via a surrounding vacuum, a process appealed to in sections 1.1.3E and F.

1.2.3 The Perfect Conductor

Alfvén's theorem. It is occasionally useful to believe the fiction that the working fluid is a perfect conductor. When $\sigma = \infty$ (i.e., $\eta = 0$), (2.31) and (2.34) become

$$\mathbf{E} = -\mathbf{V} \times \mathbf{B}, \qquad \partial_t \mathbf{B} = \nabla \times (\mathbf{V} \times \mathbf{B}). \qquad (1.2.48, 49)$$

If the fluid is in addition inviscid, we have

$$\text{if } \mathbf{n} \cdot \mathbf{B} \neq 0, \qquad [\![\mathbf{B}]\!] = 0, \qquad [\![\mathbf{V}]\!] = 0, \quad \text{on S.} \qquad (1.2.50, 51)$$

The first of these follows from the continuity of the tangential component of the total stress tensor across S, and the second is a consequence of (2.15, 48). The continuity of the normal stress implies that p is continuous. A *tangential discontinuity* of $\mathbf{B}$ and $\mathbf{V}$ can exist at a surface on which $\mathbf{n} \cdot \mathbf{B} = 0$. Since P is continuous at such a discontinuity, and since $\mathbf{n} \times \mathbf{B}$ is not, it follows that p is then discontinuous in general.

It can be shown from (2.49) that magnetic field lines are material curves. This is Alfvén's theorem, which is often more picturesquely phrased by saying that the 'lines of force are frozen to a perfectly conducting fluid in its motion'. Proofs are to be found in almost any MHD text; see for example Roberts (1967). We give six examples of the theorem to illustrate its usefulness.

Example 1. We suppose that an inviscid, perfectly conducting, incompressible fluid of uniform density ρ fills all space, as does a uniform field $\mathbf{B}_0$. When this equilibrium is disturbed, the motion of the fluid causes the straight field lines of $\mathbf{B}_0$ to bend. As a result, electric currents are created together with a Lorentz force and changed magnetic stresses. If we focus on a tube of force (i.e., a bundle of field lines) of cross-sectional area A_0, we see that its mass per unit length is $m = \rho A_0$ and the tension along it is $T = A_0 B_0^2/\mu_0$. It follows that, as for an elastic string, the disturbance will be carried as a wave in each direction along the field lines with speed $\sqrt{(T/m)}$, i.e., with the *Alfvén velocity*

$$\mathbf{V}_A = \frac{\mathbf{B}}{\sqrt{(\mu_0 \rho)}}. \qquad (1.2.52)$$

These are *Alfvén waves*. The isotropic magnetic pressure, $B^2/2\mu_0$, plays no part in this argument, and it is unchanged by the presence of the waves. Although these results rest on an analogy, it is worth remarking that a proper mathematical analysis fully substantiates them.

Example 2 illustrates how motion can enhance the magnetic energy of a system. Suppose a section of length L_0 of the tube of force of example 1 is stretched to a length $L > L_0$. To conserve mass, the cross-sectional area will reduce to $A = A_0 L_0/L$ and at the same time the field lines within it will be brought closer together, so that the field will intensify to $B = B_0 A_0/A = B_0 L/L_0$. The magnetic energy within the tube therefore increases from $M_0 = (B_0^2/2\mu_0)A_0 L_0$ to $M = (B^2/2\mu_0)AL = M_0(L/L_0)^2$. This can be thought of as the work that must be done in stretching the tube against the tension of the field lines it contains.

In the remaining examples, we shall allow diffusion to operate, i.e., we shall assume that R, though very large, is not infinite. It is evident from (2.39) that diffusion is most effective where the gradients of $\mathbf{B}$ are large, and in the next two examples such gradients occur only in small regions. It is only in these regions that diffusion is effective, and in which field lines sever and coalesce, so creating a new field line topology. Elsewhere, the frozen field picture works well or, as we shall suppose, perfectly.

Example 3 concerns the Vainshtein–Zel'dovich (1972) rope dynamo. Consider a closed flux tube (see figure 1.2a) which is *stretched* to twice its length, its cross-sectional area being halved and the field within it being doubled, as in example 2. The tube is then *twisted* to form a figure eight (figure 1.2b), and is *folded* so as to merge with itself (figure 1.2c). By this device, the total cross-sectional area, which now consists of two cross-sections of the original rope, is restored to its original value, but contains twice the original flux. The region denoted by X in figure 1.2(c) is one of high field gradient in which Alfvén's theorem is a poor approximation, i.e., the final terms of (2.31) and (2.34) are influential. In fact, ohmic diffusion can rapidly reconnect the field lines near X, and so restore the tube to its original form, but with double the contained flux (figure 1.2d), i.e., the magnetic energy has been increased by the motions — dynamo action has occurred. After, or even contemporaneously with, the diffusive reconnection, the whole of the *stretch–twist–fold* (STF) process can be repeated. Conceptually, the rope dynamo has been very helpful in the search for fast dynamos (see below), but clearly such a qualitative description of the dynamo falls far short of a mathematically satisfying demonstration that fast dynamos exist.

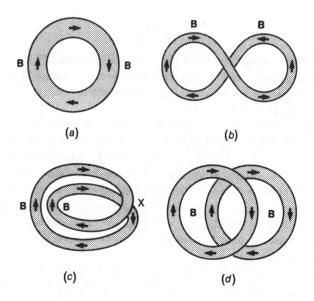

Figure 1.2 The Zel'dovich–Vainshtein dynamo.

Example 4 shows Alfvén's (1950) twisted rope dynamo. A straight flux rope is twisted in opposite directions at its ends and develops loops that resemble those that are often seen on the cord joining a telephone handset to its cradle (figure 1.3a). Because of the high field gradients in the regions where the loops leave the rope, indicated by X on figure 1.3(a), the loops detach, as indicated in figure 1.3(b). Even though, as in example 3, the time taken for one loop to separate from the rope is limited by diffusion and may be long, many new loops can start to form during this time. The mechanism is therefore also a good candidate for a fast dynamo, but again nothing definite can be said until a proper mathematical foundation has been laid.

Example 5. Diffusion plays an even more significant role in the next example, which aims to show that the *tendency* for field lines to move with a finitely conducting fluid is a useful aid to thought. The example also points to the dangers attendant on taking the approximation of perfect conductivity too literally. It further illustrates the so-called 'ω-effect', the general case of which is thought to be significant for dynamo action (see section 1.4.3). The fluid fills a sphere, V, surrounded by a vacuum $\hat{V}$. An externally-generated uniform field, $\mathbf{B}_0 = B_0 \mathbf{1}_z$, is maintained everywhere. (The unit vector in the direction of increasing coordinate c is denoted by $\mathbf{1}_c$.) At time $t = 0$ a steady motion, $\mathbf{V} = V_0(r/a)^3 \sin\theta\, \mathbf{1}_\phi$, is initiated, where V_0 is a positive constant

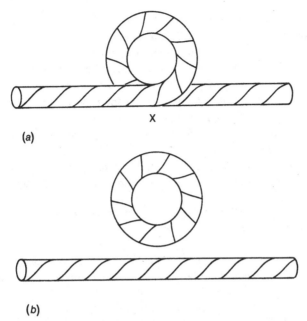

(a)

(b)

Figure 1.3 Alfvén's twisted rope dynamo.

and (r, θ, ϕ) are spherical coordinates with the z-axis as $\theta = 0$. Alfvén's theorem indicates correctly that the motion stretches out the field lines of $\mathbf{B}_0$ within V into the ϕ-direction in the Northern hemisphere of V and into the $-\phi$-direction in the Southern hemisphere of V. A final steady state is reached in which B_ϕ is given by

$$B_\phi \to \frac{B_0 V_0}{7 \eta a^3} r^2 (a^2 - r^2) \sin \theta \cos \theta, \quad t \to \infty. \tag{1.2.53}$$

It may be noted that (2.53) is continuous on S (i.e., $r = a$) with the field $\hat{B}_\phi$ in $\hat{V}$, which must be zero because $\hat{\mathbf{J}} = \mathbf{0}$. If the magnetic Reynolds number, $R = V_0 a / \eta$ is large, the field (2.53) greatly exceeds the field $\mathbf{B}_0$ from which it was induced. In this case it may be shown that, for $t \ll \tau_\eta \equiv a^2 / \eta$ (and beneath a thin diffusive boundary layer of thickness $O((\eta t)^{1/2})$ on S, within which the solution adjusts to the condition $B_\phi = 0$ on S),

$$B_\phi \sim \frac{2 B_0 V_0 t}{a^3} r^2 \sin \theta \cos \theta, \tag{1.2.54}$$

which is in fact the appropriate solution of (2.49), the equation from which Alfvén's theorem followed.

This example illustrates an important point: the assumption of perfect conductivity can be good only for $t \ll \tau_\eta$. The limits $R \to \infty$ and $t \to \infty$

are not interchangeable. Ultimately $(t \gg \tau_\eta)$, diffusive processes cannot be ignored. The initial stretching of field lines is correctly represented by (2.54), but in (2.53) the rate of stretching is exactly balanced by the rate at which (in defiance of Alfvén's theorem) the field lines diffuse through the conductor in the opposite direction. In dynamo theory, we are always interested in whether, and what kind of, a field can exist for $t \gg \tau_\eta$, and there is strictly no place for the perfectly conducting fluid approximation. Nevertheless, R is usually large, and the tendency for the field lines to move with the fluid helps one to picture the evolution of field.

Example 6 is example 5, slightly modified to bring it closer to the dynamo question. Suppose we replace the uniform inducing field, $\mathbf{B}_0$, by

$$\mathbf{B}_M = \begin{cases} (m/a^5)[(5a^2 - 3r^2)\mathbf{1}_r \cos\theta + (6r^2 - 5a^2)\mathbf{1}_\theta \sin\theta], & r \le a, \\ (m/a^3)[2\mathbf{1}_r \cos\theta + \mathbf{1}_\theta \sin\theta], & r \ge a, \end{cases}$$

$$(1.2.55)$$

which is a dipole of strength $m\mathbf{1}_z$ in $\widehat{\mathrm{V}}$. If we solve the ϕ-component of the induction equation, we find in place of (2.53, 54) the more complicated results

$$B_\phi \to \frac{mV_0}{42\eta a^8} r^2(a^2 - r^2)(23a^2 - 7r^2)\sin\theta\cos\theta, \quad t \to \infty, \qquad (1.2.56)$$

$$B_\phi \sim \frac{2mV_0 t}{a^8} r^2(5a^2 - 3r^2)\sin\theta\cos\theta, \qquad (1.2.57)$$

which can be interpreted much as in example 5. There is, however, one essential difference: the field $\mathbf{B}_M$ vanishes at infinity in $\widehat{\mathrm{V}}$ and is continuous across S; no finite $\mathbf{V}$ can be found for which the r- and θ-components of the steady induction equation can be satisfied. Unless some other sources are postulated, $\mathbf{B}_M$ will disappear in a time of order τ_η taking B_ϕ with it, i.e., (2.56) will then not hold. Nevertheless, while $t \ll \tau_\eta$, there will be little diminution of $\mathbf{B}_M$, and (2.57) will apply if R is large, so showing that B_ϕ will grow to strengths large compared with $\mathbf{B}_M$ before it vanishes with it. We return to this topic in section 1.4.3 below.

An exact integral. Finally, returning again to the case $R = \infty$, it may be shown that (2.49) possesses a Cauchy integral. Let the initial field configuration be $\mathbf{B}^{(0)}(\mathbf{x}^{(0)})$, and let $\mathbf{x}^{(0)}$ (the Lagrangian coordinate) be displaced to $\mathbf{x} = \mathbf{x}(\mathbf{x}^{(0)})$ as a result of incompressible fluid motion, the field becoming at the same time $\mathbf{B}(\mathbf{x})$. then

$$B_i = \frac{\partial x_i}{\partial x_j^{(0)}} B_j^{(0)}. \qquad (1.2.58)$$

This result provides the basis of a mapping technique much used in fast dynamo theory; see chapter 6. It may be generalised for a compressible fluid, in which the initial density $\rho^{(0)}(\mathbf{x}^{(0)})$ becomes $\rho(\mathbf{x})$:

$$\frac{B_i}{\rho} = \frac{\partial x_i}{\partial x_j^{(0)}} \frac{B_j^{(0)}}{\rho^{(0)}}. \tag{1.2.59}$$

1.3 GENERAL CONSIDERATIONS

1.3.1 Dynamo Conditions

Self-excitation. In this section we shall work with the induction equation (2.34) in the dimensionless form (2.38), i.e., we will require that

$$\partial_t \mathbf{B} = R\nabla \times (\mathbf{V} \times \mathbf{B}) + \nabla^2 \mathbf{B}, \qquad \nabla \cdot \mathbf{B} = 0. \tag{1.3.1, 2}$$

The induction equation is sometimes called 'the dynamo equation'. Whilst it is true that a dynamo field must satisfy (3.1, 2) throughout the volume V of the conductor, such a terminology is a misnomer since the field must also satisfy two crucially important requirements that solutions to (3.1, 2) are not in general required to obey:

First dynamo condition: All fields and currents must be created by fluid motion; none must be supplied by other sources external or internal (for example, thermoelectric and electrochemical sources, i.e., 'batteries', must be excluded).

Second dynamo condition: The fields and currents must persist indefinitely (or for a time long compared with τ_η, where it is assumed that the energy sources powering the motions are maintained during that time). The dynamo itself need not be steady; $\mathbf{B}$ may be oscillatory, or even chaotic if $\mathbf{V}$ is chaotic.

The mathematical expression of the first dynamo condition depends on the geometry of the system. We consider five possibilities.

System I. The conductor fills all space, but the fluid velocity is nonzero only within a bounded volume V. The field $\hat{\mathbf{B}}$ in the exterior $\hat{V}$ of V must satisfy (3.1, 2) with $\mathbf{V}$ set zero. The dynamo condition requires that $\hat{\mathbf{B}}$ and $\hat{\mathbf{E}}$ are $O(r^{-2})$ as $r \to \infty$, where r is the distance from some origin in V. Across the surface, S, of V we require that

$$\mathbf{B} = \hat{\mathbf{B}}, \qquad \mathbf{n} \times \mathbf{E} = \mathbf{n} \times \hat{\mathbf{E}}, \quad \text{on S}, \tag{1.3.3, 4}$$

where

$$\mathbf{E} = -R\mathbf{V} \times \mathbf{B} + \nabla \times \mathbf{B}, \qquad \hat{\mathbf{E}} = \nabla \times \hat{\mathbf{B}}. \tag{1.3.5, 6}$$

System II. The conductor fills all space, but the fluid velocity is nonzero only within an *un*bounded volume, V, *not* filling all space. Again the fields $\hat{\mathbf{B}}$ and $\hat{\mathbf{E}}$ in the stagnant surroundings, $\hat{V}$, of V must vanish at large distances from V, but in addition conditions must apply at infinity within V. For example, the potential difference between the ends $z = \pm\infty$ of the cylindrical V considered in section 1.4.1 must vanish since otherwise the resulting J_z would create an applied field, B_ϕ. Again (3.3, 4) must be obeyed.

System III. The conductor fills all space and the fluid motion is spatially periodic, as for the models (4.6) and (4.7) below. In this case, the components of **B** may be expanded in Fourier series based on the periodicities of the assumed **V**, and the dynamo condition requires that the zero wave number components of those series vanish since, if present, they would correspond to a uniform nonzero applied field. Similar restrictions apply when the dynamo is driven by homogeneous turbulence filling all space.

System IV. The conductor fills a bounded volume V, the exterior $\hat{V}$ of which is an insulator in which $\hat{\mathbf{J}} = \mathbf{0}$ and $\hat{\jmath} = 0$

$$\nabla \times \hat{\mathbf{B}} = \mathbf{0}, \qquad \nabla \times \hat{\mathbf{E}} = -\partial_t \hat{\mathbf{B}}, \qquad (1.3.7, 8)$$

$$\nabla \cdot \hat{\mathbf{B}} = 0, \qquad \nabla \cdot \hat{\mathbf{E}} = 0. \qquad (1.3.9, 10)$$

The first dynamo condition now reduces to

$$\hat{\mathbf{B}} = O(r^{-3}), \qquad \hat{\mathbf{E}} = O(r^{-2}), \quad r \to \infty, \qquad (1.3.11, 12)$$

i.e., $\hat{\mathbf{B}}$ is dipolar at great distances, monopoles being excluded on physical grounds. Again (3.3, 4) must hold, but usually we need not trouble ourselves with (3.4, 8, 10, 12). Once **B** in V has been derived, **E** can be deduced from (3.5), and $\hat{\mathbf{E}}$ follows from (3.8, 10, 12), up to an unknown gradient, which can however be found from (3.4). The crucial difference between **E** and **B** is that (3.3) requires that $[\![\mathbf{n} \cdot \mathbf{B}]\!] = 0$ on S, but a discontinuity in $[\![\mathbf{n} \cdot \mathbf{E}]\!]$ is allowed, according to (2.17) and (2.26), and merely determines Σ on S.

System V. The source of dynamo action lies in a small domain, D, within V and far from the boundaries, S, of the conductor; there are no sources of **B** in the exterior of D. This variant of possibility I arises in the so-called MEGA method (standing for 'Maximally Efficient Generation Approach'), an asymptotic method devised by Ruzmaikin *et al.* (1990). The dynamo condition becomes $\mathbf{B} \to \mathbf{0}$ as $d \to \infty$ in *all* directions from D; here d is distance from an origin in D.

The second dynamo condition merely requires that

$$\text{\textbf{B} does not tend to 0 for all \textbf{x} as } t \to \infty. \qquad (1.3.13)$$

Eligible velocities. In kinematic theory, all reasonable choices of **V** are allowed. But what is 'reasonable'? It is usually supposed that **V** is continuously differentiable.[2] Moreover, it is usually sufficient for a good understanding of

---

[2] This is not true in the case of the model discussed in section 1.4.1 below, where **V** is discontinuous at the surface S of a cylinder. At first sight this does not seem to be a very serious matter: one could add a thin film of lubricating conductor between the cylinder and its stagnant surroundings, which would restore the continuity of **V** without, apparently, significantly modifying the model. Nevertheless, this seemingly trifling change matters when fast dynamos are sought, as will be mentioned later in this book. See also Gilbert (1988).

dynamo mechanisms to suppose that the motions are isochoric:

$$\nabla \cdot \mathbf{V} = 0; \qquad (1.3.14)$$

in the usual terminology the fluid is 'incompressible'. When the moving fluid is confined to a volume V with a fixed surface S, we should always insist that

$$\mathbf{n} \cdot \mathbf{V} = 0 \qquad \text{on S.} \qquad (1.3.15)$$

Many theoretical models postulate time-dependent $\mathbf{V}$, but it would be physically unrealistic to allow $\mathbf{V}$ to become unbounded for any $\mathbf{x}$, and we shall ignore that possibility, thereby excluding point sources or sinks of matter. Ultimately, the correct choice of $\mathbf{V}$ is a solution of the equations of fluid motion, including the Lorentz force, but this is a considerably more difficult (nonlinear) undertaking than solving the (linear) kinematic dynamo problem. Kinematic theory may be regarded as a crucially important 'black box' embedded in the full MHD theory.

1.3.2 Kinematic Dynamos

The eigenvalue problem. If, as we shall suppose initially, $\mathbf{V}$ is steady (i.e., t-independent), (3.1–12) admit solutions of the form

$$\mathbf{B} \propto e^{\lambda t}, \qquad (1.3.16)$$

where the growth rate λ is an eigenvalue. Equation (3.1) becomes

$$\lambda \mathbf{B} = R\nabla \times (\mathbf{V} \times \mathbf{B}) + \nabla^2 \mathbf{B}, \qquad (1.3.17)$$

and the remaining essential equations are unaltered.

We shall mainly focus here on case IV of section 1.3.1, the case of greatest interest when modelling planetary, stellar and galactic dynamos. The spectrum of λ is then discrete, with limit point at $-\infty$. Greatest interest attaches to the λ (or λs) with greatest real part; supposing for simplicity only one such (although a complex λ must be one of a pair), we shall denote it by $\lambda_{\max}$. The eigenfunctions B_α form a complete set, i.e., an arbitrary field obeying (3.1–3, 7, 9, 11) can be decomposed into a sum of the form

$$\mathbf{B} = \sum_\alpha \mathbf{B}_\alpha(\mathbf{x}) \exp(\lambda_\alpha t). \qquad (1.3.18)$$

Non-self-adjointness. The eigenvalue problem is not in general self-adjoint. The growth rates, λ_α, are not therefore necessarily real, although they must occur in complex conjugate pairs, and the corresponding eigenfunctions must ensure the reality of the sum (3.18). In infinite domains, such as those of types I–III in section 1.3.1, the adjoint problem is also a dynamo problem, obtained by replacing $\mathbf{V}$ by $-\mathbf{V}$ (Roberts 1960). Such a motion may be very different in character. For example the STF flow described in example 3 of section 1.2.3 becomes a UTC motion, i.e., unfold-twist-compress! Not surprisingly the adjoint eigenfunctions may differ vastly from the eigenfunctions of the original system, even though the eigenvalues are identical. And, from a numerical point of view, it may be simpler to solve the adjoint problem than the original one, a point made long ago by Chandrasekhar (1961) in a different context, and made again by Bayly in chapter 10. In a finite domain (type IV), the adjoint problem corresponds to no physically attainable system although, again, it does have some practical uses; see Kono & Roberts (1992). In all cases, it is only when $R = 0$ that we recover a self-adjoint problem, namely that of finding the *decay modes*, and are assured of a real spectrum, even though it is entirely negative. When V is the unit sphere, $\lambda_{\max}(0) = -\pi^2$.

Another significant consequence of the non-self-adjointness of the eigenvalue problem is that a demonstration that the total magnetic energy, M, increases (i.e., the integral of $U_M = B^2/2\mu_0$ over all space) does *not* prove that the motion regenerates field. Consider the last example of section 1.2.3 as a putative dynamo and assume that $R \equiv V_0 a/\eta \gg 1$. After a time of order η/V_0^2, M starts to increase and, in a time of order $\tau_\eta = a^2/\eta$, it reaches a maximum value that is of order R^2 times its initial value, but thereafter it declines monotonically to zero with the e-folding time, τ_η/π^2 of the dipolar part of the $\mathbf{B}_M$ field given by (2.55). The zonal field owes its existence to $\mathbf{B}_M$, but this has nothing to sustain it; therefore B_ϕ must ultimately die with $\mathbf{B}_M$. Because of the high cost of numerical integrations of the dynamo problem, it is important to know how long a dynamo model should be integrated before it can be safely concluded from the persistence of a global quantity such as M that the model is self-sustaining. This point is addressed in Hughes (1993) and St. Pierre (1993).

AC and DC dynamos. The eigenvalue problem is reminiscent of many that arise in linear stability theory. The eigenvalues are continuous functions of R, which plays the role of the 'control parameter'. As R varies, mode crossing of real eigenvalues may occur, and two complex eigenvalues may emerge at that crossing. The determination of the eigenvalue spectrum, and especially of $\lambda_{\max}$, is of central importance. If

$$\Re(\lambda_{\max}) \geq 0, \qquad (1.3.19)$$

the corresponding term in (3.18) will persist for all time and the second dynamo condition (3.13) is satisfied. In fact, if $\Re(\lambda_{\max}) > 0$, $\mathbf{B}$ apparently grows without bound, a physical absurdity that arises from the linear nature of the kinematic dynamo problem, and which is removed by the nonlinear Lorentz force in the full MHD dynamo theory. If (3.19) does not hold, all terms in (3.18) disappear as $t \to \infty$, and the motion $\mathbf{V}$ does not regenerate field, at least for the value of R concerned. As in linear stability theory, the location of a marginal state is of particular interest. This is the smallest value, R_c, of $|R|$ for which

$$\Re(\lambda_{\max}) = 0. \tag{1.3.20}$$

There are two possibilities:

$$\begin{cases} \text{DC} \quad \text{(direct bifurcation)}: & \Im(\lambda_{\max}) = 0, \\ \text{AC} \quad \text{(Hopf bifurcation)}: & \Im(\lambda_{\max}) \neq 0. \end{cases} \tag{1.3.21}$$

The terms commonly used in modern stability theory are shown in parentheses, but the acronyms DC and AC are shorter and more commonly employed in other electrical contexts! When DC dynamos are preferred, there is a short cut to locating them. One simply sets $\lambda = 0$ in (3.17) and solves the resulting equation

$$0 = R_c \nabla \times (\mathbf{V} \times \mathbf{B}) + \nabla^2 \mathbf{B}, \tag{1.3.22}$$

in conjunction with (3.2, 3, 7, 9, 11), seeking a real eigenvalue R_c. Cowling (1934) noticed that the DC dynamo problem could be formulated as an 'inverse problem', in which $\mathbf{B}$ is given and a $\mathbf{V}$ that maintains it is sought; see also Fearn, Roberts & Soward (1988) and Lortz (1990).

Symmetries. As R is increased from zero, $\lambda_{\max}$ moves away from its negative decay value and may, or may not, become complex. In either case, the crucial question is whether a marginal state (3.20) is ever attained. And if not, why not try *decreasing* R from zero and renewing the search? As the example of the homopolar dynamo of section 1.1 shows, there is no reason why, in general, $\lambda_\alpha(R)$ and $\lambda_\alpha(-R)$ should coincide, even though this may be so in special cases.

Suppose for example that $\mathbf{V}$ is unchanged under the reflection $\mathbf{x} \to -\mathbf{x}$ about a suitably chosen origin. Such in fact is the case when $\mathbf{V}$ is a spherical model of a planetary core, or when $\mathbf{V}$ is a flat ellipsoidal model of a galactic disk. If also $\mathbf{V}(-\mathbf{x}) = \mathbf{V}(\mathbf{x})$, then $\lambda_\alpha(-R) = \lambda_\alpha(R)$, for all α. We may also note that flows $-\mathbf{V}(-\mathbf{x})$ and $\mathbf{V}(\mathbf{x})$ share the same eigenvalue spectrum. An interesting case arises when

$$-\mathbf{V}(-\mathbf{x}) \equiv \mathbf{V}(\mathbf{x}). \tag{1.3.23}$$

Then every eigenfunction is either of dipolar (D) type or of quadrupolar (Q) type:

$$\mathbf{B}_\alpha^{\mathrm{D}}(-\mathbf{x}) \equiv \mathbf{B}_\alpha^{\mathrm{D}}(\mathbf{x}), \qquad \mathbf{B}_\alpha^{\mathrm{Q}}(-\mathbf{x}) \equiv -\mathbf{B}_\alpha^{\mathrm{Q}}(\mathbf{x}). \tag{1.3.24, 25}$$

Similar statements can be made when $\mathbf{V}$ is unchanged by reflection in a suitably chosen plane, $z = 0$ (say) and $\mathbf{x}_R = (x, y, -z)$. Define $\mathbf{F}_R$ from any field $\mathbf{F}$ by $\mathbf{F}_R(\mathbf{x}) \equiv \mathbf{F}(\mathbf{x}_R) - 2F_z(\mathbf{x}_R)\mathbf{1}_z$. When $\mathbf{V}_R \equiv \mathbf{V}$, the eigenfunctions are either symmetric (S) or antisymmetric (A), in the sense that

$$\mathbf{B}_{R\alpha}^{\mathrm{S}} \equiv \mathbf{B}_\alpha^{\mathrm{S}}, \qquad \mathbf{B}_{R\alpha}^{\mathrm{A}} \equiv -\mathbf{B}_\alpha^{\mathrm{A}}. \tag{1.3.26, 27}$$

Thus, a dipole of moment **m** is of A-type when axial ($\mathbf{m} \times \mathbf{1}_z = 0$) but is of S-type when equatorial ($\mathbf{m} \cdot \mathbf{1}_z = 0$). The designations A and S are common in dynamo theory, although the D and Q labelling is more natural.

There is no loss of generality if, as we shall now do, we suppose that $R \geq 0$; in non-dimensionalising, we have essentially replaced **V** by $R\mathbf{V}$, but we could equally well have replaced it by $(-R)(-\mathbf{V})$.

Unsteady flows. Few naturally-occurring flows are steady, and many theoretical studies postulate time-dependent **V**, for which (3.16) is incorrect. Many investigations assume periodic **V**, of period P (say), i.e.,

$$\mathbf{V}(\mathbf{x}, t + P) \equiv \mathbf{V}(\mathbf{x}, t), \qquad \text{for all } \mathbf{x} \text{ and } t. \tag{1.3.28}$$

In this case, **B** behaves in the same way as $t \to \infty$, and the dynamo condition simply requires that $\lambda \geq 0$, where

$$\lambda = \frac{1}{P} \ln \left[\lim_{t \to \infty} \frac{B(t + P)}{B(t)} \right], \tag{1.3.29}$$

and B is any component of **B** at any **x**. More generally **V** may be a chaotic solution of the full nonlinear MHD equations. In that case, a generalization of (3.29) is necessary.

1.3.3 Necessary Conditions; Bounds

As noted below (2.45), a necessary condition for dynamo action is that R is sufficiently large. In this section we sketch the proofs of three lower bounds on R, i.e., three values of R such that, if any is applicable but is not exceeded, the dynamo fails.

First bound. Consider a steady model of type I (section 1.3.1), and let **V** be zero on S. The Biot-Savart solution of (3.2–6, 22) shows that **B** is a nontrivial solution of a homogeneous Fredholm equation,

$$\mathbf{B}(\mathbf{x}) = \frac{R}{4\pi} \int \frac{(\mathbf{V}_1 \times \mathbf{B}_1) \times (\mathbf{x} - \mathbf{x}_1)}{|\mathbf{x} - \mathbf{x}_1|^3} \, dV_1, \tag{1.3.30}$$

where $\mathbf{B}_1 = \mathbf{B}(\mathbf{x}_1)$, $dV_1 = dx_1 dy_1 dz_1$, etc. Let $B_{\max}$, the largest value of $|\mathbf{B}|$ in V or on S, occur at $\mathbf{x} = \mathbf{x}_m$. Let $V_{\max}$ be the largest value of $|\mathbf{V}|$ in V. Applying (3.30) at $\mathbf{x}_m$ we obtain

$$B_{\max} \leq \frac{R}{4\pi} V_{\max} B_{\max} \left[\int \frac{dV_1}{|\mathbf{x} - \mathbf{x}_1|^2} \right]_{\max} \leq \frac{R}{4} V_{\max} B_{\max} L_{\max}, \tag{1.3.31}$$

where $L_{\max}$ is the diameter of the smallest sphere that can contain V. If we redefine R to be $R_1 = V_{\max} L_{\max} R$, we see that a necessary condition for steady dynamo action is $R_1 \geq 4$ (Roberts 1967, p. 74).

Second bound. Consider a model of type IV. The integral of (2.43) over the whole system, i.e., over $V_\infty = V + \hat{V}$ is, in dimensionless units,

$$\frac{dM}{dt} = -\int_V (\nabla \times \mathbf{B})^2 \, dV - R \int_V \mathbf{V} \cdot [(\nabla \times \mathbf{B}) \times \mathbf{B}] \, dV, \qquad (1.3.32)$$

where $M = \frac{1}{2} \int_{V_\infty} B^2 \, dV$ is the nondimensional total magnetic energy. A variational inequality (see, for example, section 3.2 of Roberts 1967) shows that

$$\int_V (\nabla \times \mathbf{B})^2 \, dV \geq k \int_{V_\infty} B^2 \, dV, \qquad (1.3.33)$$

where we have used $-k$ instead of the principal decay rate, $\lambda_{max}(0)$, to avoid notational confusion. By using Schwarz's inequality, we may now show that

$$-\int_V \mathbf{V} \cdot [(\nabla \times \mathbf{B}) \times \mathbf{B}] \, dV \leq \frac{V_{max}}{k^{1/2}} \int_V (\nabla \times \mathbf{B})^2 \, dV. \qquad (1.3.34)$$

Combining (3.32–34), we now see that

$$\frac{dM}{dt} \leq -\left[1 - \frac{RV_{max}}{k^{1/2}}\right] \int_V (\nabla \times \mathbf{B})^2 \, dV. \qquad (1.3.35)$$

Clearly a necessary condition for dynamo action is $RV_{max} \geq k^{1/2}$ (Childress 1969). If we redefine R to be $R_2 \equiv V_{max}R$, we see that the condition is $R_2 \geq k^{1/2}$ (which is $R_2 \geq \pi$ in the case of a sphere of radius L). Equation (3.35) also shows that, if $R_2 \geq k^{1/2}$, then

$$\lambda_{max} \leq \frac{(R_2 - k^{1/2})}{2k^{1/2}}. \qquad (1.3.36)$$

In any given situation, we may be able to capitalise on the rotational invariance of the induction equation to extract the best possible result from (3.35). Thus, if V is axisymmetric, we may transform to the frame rotating about that axis for which V_{max} is least.

Third bound. This also applies to models of type IV, but only those for which $\mathbf{V} = \mathbf{0}$ on S. With the help of (2.46) and (2.47), a preliminary transformation is made of the last integral in (3.32):

$$-\int_V \mathbf{V} \cdot [(\nabla \times \mathbf{B}) \times \mathbf{B}] \, dV = \int_V B_i B_j \nabla_i V_j \, dV = \int_V B_i B_j d_{ij} \, dV, \qquad (1.3.37)$$

where $d_{ij} = \frac{1}{2}(\nabla_i V_j + \nabla_j V_i) = d_{ji}$ is the rate of deformation tensor. It is easily seen that

$$\int_V B_i B_j d_{ij} \, dV \leq \zeta_{max} \int_V B^2 \, dV \leq 2M\zeta_{max}, \qquad (1.3.38)$$

where ζ_{max} is the largest value taken by any eigenvalue of $d_{ij}(\mathbf{x})$ anywhere in V. From (3.32, 33, 37, 38), it now follows that

$$\frac{dM}{dt} \leq -2(k - R\zeta_{max})M. \qquad (1.3.39)$$

Thus, a necessary condition for dynamo action is $R\zeta_{max} \geq k$. If we redefine R to be $R_3 \equiv \zeta_{max} R$, we see that the condition is $R_3 \geq k$.

Inequality (3.39) also places a bound on the growth rate of the field:

$$\lambda_{max} \leq R_3 - k, \qquad (1.3.40)$$

which in unscaled units shows that

$$\lambda_{max} \leq \zeta_{max}, \qquad (1.3.41)$$

(Backus 1958, see also Proctor 1977). The aim of fast dynamo theory, which is expounded in later chapters, is to find motions for which λ_{max} is of order ζ_{max} as $R \to \infty$. Such growth rates are, remarkably, independent of η!

Another bound of some interest has been derived by Busse (1975); see also section 3.2 of Roberts (1987).

1.3.4 Axisymmetry; Antidynamo Theorems

Axisymmetric representation. Rotational forces in a naturally-occurring body provide a preferred axis that may be evident in its MHD structure. Axisymmetry has therefore a particular significance to both kinematic and MHD dynamo theory. It is convenient to establish a notation at the outset. We shall consider type IV systems for axisymmetric V only.

Let (s, ϕ, z) be cylindrical coordinates with $s = 0$ as axis of symmetry. Separate **B** and **V** into their axisymmetric and asymmetric parts by first averaging each cylindrical component over ϕ to obtain its axisymmetric part, e.g.,

$$\overline{B}_s(s, z, t) = \frac{1}{2\pi} \int_0^{2\pi} B_s(s, \phi, z, t) \, d\phi, \qquad (1.3.42)$$

and then defining the asymmetric parts by subtraction, e.g., $\mathbf{B}' = \mathbf{B} - \overline{\mathbf{B}}$. Thus we have

$$\mathbf{B} = \overline{\mathbf{B}} + \mathbf{B}', \qquad \mathbf{V} = \overline{\mathbf{V}} + \mathbf{V}'. \qquad (1.3.43, 44)$$

By (3.2), we may further write $\overline{\mathbf{B}}$ as the sum of meridional (M) and zonal (ϕ) fields:

$$\overline{\mathbf{B}} = \overline{\mathbf{B}}_M + \overline{\mathbf{B}}_\phi, \qquad (1.3.45)$$

where

$$\overline{\mathbf{B}}_M = \nabla \times [A(s, z, t)\mathbf{1}_\phi], \qquad \overline{\mathbf{B}}_\phi = B(s, z, t)\mathbf{1}_\phi. \qquad (1.3.46, 47)$$

The function sA is constant on the lines of force of the meridionally projected part, $\overline{\mathbf{B}}_M$, of $\overline{\mathbf{B}}$.

In a similar way, we may write

$$\mathbf{V} = \overline{\mathbf{V}}_M + \overline{\mathbf{V}}_\phi, \qquad (1.3.48)$$

and, when the fluid is incompressible so that (3.14) holds,

$$\mathbf{V}_M = \nabla \times [s^{-1}\chi(s,z,t)\mathbf{1}_\phi], \qquad \mathbf{V}_\phi = s\zeta(s,z,t)\mathbf{1}_\phi. \qquad (1.3.49,50)$$

Here χ is constant on the streamlines of the meridian flow, and ζ is the zonal shear.

The induction equation (3.1) may be similarly divided into axisymmetric and asymmetric parts:

$$\partial_t \overline{\mathbf{B}} = R\nabla \times (\overline{\mathbf{V}} \times \overline{\mathbf{B}} + \overline{\mathcal{E}}) + \nabla^2 \overline{\mathbf{B}}, \qquad (1.3.51)$$

$$\partial_t \mathbf{B}' = R\nabla \times (\overline{\mathbf{V}} \times \mathbf{B}' + \mathbf{V}' \times \overline{\mathbf{B}} + \mathcal{E}') + \nabla^2 \mathbf{B}', \qquad (1.3.52)$$

where

$$\overline{\mathcal{E}} = \overline{\mathbf{V}' \times \mathbf{B}'}. \qquad (1.3.53)$$

Moreover (3.51) may be separated into its meridional and zonal parts:

$$\partial_t A + Rs^{-1}\mathbf{V}_M \cdot \nabla(sA) = \Delta A + R\overline{\mathcal{E}}_\phi, \qquad (1.3.54)$$

$$\partial_t B + Rs\mathbf{V}_M \cdot \nabla(s^{-1}B) = \Delta B + Rs\overline{\mathbf{B}}_M \cdot \nabla\zeta + R(\nabla \times \overline{\mathcal{E}})_\phi, \qquad (1.3.55)$$

where $\Delta \equiv \nabla^2 - s^{-2}$. In the insulating exterior, $\widehat{V}$, of V, we have by (3.7, 9, 11)

$$\Delta\widehat{A} = 0, \qquad \widehat{B} = 0, \qquad (1.3.56,57)$$

$$\widehat{A} = O(r^{-2}), \qquad r \to \infty, \qquad (1.3.58)$$

while (3.3) demands that

$$A = \widehat{A}, \qquad \mathbf{n} \cdot \nabla A = \mathbf{n} \cdot \nabla\widehat{A}, \qquad B = \widehat{B}, \quad \text{on S.} \qquad (1.3.59,60,61)$$

Cowling's theorem. Suppose now that $\overline{\mathcal{E}} \equiv 0$; e.g., assume that $\mathbf{V}' \equiv \mathbf{0}$ or $\mathbf{B}' \equiv \mathbf{0}$. Equation (3.54) is then essentially reduced to the heat conduction equation for sA, the $\mathbf{V}_M$ term on its left-hand side being part of the motional derivative which convects the 'heat' but is not a source.

According to (3.58), the 'temperature' is held zero at infinity. Not surprisingly therefore, A tends to zero with an e-folding time of order τ_η. This argument may be formalised by multiplying (3.54) by s^2A, integrating the result over V, and using (3.14, 15, 58–60) to obtain

$$\frac{d}{dt}\int_V (sA)^2 \, dV = -2\int_V [\nabla(sA)]^2 \, dV \le 0. \qquad (1.3.62)$$

Once $\overline{\mathbf{B}}_M$ has vanished, the only surviving source in (3.55), namely the ω-effect source, $Rs\overline{\mathbf{B}}_M \cdot \nabla\zeta$, also disappears and (3.55) is reduced essentially to a heat conduction equation, this time for B/s as 'temperature'. Again B disappears during a time of order τ_η. This result may be formalised by multiplying (3.55) by B/s^2, integrating over V, and using (3.14, 15, 57, 61) to obtain

$$\frac{d}{dt}\int_V \left(\frac{B}{s}\right)^2 dV = -2\int_V \left[\nabla\left(\frac{B}{s}\right)\right]^2 dV \le 0. \qquad (1.3.63)$$

In short, when $\overline{\mathcal{E}} = 0$, all components of $\overline{\mathbf{B}}$ disappear as $t \to \infty$ (Cowling 1957, Braginsky 1964 a).

If $\mathbf{V'} = \mathbf{0}$, (3.52) is satisfied for all t by $\mathbf{B'} \equiv \mathbf{0}^3$. Cowling's theorem follows: *an axisymmetric magnetic field cannot be maintained by dynamo action.* Neither of the two proofs of this theorem given by Cowling (1934) is completely satisfactory, and there have been several subsequent investigations aimed at greater rigour. None of these has cast substantial doubt on the truth of the theorem. They are reviewed in section 2.3 of Fearn *et al.* (1988), where further references may be found.

Further antidynamo theorems. Even if $\mathbf{V'} \equiv \mathbf{0}$, it is necessary to restrict attention to the $\mathbf{B'} \equiv \mathbf{0}$ solutions of (3.52). Thus Cowling's theorem shows more generally that an axisymmetric flow cannot maintain the axisymmetric part of any field. Examples exist, however, that show that an axisymmetric flow *can* maintain a purely asymmetric field (see sections 1.4.1 and 1.4.2).

Cowling's theorem was the earliest and most significant of a number of 'anti-dynamo theorems' that rule out whole classes of solutions even when the necessary conditions of section 1.3.3 are satisfied. Other theorems include:

(a) *A two-dimensional magnetic field cannot be maintained by dynamo action.* (By 'two-dimensional' we mean that, in some coordinate system (x, y, z), $\mathbf{B}$ is independent of z.)

(b) The toroidal velocity theorem: *an incompressible motion in a spherical V having a zero radial component everywhere cannot maintain a field.* (The adjective 'toroidal' will be defined in section 1.4.2.) The theorem still holds if $\eta = \eta(r)$, where (r, θ, ϕ) are spherical coordinates with the origin fixed at the centre of V. It also appears that the theorem may also be true if $\eta = \eta(r, \theta)$; see Donner & Brandenberg (1990).

(c) Zel'dovich's theorem. *An incompressible motion, in which $V_z \equiv 0$ in some Cartesian coordinate system (x, y, z), cannot maintain a magnetic field* (Cowling 1957, Zel'dovich 1957, Zel'dovich & Ruzmaikin 1980).

(d) The radial velocity theorem. *The radial motion $\mathbf{V} \equiv V(r, t)\mathbf{1}_r$ cannot maintain a magnetic field in a sphere* (Namikawa & Matsushita 1970); see also Ivers & James (1986, 1988).

For further details and speculations, see section 2.4 of Fearn *et al.* (1988). The suite of anti-dynamo theorems is often summarised by saying that working dynamos have a 'low degree of symmetry'.

---

[3] This is not true if $\mathbf{V'} \ne \mathbf{0}$, and $\mathbf{B'}$ and $\overline{\mathcal{E}}$ will become nonzero, even if they are zero initially. Thus Cowling's theorem implicitly requires that $\mathbf{V'} \equiv \mathbf{0}$.

1.4 WORKING DYNAMOS

1.4.1 Simple Models

The Ponomarenko dynamo; helicity. The first proofs that the conditions of section 1.3.1 could be met were provided by Herzenberg (1958), who assumed a steady **V**, and by Backus (1958), who used a time-dependent **V**. Today, the simplest known dynamo is that of Ponomarenko (1973). In the terminology of section 1.3.1, it is of type II: the conductor fills all space but the motions are confined to a cylinder:

$$\mathbf{V} = \begin{cases} (0, \omega s, U), & s < a, \\ 0, & s > a, \end{cases} \qquad (1.4.1)$$

where ω and U are constants. When separated into cylindrical components the induction equation (2.34) is found to admit separable solutions in which

$$\mathbf{B} \propto \exp(im\phi + ikz + \lambda t), \qquad (1.4.2)$$

where m, k and λ are constants. Despite the axisymmetry of the motion (4.1), growing solutions exist when $|\omega U|$ is large enough. These are asymmetric and not two-dimensional, so evading Cowling's theorem and theorem (a) of section 1.3.4. It transpires that dynamo action is efficient, the marginal state occurring when

$$R_{\mathrm{Po}} \equiv a(U^2 + \omega^2 a^2)^{1/2}/\eta \approx 17.7221175. \qquad (1.4.3)$$

This state arises for $m = 1$, $U/\omega a \approx 1.314$, $ka \approx -0.38754$ and $\Im(\lambda) \approx 0.41029961\eta/a^2$.

With a technological demonstration of dynamo action in mind, Gailitis & Freiberg (1976, 1980) generalised the model in a number of ways. They considered a model of finite length in which the 'fluid' of the dynamo returned in an outer sleeve. Such an experimental device was built at the MHD facility at Salaspils, near Riga; see Gailitis *et al.* (1987, 1989). Insufficiently large values of R_{Po} were attained for self-excitation to occur, but it was shown that the decay of an initially imposed field was inhibited by the motion.

A screw dynamo has been investigated by Ruzmaikin, Sokoloff & Shukurov (1988c), Ruzmaikin *et al.* (1989) and by Sokoloff, Shukurov & Shumkina (1989). This differs from motion (4.1) by having s-dependent ω and U that vanish on $s = a$, so that the infinite shear on $s = a$ is avoided. As noted in section 1.3.1, this (unlike the Ponomarenko dynamo) is therefore not a fast dynamo.

As well as being the simplest known example, the Ponomarenko model exhibits with maximum clarity the effectiveness of helical motions in generating field. The *helicity* of a flow **V** is defined by

$$H = \mathbf{V} \cdot \omega, \qquad \omega = \nabla \times \mathbf{V}, \qquad (1.4.4, 5)$$

where ω is the vorticity. In the case of (4.1), $H = 2\omega U$. Helicity is reversed under coordinate reflections such as $x \to -x$ or $x \to x_R$ (see section 1.3.2). In the terminology of vector calculus, V is a polar vector but ω is a skew vector (or pseudo-vector); their scalar product is therefore a pseudoscalar which reverses sign on coordinate reflection; see section 1.5.2. Almost all successful homogeneous dynamo models operate through the helicity of their motions, which introduces the skewness lacked by the fluid container (section 1.1). The significance of helicity was first recognised in a qualitative way by Parker (1955), in a mathematical but disguised way by Braginsky (1964a), and in a transparent way by Steenbeck, Krause & Rädler (1966). Steenbeck & Krause (1966) christened it 'Schraubensinn', but Moffatt (1969) later proposed the more felicitous term helicity, which has stuck.

ABC flows. The effectiveness of helical motions is also clearly demonstrated by spatially-periodic motions (type III of section 1.3.1), such as that of G. O. Roberts (1972) for which, in scaled units,

$$V = (\sin y, \sin x, \cos x - \cos y). \qquad (1.4.6)$$

The flow in each cell of (4.6) resembles the Ponomarenko motion, the directions of V being opposite in adjacent cells, i.e., the Ponomarenko-like dynamos are aligned alternately in the $+z$ and $-z$ directions. When R is large enough, the two-dimensional motion (4.6) regenerates a three-dimensional field, with components proportional to $\exp(ikz)$, where $k \neq 0$.

Motion (4.6) is a *Beltrami flow*, i.e., a flow in which V and ω are everywhere parallel. It is clear from (4.4) that generally $|H| \leq |V||\omega|$ and Beltrami flows are maximally helical in the sense that $|H| = |V||\omega|$. The motion (4.6) is a particular case of the more general Beltrami flow

$$V = (C \sin z + B \cos y, A \sin x + C \cos z, B \sin y + A \cos x), \qquad (1.4.7)$$

which is now commonly called an 'ABC-flow' (A for Arnol'd, B for Beltrami and C for Childress). Both (4.6) and (4.7) have been intensively studied in the fast dynamo context.

Another cylindrical dynamo, so simple that its action can be demonstrated by purely analytic arguments, is that of Lortz (1968). See Eltayeb & Loper (1988) for a more recent discussion.

1.4.2 Numerical Models

Toroidal-poloidal decomposition. None of the models so far considered are, in the terminology of section 1.3.1, of class IV, but it precisely those models, operating in a simple domain such as a sphere, that are most closely related to naturally-occurring dynamos. The motivation for studying spherical dynamos is strong, and we first introduce a technique that is very useful for solving the dynamo equations numerically when S is the unit sphere.

In view of (2.3), $\mathbf{B}$ may be divided into *toroidal* and *poloidal* parts:

$$\mathbf{B} = \mathbf{B}_T + \mathbf{B}_P, \qquad (1.4.8)$$

where

$$\mathbf{B}_T = \nabla \times (T\mathbf{x}) = \left[0, \frac{1}{\sin\theta}\frac{\partial T}{\partial\phi}, -\frac{\partial T}{\partial\theta}\right], \qquad (1.4.9)$$

$$\mathbf{B}_P = \nabla \times \nabla \times (S\mathbf{x}) = \left[\frac{L^2 S}{r^2}, \frac{1}{r}\frac{\partial^2(rS)}{\partial\theta\partial r}, \frac{1}{r\sin\theta}\frac{\partial^2(rS)}{\partial\phi\partial r}\right], \qquad (1.4.10)$$

give the spherical components and

$$L^2 = -\left\{\frac{1}{\sin\theta}\frac{\partial}{\partial\theta}\left[\sin\theta\frac{\partial}{\partial\theta}\right] + \frac{1}{\sin^2\theta}\frac{\partial^2}{\partial\phi^2}\right\} \qquad (1.4.11)$$

is the angular momentum operator. The functions $T(r,\theta,\phi,t)$ and $S(r,\theta,\phi,t)$ are called 'defining scalars'.

The canonical examples of toroidal and poloidal vectors arise for axisymmetric T and S. When T is independent of ϕ, the only nonzero component of $\mathbf{B}_T$ is $B_{T\phi}$ and this corresponds to B in (3.47). When S is independent of ϕ, $B_{P\phi}$ is zero and the lines of force of $\mathbf{B}_P$ therefore lie in meridian planes. This corresponds to the A term in (3.46). In short, in the axisymmetric case, $\mathbf{B}_T = \mathbf{B}_\phi$ and $\mathbf{B}_P = \mathbf{B}_M$; see section 1.3.4.

Any toroidal vector is orthogonal to any poloidal vector in the sense that $\langle\langle \mathbf{B}_T \cdot \mathbf{B}_P \rangle\rangle = 0$. Here $\langle\langle Q \rangle\rangle$ denotes the 'horizontal' average of a quantity $Q(r,\theta,\phi,t)$:

$$\langle\langle Q \rangle\rangle = \frac{1}{4\pi}\int Q\,d\Omega, \qquad (1.4.12)$$

where $d\Omega = \sin\theta\,d\theta d\phi$ is the solid angle; $\langle\langle Q \rangle\rangle$ depends on r and t alone.

Since L^2 annihilates functions that depend on r alone, there is no way of including a radially symmetric $\mathbf{B}$ into (4.8); such a monopole field is in any case inadmissible on physical grounds. Also, since $L^2 f(r) \equiv 0$, we may without loss of generality remove from T and S their horizontally averaged parts $\langle\langle T \rangle\rangle$ and $\langle\langle S \rangle\rangle$, i.e., we may assume that

$$\langle\langle T \rangle\rangle = 0, \qquad \langle\langle S \rangle\rangle = 0. \qquad (1.4.13, 14)$$

It is then easy to invert $L^2 T$ and $L^2 S$ for such functions. Let $Y_n^m(\theta,\phi) = P_n^m(\theta)\exp(im\phi)$ be a surface harmonic, where $P_n^m(\theta)$ is the associated Legendre function. After expanding $L^2 T$ as

$$L^2 T = \sum_{n=1}^{\infty}\sum_{m=-n}^{n} F_n^m(r,t) Y_n^m(\theta,\phi), \qquad (1.4.15)$$

we may use the fact that $L^2 Y_n^m = n(n+1) Y_n^m$ to obtain

$$T = \sum_{n=1}^{\infty} \sum_{m=-n}^{n} T_n^m(r,t) Y_n^m(\theta,\phi), \qquad (1.4.16)$$

where $T_n^m = F_n^m/n(n+1)$.

The curl of a toroidal vector is a poloidal vector and, more unexpectedly, *vice versa*. Consequently a toroidal field is produced by a poloidal current and *vice versa*: by (2.1) and (2.28) we have

$$\mathbf{J} = \mathbf{J}_T + \mathbf{J}_P, \qquad (1.4.17)$$

where

$$\mu_0 \mathbf{J}_T = \nabla \times (-\nabla^2 S \mathbf{x}), \qquad \mu_0 \mathbf{J}_P = \nabla \times \nabla \times (T \mathbf{x}). \qquad (1.4.18,19)$$

The generalization of (4.8–10) to other coordinate systems fails (except in the case of Cartesian coordinates) to produce significant simplifications, largely because in such systems the curl of a poloidal vector has a poloidal part. The following way of extracting the defining scalars and the equations that govern them (see (4.41, 42) below) then requires complicating amendments.

Toroidal fields possess no radial component according to (4.9), and the poloidal defining scalar of a divergenceless field can therefore easily be obtained from its radial component; see (4.10). Thus

$$\mathbf{x} \cdot \mathbf{B} = r B_r = r B_{Pr} = L^2 S, \qquad \mathbf{x} \cdot \mathbf{J} = r J_r = r J_{Pr} = L^2 T/\mu_0, \quad (1.4.20,21)$$

or equivalently

$$T = L^{-2}(\mu_0 \mathbf{x} \cdot \mathbf{J}), \qquad S = L^{-2}(\mathbf{x} \cdot \mathbf{B}). \qquad (1.4.22,23)$$

The required inversions of L^2 are easily performed, as described above.

It is clear from (4.18, 19) that, in the vacuum $\hat{V}$ surrounding V,

$$\hat{T} = 0, \qquad \nabla^2 \hat{S} = 0. \qquad (1.4.24,25)$$

According to (4.10, 25),

$$\hat{\mathbf{B}} = \nabla \hat{\Phi}, \qquad \hat{\Phi} = \partial(r\hat{S})/\partial r. \qquad (1.4.26,27)$$

According to section 1.3.1, the relevant solution of (4.25) is

$$\hat{S}(r,\theta,\phi,t) = \sum_{n=1}^{\infty} \sum_{m=-n}^{n} \hat{S}_n^m(r,t) Y_n^m(\theta,\phi), \qquad (1.4.28)$$

where, for some $A_n^m(t)$,

$$\hat{S}_n^m(r,t) = A_n^m(t) r^{-(n+1)}. \qquad (1.4.29)$$

It follows that, throughout $\hat{V}$ and on S,

$$\frac{\partial \hat{S}_n^m(r,t)}{\partial r} + (n+1)\hat{S}_n^m(r,t) = 0. \qquad (1.4.30)$$

According to (3.3)

$$T = \hat{T}, \quad S = \hat{S}, \quad \partial S/\partial r = \partial \hat{S}/\partial r, \quad \text{on } r = 1. \qquad (1.4.31,32,33)$$

It follows from (3.24, 30) that

$$T_n^m = 0, \quad \frac{\partial S_n^m}{\partial r} + (n+1)S_n^m = 0, \quad \text{at } r = 1. \qquad (1.4.34,35)$$

By applying (3.34, 35) to solutions, **B**, of the induction equation, we satisfy the first dynamo condition without having to determine the solution in $\hat{V}$ explicitly.

Two further points should be made. First, if (as is often the case) the fluid is supposed to be incompressible, so that (3.14) is obeyed[4], then **V** too can be divided as in (4.8):

$$\mathbf{V} = \mathbf{V}_T + \mathbf{V}_P, \qquad (1.4.36)$$

where

$$\mathbf{V}_T = \nabla \times (T\mathbf{x}), \quad \mathbf{V}_P = \nabla \times \nabla \times (S\mathbf{x}), \qquad (1.4.37,38)$$
$$T = L^{-2}(\mathbf{x} \cdot \boldsymbol{\omega}), \quad S = L^{-2}(\mathbf{x} \cdot \mathbf{V}); \qquad (1.4.39,40)$$

T and S can be expanded as in (4.16).

Spectral equations and their solutions. One of the first attempts to solve the dynamo equations in a sphere was due to Bullard & Gellman (1954). Assuming steady flow, they expanded **B** and **V** in toroidal and poloidal harmonics, as in (4.16). They substituted into (2.38) and extracted equations for each T_n^m and S_n^m by multiplying (2.38) by Y_n^{m*} and forming the horizontal average (4.12) of the resulting equation. In this way they obtained an infinite set of ordinary differential equations of the form

$$\lambda T_\gamma = R \sum_{\alpha,\beta}[(S_\alpha T_\beta T_\gamma) + (T_\alpha S_\beta T_\gamma) + (S_\alpha S_\beta T_\gamma) + (T_\alpha T_\beta T_\gamma)] + \nabla_\gamma^2 T_\gamma, \quad (1.4.41)$$
$$\lambda S_\gamma = R \sum_{\alpha,\beta}[(S_\alpha S_\beta S_\gamma) + (T_\alpha S_\beta S_\gamma) + (S_\alpha T_\beta S_\gamma)] + \nabla_\gamma^2 S_\gamma, \qquad (1.4.42)$$

where

$$\nabla_\gamma^2 = \frac{d^2}{dr^2} - \frac{n_\gamma(n_\gamma+1)}{r^2}. \qquad (1.4.43)$$

---

[4] We must also exclude purely radial flows such as stellar winds, or winds in accretion disks.

Here α is an abbreviation for (n_α, m_α) and similarly for β and γ.

The coupling coefficient $(\mathcal{S}_\alpha T_\beta T_\gamma)$ represents the creation of the T_γ field from the T_β field by the inductive action of the $\mathcal{S}_\alpha$ motion, through the term $\nabla \times (\mathbf{V} \times \mathbf{B})$ in (2.38), and similarly for the other coupling coefficients, which are too complicated to set down explicitly here; see, for example, Kono & Roberts (1991). The evaluation of these coefficients is arduous and can lead to error. It was therefore partially automated by Pekeris, Accad & Shkoller (1973) and by Kumar & Roberts (1975); it was completely automated, using computer algebra, by Dudley (1988) and by Nakajima & Kono (1991).

In practice, the eigenvalue problem posed by (4.34, 35, 41, 42) is truncated in γ, and is integrated over r by standard methods, usually by finite differencing in r and solving the resulting algebraic eigenvalue problem by the QR algorithm or by inverse iteration. Dynamo action is claimed when $\Re(\lambda_{\max}) > 0$, where $\lambda_{\max}$ converges convincingly as the truncation level is raised.

The first successful demonstration of dynamo action by this method was due to Pekeris *et al.* (1973), who assumed an asymmetric Beltrami flow. Dudley (1988) demonstrated that some very simple flows, such as the spherical Ponomarenko dynamo (or 'spheromac')

$$\mathbf{V} = T_1^0 + \epsilon \mathcal{S}_1^0, \qquad (1.4.44)$$

where

$$T_1^0(r) = \mathcal{S}_1^0(r) = \sin \pi r, \qquad (1.4.45)$$

could generate asymmetric $\mathbf{B}$; see also Dudley & James (1989). For $\epsilon = 0.17$, they obtained $R_c \approx 155$.

Although the spectral method devised by Bullard and Gellman provides a very convenient way of dealing with the first dynamo condition, it does not necessarily provide the most economical approach. At high truncation levels, or when $\mathbf{V}$ contains many harmonic components (and of course, in the full MHD dynamo problem, there will be as many $\mathbf{V}$ coefficients to contend with as $\mathbf{B}$ coefficients), very many interaction coefficients must be evaluated. Pseudo-spectral methods are then more efficient, in which the $\mathbf{V} \times \mathbf{B}$ products are evaluated in physical space.

Translated back into physical space, condition (4.35) is non-local, i.e., it connects together S at different points on $r = 1$. There are other, equivalent and efficient methods of implementing this condition, and it is quite unnecessary to use spherical harmonics at all in integrating the equations forward in time; finite difference methods can be used, although it still makes sense to take advantage of fast Fourier methods, by expanding in Fourier coefficients in ϕ and in Tchebychev series in r. The first attempt in this direction seems to have been made by Thirlby (reported in Roberts 1971), but the method was later developed independently by Ivanova & Ruzmaikin (1977), Braginsky (1978) and others.

In a highly rotating body, such as a planetary core, Coriolis forces impart a high degree of two-dimensionality to the dynamics, and cylindrical coordinates are much easier to work with, both analytically and computationally. It is then of interest to work with a coordinate system in V that combines the virtues of both cylindrical and spherical coordinates (Roberts 1992, Nakajima & Roberts 1994).

The incorporation of the first dynamo condition for a spherical V is so simple that it has even been used in modelling dynamos operating in galactic disks, the inducing motions being then confined to the neighbourhood of the equatorial plane ($\theta = \pi/2$), the remainder of the sphere being taken to be the galactic halo; see Brandenburg, Tuominen & Krause (1990).

1.4.3 Asymptotic Methods; the $\alpha\omega$-Dynamo

The nearly symmetric dynamo. It is always profitable, when faced with a situation as difficult as the kinematic dynamo problem, to attempt to find solutions by expansion, but the results of section 1.3.3, all of which demonstrate that R must be O(1) for regeneration, show that (regular) expansion of the solution in R is a lost cause. Braginsky (1964a, b) wondered whether (singular) expansion in R^{-1} might fare better. Since the dynamo should be marginal for small values of R^{-1}, one should expand about a state which is non-regenerative for $R^{-1} = 0$, i.e., for $R = \infty$, and Braginsky chose that state to be one of purely zonal V and B, which (like all states of aligned field and flow) creates no emf. The choice of such a starting point is partly motivated by dynamical considerations: the dynamics of rapidly rotating bodies of fluid are strongly influenced by rotation (the basic reason why a magnetic compass needle points approximately North on Earth's surface and why solar activity is approximately symmetric about the Sun's equator), and in particular Coriolis forces tend to deflect motions into the zonal direction, shearing the magnetic field into the same direction as they do so.

In the notation of section 1.3.4, Braginsky (1964a) supposed steady flows of the form

$$\mathbf{V} = \overline{\mathbf{V}}_\phi + R^{-1/2}\mathbf{V}' + R^{-1}\overline{\mathbf{V}}_M. \qquad (1.4.46)$$

Here the scaling (2.39) is envisaged, and R is based on a typical magnitude of the zonal flow; $\overline{\mathbf{V}}_\phi$, $\mathbf{V}'$ and $\overline{\mathbf{V}}_M$ are all O(1). When he substituted (4.46) into (2.39), he found that B must be an infinite series in powers of $R^{-1/2}$, which can be conveniently written as

$$\mathbf{B} = \overline{\mathbf{B}}_\phi + R^{-1/2}\mathbf{B}' + R^{-1}\overline{\mathbf{B}}_M, \qquad (1.4.47)$$

where $\overline{\mathbf{B}}_\phi$, $\mathbf{B}'$ and $\overline{\mathbf{B}}_M$ are themselves infinite series in R^{-1} but are dominantly O(1). He evaluated the $\overline{\mathcal{E}}_\phi$ and $(\nabla \times \overline{\mathcal{E}})_\phi$ sources appearing in (3.54) and

(3.55) and discovered a remarkable and (at first) a mysterious fact: if certain 'effective' fields, A_e and χ_e, were introduced to replace A and χ, the equations simplified enormously, and became in fact

$$\partial_t A + s^{-1}\overline{\mathbf{V}}_M \cdot \nabla(sA) = \Delta A + \alpha B, \qquad (1.4.48)$$

$$\partial_t B + s\overline{\mathbf{V}}_M \cdot \nabla(s^{-1}B) = \Delta B + s\overline{\mathbf{B}}_M \cdot \nabla\zeta, \qquad (1.4.49)$$

where we have dropped the suffix e on A and $\overline{\mathbf{V}}_M$ (see (3.49)). It is the source αB in (4.48) that is the vital new ingredient, lacking which the dynamo must fail (Cowling's theorem). This has become known as the 'α-effect', for no better reason than the accidental choice of α for the coefficient of B. The B field is supported by the ω-effect term $s\overline{\mathbf{B}}_M \cdot \nabla\zeta$ in (4.49). Together, (4.48) and (4.49) define the $\alpha\omega$-dynamo, of which more will be said in section 1.5.4.

Braginsky derived the explicit expression for $\alpha(s,z)$ as a functional of $\mathbf{V}'$. Expanding the meridional components of the asymmetric velocity in the form

$$\mathbf{V}'_M = \overline{V}_\phi \sum_{m=1}^{\infty}(\mathbf{u}_M^{mc}\cos m\phi + \mathbf{u}_M^{ms}\sin m\phi), \qquad (1.4.50)$$

he showed that

$$\alpha = \sum_{m=1}^{\infty}m^{-1}[(1-m^2)(s^{-1}(\mathbf{u}^{ms}\times\mathbf{u}^{mc})_\phi + \nabla u_z^{mc}\cdot\nabla(ru_r^{ms}) - \nabla u_z^{ms}\cdot\nabla(ru_r^{mc})].$$
$$(1.4.51)$$

In his first derivation, Braginsky (1964a) established this result for a single wavelike $\mathbf{V}'$ that he could reduce to a stationary velocity by transforming to a rotating frame, but later (Braginsky 1964b) he extended the result to $\mathbf{V}'$ that consisted of a sum of waves, each travelling zonally with a different angular velocity. The resulting expression for α consisted of a sum of expressions (4.51), one term for each wave.

Two remarkable features of his results should be noted. First, it might seem at first sight that the theory has little to do with (say) the geodynamo since, according to (4.47), $R^{-1/2}\mathbf{B}'/R^{-1}\overline{\mathbf{B}}_M = O(R^{1/2}) \gg 1$, suggesting that, if the theory were applicable to Earth, the principal component of the geomagnetic field would be an equatorial dipole rather than the observed axial dipole. In fact, however, diffusive effects do not enter the asymptotic expansion of the solution to the first *two* orders, i.e., the fluid might just as well be perfectly conducting to order $R^{-1/2}$, and Alfvén's theorem shows to that order that, since $\mathbf{V}$ cannot cross the core surface, S, neither can $\mathbf{B}$. Thus, to the leading $R^{-1/2}$ order, $R^{-1/2}\mathbf{B}'$ (like the toroidal field $\mathbf{B}_\phi$) is zero in the exterior $\hat{V}$ of V, including Earth's surface. The field $R^{-1/2}\hat{\mathbf{B}}'$ is in fact $O(R^{-3/2})$, and so $R^{-1/2}\hat{\mathbf{B}}'/R^{-1}\hat{\mathbf{B}}_M = O(R^{-1/2}) \ll 1$, consistent with the slight tilt of the geomagnetic axis of Earth relative to its geographical axis.

Second, the expression (4.51) for α vanishes if, for any choice of origin for ϕ, the cosine or the sine terms are missing. In meteorological parlance, the wave (4.50) must be 'tilted'. In a subsequent notable advance, Soward (1972) — see also Soward (1990) — recovered Braginsky's results and interpreted the effective variables through a pseudo-Lagrangian treatment. In these terms, he was able to simplify (4.51) very considerably. He found, in the case of steady $\mathbf{V}$, that

$$\alpha = - \oint_C d\mathbf{x} \cdot \nabla \times \widetilde{\mathbf{V}} \Big/ \oint_C d\mathbf{x} \cdot \widetilde{\mathbf{V}}, \tag{1.4.52}$$

where $\widetilde{\mathbf{V}} = \mathbf{V} - \mathbf{V}_{eM}$. It is difficult to define the curve C in a few words. One notes first that, although the field lines of the flow $\overline{\mathbf{V}}_\phi + R^{-1/2}\mathbf{V}'$ do not in general close on themselves, it is possible to add to $\overline{\mathbf{V}}_\phi + R^{-1/2}\mathbf{V}'$ an $O(R^{-1})$ part (which is in fact the difference between $R^{-1}\overline{\mathbf{V}}_{eM}$ and $R^{-1}\overline{\mathbf{V}}_M$) such that the streamlines of the resulting flow (namely $\widetilde{\mathbf{V}}$) do close; these streamlines provide a family of curves, C. To find α for a particular latitude circle, one must determine the particular curve C whose average displacement from that latitude circle is zero. This is the C used in (4.52). Since $d\mathbf{x} = \mathbf{V}dt$, (4.52) displays an intimate connection between α and the helicity (4.4) of the motion. This will be considered further in section 1.5.2.

It is perhaps also surprising that, if we adopt the 'natural symmetry' (3.23) for $\mathbf{V}$ and seek dipole-type solutions for which (3.24) holds then, according to (4.52), α vanishes on the equatorial plane, as does B, so that $\overline{\mathcal{E}}_\phi = \alpha B$ vanishes quadratically. This is a consequence of the asymptotic approach, which actually holds in the next approximation also (Tough 1967). Fearn & Proctor (1987) observe however that this is not generally the case: for solutions of dipole parity, B vanishes on the equatorial plane but $\overline{\mathcal{E}}_\phi$ does not.

$\alpha\omega$-solutions; the dynamo wave. Braginsky (1964c) integrated a number of $\alpha\omega$-dynamo models numerically, and since that time so many further models have been studied that it would be absurd to undertake a review of the relevant literature here. As one may surmise from the form of (4.48, 49), the success or failure of an $\alpha\omega$-dynamo depends on the size of the *dynamo number*,

$$D = R_\alpha R_\omega, \tag{1.4.53}$$

where $R_\alpha = \tilde{\alpha}\mathcal{L}/\eta$ is the α-effect Reynolds number, based on a characteristic magnitude, $\tilde{\alpha}$, of (the now dimensioned) α; $R_\omega = \tilde{\zeta}\mathcal{L}^2/\eta$ is an ω-effect Reynolds number based on a characteristic magnitude, $\tilde{\zeta}$, of (the now dimensioned) ζ. Of interest is the critical value, D_c, at which dynamo action becomes marginally possible. This depends in general on the sign of D. When the sources α and ω are smoothly distributed over V, the dynamo tends to be of AC type and, starting with Steenbeck & Krause (1969), spherical models have been constructed in which magnetic activity moves continually from poles to the equator in a way reminiscent of the solar cycle. This applies when $\mathbf{V}_M$ is small. A sufficiently large meridional flow can turn an AC-dynamo into a DC-dynamo, and can even allow it to operate more efficiently, i.e., at a smaller value of D (Roberts 1972, 1988, 1989).

The nature of the oscillatory $\alpha\omega$-dynamo was clarified early by Parker (1955), using a very simple planar model in which (x, y, z) corresponded to (r, θ, ϕ). Thus the shear responsible for the ω-effect became $V_z 1_z$, which, acting on the meridional field, now $B_x 1_x + B_y 1_y$, gave rise to a zonal field $B_z 1_z$. Simplifying by assuming that $B_y = 0$ and that α and $\omega = \partial V_z / \partial x$ are constants, the analogues of (4.48) and (4.49) became

$$\frac{\partial B_z}{\partial t} = \frac{\partial^2 B_z}{\partial y^2} + R_\omega B_x, \tag{1.4.54}$$

$$\frac{\partial B_x}{\partial t} = \frac{\partial^2 B_x}{\partial y^2} + R_\alpha \frac{\partial B_z}{\partial y}. \tag{1.4.55}$$

These admit solutions of the form

$$\mathbf{B} \propto \exp(iky + \lambda t), \tag{1.4.56}$$

provided that

$$(\lambda + k^2)^2 = ikD. \tag{1.4.57}$$

One root of (4.57) may have a positive real part:

$$\lambda = -k^2 + [1 + i\operatorname{sgn}(kD)](kD/2)^{1/2}. \tag{1.4.58}$$

This is marginal for wave number k if $D = D_c$, where

$$|D_c| = 2|k|^3. \tag{1.4.59}$$

This marginal state is clearly oscillatory, and the field (4.56) progresses in the y-direction as a regenerating *dynamo wave*, with a phase velocity

$$c_p = -|k|(\operatorname{sgn}D_c), \tag{1.4.60}$$

that depends on the sign of D. The way that R_α and R_ω contribute individually to the sign of D_c decides on the relative phase of B_x and B_z. We have

$$\frac{B_x}{B_z} = [1 + i\operatorname{sgn}(kD)]\frac{k^2}{R_\omega}, \tag{1.4.61}$$

so that

$$\arg(B_x/B_z) > 0, \quad \text{if } \operatorname{sgn}(kR_\alpha) > 0, \tag{1.4.62}$$

$$\arg(B_x/B_z) < 0, \quad \text{if } \operatorname{sgn}(kR_\alpha) < 0. \tag{1.4.63}$$

Parker (1957) provided a simple qualitative explanation of a dynamo wave, and suggested (Parker 1957) that the solar cycle is a dynamo wave, but one operating in spherical geometry. He argued that $R_\omega = (\mathcal{L}^3/\eta)(\partial\omega/\partial r)$ is probably negative in the Sun, and that R_α is positive in its northern hemisphere and negative in its southern hemisphere. Using the correspondence $(x, y, z) \sim (r, \theta, \phi)$ noted above, he deduced that magnetic activity on the Sun should progress from its poles to its equator, as is in fact observed. Steenbeck & Krause (1969), and many subsequent authors, performed integrations of spherical models that supported Parker's proposal, but integrations of the full MHD equations, and also results from helioseismology, later cast substantial doubt on the details of Parker's picture. It is now believed that $\partial\omega/\partial r$ is small and perhaps positive over much of the convection zone. The origin of magnetic activity, and the progression of solar activity, are currently thought to have their origins at the base of the convection zone, in a layer in which $\partial\omega/\partial r$ is very large. This layer, rather than the bulk of the convection zone, is seen as controlling the dynamo wave; see Berthomieu & Cribier (1990), Brandenberg & Tuominen (1991), Cox *et al.* (1991) and Rosner & Weiss (1992). Although we will not take this subject further here, the case of the solar dynamo well illustrates the dangers of kinematic theory: a bad guess of **V** can be seriously misleading. In the final analysis, there is no substitute for solving the full MHD equations.

Kumar & Roberts (1975) integrated three-dimensional models in which the motions became axisymmetric as a certain parameter tended to zero. This allowed them to compare explicitly their results in that limit with those of the corresponding two-dimensional model with α given by (4.51). They found the agreement very satisfactory.

Ruzmaikin *et al.* (1990) argue that, when D is large, only the regions where $\alpha\omega$ is greatest are important, and that the solution will die away with distance from those regions, i.e., the dynamo is, in the terminology of section 1.3.1, of type V. With this in mind they developed an asymptotic technique (MEGA; see section 1.3.1) to determine D_c and the form of the solution. They used MEGA to explain the magnetism of the Sun, Earth and galaxies. See also section 2.4 of Roberts & Soward (1992).

With galactic dynamos or accretion disks in mind, there have been a number of integrations of the $\alpha\omega$-equations in appropriately flattened domains, V (see Stepinski & Levy 1990 and also Soward 1992*a, b*, where earlier references are given) and even in toroidal V (e.g., Schmitt 1990)!

1.5 MEAN FIELD ELECTRODYNAMICS

1.5.1 Basic Ideas Applied in a Simple Case

Mean field thermodynamics. The basic ideas of mean field electrodynamics may be most easily appreciated in a simpler context than electromagnetic theory, namely that of determining the evolution of the mean temperature in a turbulent incompressible fluid. The governing equation is

$$\partial_t T = -\nabla \cdot (\mathbf{q} + T\mathbf{V}), \qquad \mathbf{q} = -\kappa \nabla T, \qquad (1.5.1, 2)$$

where T is (from now onwards) temperature and $\mathbf{q}$, the heat conduction vector, is given by Fourier's law; here κ is the thermal diffusivity. (Strictly $\mathbf{q}$ is the heat conduction vector divided by ρC_p, where ρ is density and C_p is the specific heat at constant pressure.)

Suppose that the turbulence is statistically 'homogeneous' and 'steady', i.e., that the statistical properties of $\mathbf{V}$ are invariant to translations in space and time, so that they are the same for all $\mathbf{x}$ and t. In the same spirit as kinematic dynamo theory, we suppose that these statistical properties are known. For every variable, we define a mean (over ensembles), e.g., $\overline{T}$ is the average of T over ensembles, and we can then derive the fluctuating remnant of that variable, e.g., $T' = T - \overline{T}$. From now on the notation of section 1.3.4, in which $\overline{T}$ is an average of T over ϕ, is abandoned. We have

$$T = \overline{T} + T', \qquad \mathbf{V} = \mathbf{V}', \qquad (1.5.3, 4)$$

the latter recognizing that $\mathbf{V}$ is isotropic, i.e., has a zero mean.

When we substitute (5.3, 4) into (5.1) and average, we obtain

$$\partial_t \overline{T} = -\nabla \cdot (\overline{\mathbf{q}} + \overline{\mathcal{Q}}), \qquad \partial_t T' = -\nabla \cdot (\mathbf{q}' + \overline{T}\mathbf{V}' + \mathcal{Q}'), \qquad (1.5.5, 6)$$

where

$$\mathcal{Q} = T'\mathbf{V}', \qquad \overline{\mathbf{q}} = -\kappa \nabla \overline{T}, \qquad \mathbf{q}' = -\kappa \nabla T'. \qquad (1.5.7, 8, 9)$$

First order smoothing. There are two routes we can follow: the analytic and the heuristic. In the former, we note first that (5.6) and (5.9) imply that T' and therefore $\overline{\mathcal{Q}}$ are linear functionals of $\overline{T}$ whose precise forms depend on the statistical properties of the turbulence:

$$T' = \mathcal{F}(\overline{T}), \qquad \overline{\mathcal{Q}} = \mathcal{H}(\overline{T}), \qquad (1.5.10, 11)$$

where (5.11) is obtained by substituting (5.10) into (5.7) and averaging. The actual evaluation of $\mathcal{F}$ and $\mathcal{H}$ is in general next to impossible. It can be accomplished however after imposing some closure approximation, the crudest being 'first order smoothing', sometimes also called the 'second order correlation approximation'. This supposes that

$$\text{either} \quad Pe \equiv ul/\kappa \ll 1, \quad \text{or} \quad St \equiv u\tau/l \ll 1, \quad (1.5.12, 13)$$

where u is the rms turbulent velocity; l is the correlation length and τ is the correlation time of the turbulence; Pe and St are the microscale Péclet and Strouhal numbers. Under these conditions, (5.7) shows that

$$\text{either} \quad \frac{|\boldsymbol{Q}'|}{|\mathbf{q}'|} = O(Pe) \ll 1, \quad \text{or} \quad \frac{|\nabla \cdot \boldsymbol{Q}'|}{|\partial_t T'|} = O(St) \ll 1, \quad (1.5.14, 15)$$

so that to a good approximation, we may set

$$\boldsymbol{Q}' = 0, \quad (1.5.16)$$

in (5.6) to obtain

$$\partial_t T' - \kappa \nabla^2 T' = -\mathbf{V}' \cdot \nabla \overline{T}. \quad (1.5.17)$$

We may regard the right-hand side of (5.17) as known, and solve it in the form

$$\begin{aligned} T'(\mathbf{x}, t) = \int T'(\mathbf{x}_1, 0) G(\mathbf{x} - \mathbf{x}_1, t)\, dV_1 \\ - \int_0^t dt_1 \int \mathbf{V}'(\mathbf{x}_1, t_1) \cdot \nabla_1 \overline{T}(\mathbf{x}_1, t_1) G(\mathbf{x} - \mathbf{x}_1, t - t_1)\, dV_1, \end{aligned} \quad (1.5.18)$$

where G denotes Green's function

$$G(\mathbf{x}, t) = \frac{1}{(4\pi\kappa t)^{3/2}} \exp\left(-\frac{\mathbf{x}^2}{4\kappa t}\right). \quad (1.5.19)$$

The form (5.18) is appropriate to an unbounded volume V, and should be a good approximation even for bounded V when $l \ll \mathcal{L}$, except for points situated within a distance l of the surface S of V. When we substitute (5.18), which is the required functional (5.10), into (5.7) and average, we obtain the required functional (5.11):

$$\begin{aligned} \overline{Q}_i(\mathbf{x}, t) = \int \overline{T'(\mathbf{x}_1, 0) V_i'(\mathbf{x}, t)}\, G(\mathbf{x} - \mathbf{x}_1, t)\, dV_1 \\ - \nabla_j \int_0^t dt_1 \int \overline{T}(\mathbf{x} - \mathbf{x}_1, t - t_1) Q_{ij}(\mathbf{x}_1, t_1) G(\mathbf{x}_1, t_1)\, dV_1, \end{aligned} \quad (1.5.20)$$

where

$$Q_{ij}(\mathbf{x}_1, t_1) = \overline{V_i'(\mathbf{x}, t) V_j'(\mathbf{x} + \mathbf{x}_1, t + t_1)} \quad (1.5.21)$$

is the two-point two-time velocity correlation, which is independent of $\mathbf{x}$ and t because of the assumed homogeneity and steadiness of the turbulence. We may note for future reference that, by definition,

$$Q_{ij}(\mathbf{x}_1, t_1) = Q_{ji}(-\mathbf{x}_1, -t_1), \quad (1.5.22)$$

but that Q_{ij} is not, in general, symmetric. In fact, we shall later find it convenient to divide Q_{ij} into symmetric (S) and antisymmetric (A) parts:

$$Q_{ij} = Q_{ij}^{\mathrm{S}} + Q_{ij}^{\mathrm{A}}, \qquad (1.5.23)$$

where

$$Q_{ij}^{\mathrm{S}}(-\mathbf{x}_1, -t_1) = Q_{ij}^{\mathrm{S}}(\mathbf{x}_1, t_1) = Q_{ji}^{\mathrm{S}}(\mathbf{x}_1, t_1), \qquad (1.5.24)$$

$$Q_{ij}^{\mathrm{A}}(-\mathbf{x}_1, -t_1) = -Q_{ij}^{\mathrm{A}}(\mathbf{x}_1, t_1) = Q_{ji}^{\mathrm{A}}(\mathbf{x}_1, t_1). \qquad (1.5.25)$$

Because T' and $\mathbf{V}'$ should be uncorrelated for $t \gg \tau$, we may ignore the first integral on the right-hand side of (5.20). Since $Q_{ij}(\mathbf{x}_1, t_1)$ is negligibly small if $|\mathbf{x}_1| > l$ or if $t_1 > \tau$, the arguments of $\overline{T}$ in the second integral may, in the first approximation, be replaced by $\mathbf{x}$ and t; for the same reason we may replace the lower limit in the second integral by $-\infty$. We may therefore replace (5.20) by

$$\overline{Q}_i = -\kappa_{ij}^{\mathrm{turb}} \nabla_j \overline{T}, \qquad (1.5.26)$$

where

$$\kappa_{ij}^{\mathrm{turb}} = \int_0^\infty dt_1 \int Q_{ij}^{\mathrm{S}}(\mathbf{x}_1, t_1) G(\mathbf{x}_1, t_1) \, dV_1 \qquad (1.5.27)$$

is a 'turbulent thermal diffusivity tensor'. It may be verified that Q_{ij}^{A} makes no contribution to (5.5), so we have omitted it from (5.27). Suppose now that the turbulence is 'isotropic', i.e., that its statistical properties are unchanged under rotation and reflection of axes. Then

$$\kappa_{ij}^{\mathrm{turb}} = \kappa^{\mathrm{turb}} \delta_{ij}, \qquad \kappa^{\mathrm{turb}} = \frac{1}{3} \kappa_{kk}^{\mathrm{turb}}, \qquad (1.5.28, 29)$$

and (5.26) becomes

$$\overline{Q} = -\kappa^{\mathrm{turb}} \nabla \overline{T}. \qquad (1.5.30)$$

This has exactly the same form as (5.8): the effect of the turbulence is exactly the same as molecular diffusion. We may write (5.5) simply as

$$\partial_t \overline{T} = -\nabla \cdot \overline{\mathbf{q}}^{\mathrm{T}}, \qquad \overline{\mathbf{q}}^{\mathrm{T}} = -\kappa^{\mathrm{T}} \nabla \overline{T}, \qquad (1.5.31, 32)$$

where κ^{T} is the total thermal diffusivity:

$$\kappa^{\mathrm{T}} = \kappa + \kappa^{\mathrm{turb}}. \qquad (1.5.33)$$

In the limit $Pe \to 0$ with St held fixed, (5.27) becomes

$$\kappa_{ij}^{\mathrm{turb}} \approx \int_0^\infty dt_1 \int Q_{ij}^{\mathrm{S}}(\mathbf{x}_1, 0) G(\mathbf{x}_1, t_1) \, dV_1 = \frac{1}{4\pi\kappa} \int \frac{Q_{ij}^{\mathrm{S}}(\mathbf{x}_1, 0)}{|\mathbf{x}_1|} \, dV_1 = \mathrm{O}(Pe^2)\kappa. \qquad (1.5.34)$$

In the limit $St \to 0$ with Pe held fixed, (5.27) becomes

$$\kappa_{ij}^{\text{turb}} \approx \int_0^\infty dt_1 \int Q_{ij}^S(0, t_1) G(\mathbf{x}_1, t_1) \, dV_1 = \int_0^\infty Q_{ij}^S(0, t_1) \, dt_1 = \mathrm{O}(St \, Pe)\kappa.$$

$$(1.5.35)$$

In either limit, we see from (5.12, 13), that $|\kappa_{ij}| \ll \kappa$, and turbulence has almost no effect on the diffusion of $\overline{T}$. This is not true in the case $St \to 0$ with $St \, Pe \gg 1$ held fixed, in which (5.35) holds and $\kappa^{\text{turb}} \gg \kappa$. But this also is not very relevant, since for naturally-occurring turbulence $St = \mathrm{O}(1)$ and Pe is often large compared with 1. In these circumstances (5.27) is untenable.

Reynolds-type argument. Turning now to the heuristic argument, Reynolds was struck by the analogy between molecular transport and turbulent transport. Fourier's law (5.2) follows from a very well-known argument from the kinetic theory of gases. One pictures molecules moving randomly and a fictional surface perpendicular to ∇T. The molecules crossing this surface from the colder to the hotter side carry on average more kinetic energy than the ones crossing in the reverse direction. The flux of heat is roughly $\frac{1}{3}\rho C_p \mathbf{v}^{\text{mol}} l^{\text{mol}} \cdot \nabla T$, i.e., $\mathbf{q}$ obeys (5.2) with $\kappa = \frac{1}{3}v^{\text{mol}} l^{\text{mol}}$. (Here v^{mol} and l^{mol} are the rms velocity and free path of the molecules.) Reynolds' idea was essentially that randomly moving eddies of a turbulent flow carry heat in much the same way as do the randomly moving molecules, so that by analogy $\kappa^{\text{turb}} \approx \frac{1}{3}ul$, which (on taking $\tau = l/u$) essentially agrees with (5.35). It should be emphasized that the fact that the first order smoothing approximation has yielded an expression for $\overline{\mathbf{Q}}$ proportional to $-\nabla \overline{T}$ in no way makes that approximation tenable outside its region (5.12, 13) of validity, and does not provide strong support for Reynolds's argument outside that region; of course neither (5.34) nor (5.35) can then be trusted.

1.5.2 Turbulent Diffusion of Field

Heuristic approach; first order smoothing. The first discussions of induction in a turbulent conductor followed along much the same lines as that in section 1.5.1 above, but it was later realised that the diffusion of a vector quantity like $\mathbf{B}$ is a different, and more complicated, phenomenon than the diffusion of a scalar field like T. Especially significant is the fact that magnetic field lines can be stretched by turbulent motion and so cause the magnetic field energy to increase. Thus, while it is true that turbulence introduces a diffusivity, η^{turb}, of mean field that may greatly exceed η, this by no means tells the whole story.

With the same re-interpretation of the average denoted by the overbar, (3.43, 44, 51–53) remain valid. We may again use the analytic and heuristic arguments approaches of section 1.5.1. Analytic progress is again very difficult unless we again employ first order smoothing, which here requires that

$$\text{either} \quad Rm \equiv ul/\eta \ll 1, \quad \text{or} \quad St \equiv u\tau/l \ll 1, \quad (1.5.36, 37)$$

where Rm is the microscale magnetic Reynolds number. Under conditions (5.36) and (5.37) we have, in a similar manner to section 1.5.1,

$$\mathcal{E}' = 0, \qquad (1.5.38)$$

to a good approximation, and (3.52) becomes (in dimensioned units)

$$\partial_t \mathbf{B}' - \nabla \times (\mathbf{V} \times \mathbf{B}') - \eta \nabla^2 \mathbf{B}' = \nabla \times (\mathbf{V}' \times \overline{\mathbf{B}}), \qquad (1.5.39)$$

where again the right-hand side may be considered as known.[5]

For simplicity, consider first the case in which $\mathbf{V} = 0$ and the turbulence is statistically homogeneous and steady. In analogy with (5.20) we find that

$$\overline{\mathcal{E}}_i = \epsilon_{ilp}\epsilon_{pnq}\epsilon_{qmj} \int_0^\infty dt_1 \int \overline{B}_j(\mathbf{x} - \mathbf{x}_1, t - t_1) Q_{lm}(\mathbf{x}_1, t_1) \nabla_{1n} G(\mathbf{x}_1, t_1) dV_1,$$
$$(1.5.40)$$

where Green's function is now redefined in the obvious way: η replaces κ in (5.19). As before, the integrand in (5.40) is essentially zero unless $|\mathbf{x} - \mathbf{x}_1|$ and $t - t_1$ are small, and we may expand $\overline{B}_j(\mathbf{x} - \mathbf{x}_1, t - t_1)$ in Taylor series to obtain

$$\overline{\mathcal{E}}_i = \alpha_{ij}\overline{B}_j + \beta_{ijk}\nabla_k\overline{B}_j + \cdots, \qquad (1.5.41)$$

where

$$\alpha_{ij} = -\epsilon_{ilp}\epsilon_{pnq}\epsilon_{qmj} \int_0^\infty dt_1 \int [\nabla_{1n}Q_{lm}^A(\mathbf{x}_1, t_1)] G(\mathbf{x}_1, t_1)\, dV_1, \qquad (1.5.42)$$

$$\beta_{ijk} = \epsilon_{ilp}\epsilon_{pkq}\epsilon_{qmj} \int_0^\infty dt_1 \int Q_{lm}^S(\mathbf{x}_1, t_1) G(\mathbf{x}_1, t_1)\, dV_1. \qquad (1.5.43)$$

The S and A parts of Q_{ij} do not contribute to (5.42) and (5.43) respectively. We have used that fact (and also that $\nabla_{1l}Q_{lm}(\mathbf{x}_1, t_1) = 0$ from the incompressibility of the fluid) to simplify (5.43).

Consider pseudo-isotropic turbulence, that is, turbulence whose statistical properties are unchanged on rotation (but not reflection) of axes.[6] In this case (5.40–43) simplify considerably. From the only isotropic tensors available, we have

$$\alpha_{ij} = \alpha\delta_{ij}, \qquad \beta_{ijk} = \beta\epsilon_{ijk}, \qquad (1.5.44, 45)$$

where

$$\alpha = \frac{1}{3}\alpha_{kk} = -\frac{1}{3}\int_0^\infty \int [\nabla_{1l}\epsilon_{lmn}Q_{mn}^A(\mathbf{x}_1, t_1)]G(\mathbf{x}_1, t_1)dV_1, \qquad (1.5.46)$$

---

[5] The possibility that $\mathbf{B}'$ could maintain itself indefinitely through turbulent induction, without the help of $\overline{\mathbf{B}}$, is not considered here, i.e., we assume that $\overline{\mathbf{B}}$ is not identically zero.

[6] The nature and difference between statistical isotropy and pseudo-isotropy may be illustrated by the example of a box of screws pointing in random directions. The direction of the thread on the screws is reversed when viewed in a mirror, showing that it is a pseudo-isotropic system. If half the screws had a left-handed thread, the system would be statistically unaffected by reflection in the mirror, and both systems would be isotropic.

$$\beta = \frac{1}{6}\epsilon_{ijk}\beta_{ijk} = \frac{1}{3}\int_0^\infty \int Q^S_{kk}(\mathbf{x}_1, t_1)G(\mathbf{x}_1, t_1)dV_1. \tag{1.5.47}$$

The successive terms in expansion (5.41) decrease in the ratio $l/\mathcal{L} \ll 1$, and there is little point in retaining terms beyond the ones shown explicitly. By (5.44, 45) we then have

$$\overline{\mathcal{E}} = \alpha\overline{\mathbf{B}} - \beta\nabla \times \overline{\mathbf{B}}. \tag{1.5.48}$$

The second of these terms has exactly the form expected from Ohm's law, and this shows that this represents a turbulent enhancement of mean field with β as turbulent diffusivity, η^{turb}. In fact, we may now rewrite (3.51) in dimensioned form as

$$\partial_t\overline{\mathbf{B}} = \nabla \times (\alpha\overline{\mathbf{B}}) + \eta^T\nabla^2\overline{\mathbf{B}}, \tag{1.5.49}$$

where η^T is the total magnetic diffusivity:

$$\eta^T = \eta + \beta. \tag{1.5.50}$$

It may be shown (Krause & Roberts 1973) that (5.47) yields a positive β. The form of (5.49) is closely related to that obtained in section 1.4.3, despite the very different nature of the averaging process. The principal differences are first that, because of the dominance of $\overline{B}_\phi$ in Braginsky's expansion, the mean emf is, in essence, nonisotropic: $\overline{\mathcal{E}} = \alpha\overline{B}_\phi\mathbf{1}_\phi$. Second, the poloidal field and velocity are not the actual ones; they are 'effective'. Third, there is no enhancement of η to leading order. In the present case of turbulent induction, β is so large in many astrophysical contexts that the molecular diffusivity, η, is ignored; see below.

It is the first terms on the right-hand sides of (5.48) and (5.49) that are totally unfamiliar in classical electromagnetic theory. The form of these terms closely parallels those found in the Braginsky theory (section 1.4.3), and the term α-effect is used here also.

Parity. The coefficient, α, is, like the helicity (4.4), a pseudo-scalar. We recall here that rotations and reflections of axes are concerned with transformations

$$x'_i = a_{ij}x_j, \tag{1.5.51}$$

where the matrix of direction cosines a_{ij} is orthogonal: $a_{ik}a_{jk} = \delta_{ij}$, so that $\det \mathbf{a} = \pm 1$. A rotation of axes involves a proper orthogonal matrix: $\det \mathbf{a} = +1$; for a reflection $\det \mathbf{a} = -1$. When both rotations and reflections are envisaged, two kinds of tensors appear, depending on their 'parity':

$$T'_{i'j'\ldots k'} = a_{i'i}a_{j'j}\ldots a_{k'k}T_{ij\ldots k}, \qquad P'_{i'j'\ldots k'} = (\det \mathbf{a})a_{i'i}a_{j'j}\ldots a_{k'k}P_{ij\ldots k}, \tag{1.5.52, 53}$$

applying respectively to the 'true' tensors and the 'pseudo-tensors'. It is easily verified that δ_{ij} is a true tensor but, from the definition of $\det \mathbf{a}$, ϵ_{ijk} is pseudo-tensor.

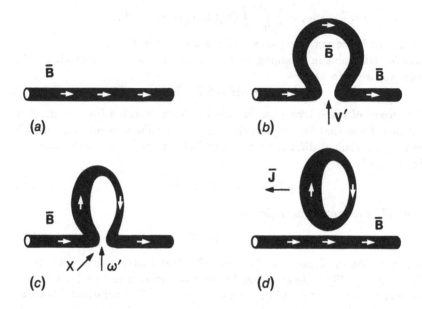

Figure 1.4. Illustration of the α-effect.

The situation may be clarified by considering the reflection $\mathbf{x}' = -\mathbf{x}$ considered in section 1.3.2. for which $a_{ij} = -\delta_{ij}$. Since $\mathbf{x}$ is a true (or 'polar') vector, the same must be true of ∇ and $\nabla\cdot$ and of the rate of change of the positions of mass and charge, i.e., $\mathbf{V}$ and $\mathbf{J}$ must also be true vectors. Because ϵ_{ijk} is a pseudo-tensor, so are the $\times$ and $\nabla\times$ operators. It follows that $\mathbf{B}$ is a pseudo- (or 'skew') vector. Therefore α_{ij} and β_{ijk} are skew tensors, from which it follows from (5.44, 45) that α is pseudo-scalar and β is a true scalar.

The helicity H and α are both pseudo-scalars. They tend to have opposite signs. The reason for this is illustrated in figure 1.4, where the action of a helical eddy on a straight line of force (figure 1.4a) of $\overline{\mathbf{B}}$ is sketched; Alfvén's theorem (section 1.2.3) helps the process to be visualised. The $\mathbf{V}'$ motion of the eddy bends the line into a giant Ω (figure 1.4b). The vorticity, ω', of the eddy gives that Ω a twist that brings it out of the plane of the paper, say perpendicular to it (figure 1.4c). Diffusion proceeds more rapidly than elsewhere at the base of the twisted Ω, indicated by X on figure 1.4(c), where the field gradients are large. Deviations from Alfvén's theorem occur there, and the field lines of the Ω detach from the parent line, to create a circular loop of field (figure 1.4d) Many of these taken together are equivalent to a mean current $\overline{\mathbf{J}}$ antiparallel to $\overline{\mathbf{B}}$. This corresponding α is therefore negative, a consequence of assuming a positively correlated $\mathbf{V}'$ and ω', i.e., a positive H. A negative H would have resulted in a positive α.

There is clearly a close connection between the process shown in figure 1.4 and the scenario of example 4 of section 1.2.3; see figure 1.3. In the former the loop turns by 90° so that it is perpendicular to the plane of the paper, whereas in figure 1.3 it is turned through 180° so that it lies in the plane of the paper. This illustrates the unsatisfactory nature of arguments such as the one just presented. Why should the loop only make a quarter turn? Why not let it turn through 270° so that it produces a positive α-effect? In reality, when approximations such as (5.36, 37) hold, the loop only turns through a small angle, and the α-effect created is therefore also small.[7]

It is easy to give expressions analogous to (5.34, 35) for α_{ij} and β_{ijk} in the low conductivity limit (5.36) or in the short correlation time limit (5.37), but generally $St = O(1)$ in naturally-occurring systems, and often Rm is large also. Heuristic arguments then give results such as

$$\alpha \simeq -\frac{1}{3} H\tau, \qquad \beta \simeq \frac{1}{3} u^2 \tau, \qquad (1.5.54, 55)$$

which are completely independent of η. One might expect however, that the turbulence would, by continually stretching, twisting and folding an initially smooth field, eventually create structures of such a short length scale that diffusion would be significant. One might therefore anticipate that, at that stage, α and β would depend on η. One motivation for the study of fast dynamos is to understand whether, and when, results such as (5.54, 55) are tenable. The above qualitative arguments suggest that they might be some-times valid. We there supposed, for clarity, that the $\mathbf{V}'$ and ω' operated in sequence; in reality steps (b) and (c) in figure 1.4 occur simultaneously. And, although diffusion at X is limited in speed by η, new loops can start forming during the detachment of the one shown. The creation of new flux may therefore not be significantly impeded by diffusion of flux through the conductor, i.e., the process may be 'fast'. The same may be true of example 4 of section 1.2.3.

The connection between H and α is not generally as close as (5.54) might suggest, and it is possible to conceive of turbulent statistics such that H is zero and α is not, or conversely (Gailitis 1974). Although dynamo models have been constructed in which $\alpha \equiv 0$ (Gailitis 1970, Gailitis & Freiberg 1974), a more typical situation is that $\alpha > 0$ in one part of the system, and $\alpha < 0$ in the remainder, with perhaps the overall integral of α zero; e.g., see (5.67) below for a spherical system.

---

[7] This explains why, unlike the corresponding step to (5.26) in section 1.5.1, two terms are retained in (5.41) despite the fact that, formally speaking, the first term is apparently $O(\mathcal{L}/l) \gg 1$ times larger than the first. In naturally-occurring systems, turbulent motions are usually far from being maximally helical in the sense defined above (4.7), and the first term exceeds the second by $O((H/u)(\mathcal{L}/l))$, which may, or may not, be large.

Finally, we should note that, in truly isotropic turbulence, in which all statistical properties are unchanged by coordinate reflection, all pseudo-scalars are zero, including α and H.

1.5.3 More General Situations

Creation of helical turbulence. Although, in the spirit of kinematic theory, we have focussed on pseudo-isotropy and its magnetic consequence (the α-effect), it is not at all clear how lack of mirror-symmetry can, in practice, survive the turbulent cascade. The very smallest scales of motion are oblivious to many effects that drastically affect the larger scales of motion. For example, when the eddy turnover time, τ, is short compared with the rotation period, $2\pi/\Omega$, the microscale motions in a system rotating with the angular velocity Ω will not be much affected by Coriolis forces. Similarly they are little affected by stratification when, as is often the case, $\tau \ll N^{-1}$, where N is the Brunt frequency. Turbulence on these scales is very close to isotropy, the inhomogeneities and the anisotropies of the largest scales being randomised away in the turbulent cascade.[8] The deviations from isotropy are, nevertheless, significant and may, in some circumstances create an α-effect. To examine this question, we can play a game, the rules of which have been adumbrated in section 1.5.2. We shall ignore anisotropies in the turbulence created by the mean flow, $\overline{\mathbf{V}}$, apart from those associated with the Coriolis force. In this subsection, we are essentially touching on dynamical questions whose answers influence the choice of kinematic dynamo models.

Effect of stratification. In a zeroth approximation, planetary cores and stellar convection zones have spherical symmetry. We expect therefore that the statistical properties of the turbulence will be neither homogeneous nor isotropic, but will depend on $\mathbf{x}$. In particular, its statistical properties at $\mathbf{x}$ will have a preferred direction, namely $\mathbf{1}_r$, the radial polar vector through that point. In the zeroth approximation, the coefficients α_{ij} and β_{ijk} in the general expression (5.41) for $\overline{\mathcal{E}}$ are those appropriate to the dominating isotropic part of the turbulence:

$$\alpha_{ij}^{(0)} = 0, \qquad \beta_{ijk} = \beta^{(0)}\epsilon_{ijk}. \qquad (1.5.56, 57)$$

Through the slight radial anisotropy, these results are modified to

$$\alpha_{ij} = \alpha_{ij}^{(0)} + \alpha_{ij}^{(1)}, \qquad \beta_{ijk} = \beta_{ijk}^{(0)} + \beta_{ijk}^{(1)}. \qquad (1.5.58, 59)$$

We now try to construct the most general pseudo-tensors, $\alpha_{ij}^{(1)}$ and $\beta_{ijk}^{(1)}$ from $\mathbf{x}, \delta_{ij}$ and ϵ_{ijk}, and which vanish when stratification is unimportant, i.e., when

---

[8] According to Braginsky & Meytlis (1990), we should make an exception of Earth's core, where the turbulence at small scales is fed directly by convective instabilities of the same spatial scale and not by cascade.

x is set zero. It is found that these are

$$\alpha_{ij}^{(1)} = \alpha_1 \epsilon_{ijk} x_k + \alpha_2 x_i x_j, \tag{1.5.60}$$

$$\beta_{ijk}^{(1)} = \beta_1 x_i \delta_{jk} + \beta_2 x_j \delta_{ki} + \beta_3 x_k \delta_{ij} + \beta_4 x_i x_l \epsilon_{ljk} + \beta_5 x_j x_l \epsilon_{lki} + \beta_6 x_k x_l \epsilon_{lij} + \beta_7 x_i x_j x_k, \tag{1.5.61}$$

where, as for β_0 in (5.57), the coefficients may depend on r. The coefficients $\alpha_1, \beta_4, \beta_5$ and β_6 are true scalars while $\alpha_2, \beta_1, \beta_2, \beta_3$ and β_7 are pseudo-scalars. The latter depend on the statistical properties of the dominating isotropic turbulence and are therefore zero. Thus (5.60, 61) reduce to

$$\alpha_{ij}^{(1)} = \alpha_1 \epsilon_{ijk} x_k, \qquad \beta_{ijk}^{(1)} = \beta_4 x_i x_l \epsilon_{ljk} + \beta_5 x_j x_l \epsilon_{lki} + \beta_6 x_k x_l \epsilon_{lij}. \tag{1.5.62, 63}$$

Slightly redefining α_1, we obtain from (5.41, 62, 63),

$$\mathcal{E} = -\alpha_1 x \times \overline{B} - \beta_0 \nabla \times \overline{B} - \beta_4 (x \cdot \nabla \times \overline{B})x + \beta_5 x \times \nabla(x \cdot \overline{B}) - \beta_6 x \times (x \cdot \nabla)\overline{B}. \tag{1.5.64}$$

On inspecting the undifferentiated $\overline{B}$ term in (5.64), we see that we do not have an α-effect: it may be interpreted as replacing the real mean velocity by an effective mean velocity, $\overline{V}_e = \overline{V} - \alpha_1 x$.

Effect of rotation. The story is very similar if we suppose that stratification is unimportant but rotation is. The analysis follows that leading to (5.64), with the essential difference that Ω, unlike the x it replaces, is a skew vector. As a result, α_1 and α_2 are zero, and we obtain instead of (5.64)

$$\mathcal{E} = -\beta_0 \nabla \times \overline{B} + (\beta_2 + \beta_3)\nabla(\Omega \cdot \overline{B}) - \mu_0 \beta_3 \Omega \times \overline{J}. \tag{1.5.65}$$

(We have here specialized to the case $\Omega\tau \ll 1$ of slow rotation, in which terms quadratic and cubic in Ω may be neglected.) The last term in (5.65) can in principle drive a dynamo, but in practice it is less significant than the dominating β_0 term. The crucial fact, evident from (5.65), is that rotation does not give rise to an α-effect.

Suppose next that the stratification and rotation act together. To simplify the discussion, let us suppose that both effects are so small that we need only retain terms linear in x and linear in Ω. We find that (with newly defined α_1 and α_2)

$$\mathcal{E} = -\beta_0 \nabla \times \overline{B} + \alpha_1 (\Omega \cdot x)\overline{B} + \alpha_2 x(\Omega \cdot \overline{B}) + \alpha_3 \Omega(x \cdot \overline{B}), \tag{1.5.66}$$

where we have not included the small terms of (5.64, 65). At last we have obtained a genuine α-effect, although it is not pure α-effect because of the last two, non-isotropic terms on the right-hand side of (5.66). In fact, there is no known naturally-occurring turbulence that produces a pure α-effect. The last two terms on the right-hand side of (5.66) are apparently as large as the first. For example, a simplified model of Steenbeck *et al.* (1966) gave $\alpha_2 = \alpha_3 = -\alpha_1/4$. Fortunately the α_2 and α_3 contributions to (5.66) do not damage the ability of the α_1 term to regenerate field, although they modify the results of numerical integrations performed in their absence. It is the general tendency in investigations of mean field dynamo models to ignore all but the pseudo-isotropic α- and β-terms in the hope/belief that the omission of the remaining, non-isotropic terms will not be damaging or misleading; this has recently been contested in the case of galactic dynamos (Rüdiger 1990).

It may be seen from (5.66) that α has the form

$$\alpha = \alpha_1(r)\cos\theta, \qquad (1.5.67)$$

so that, in terms of the symmetries considered in section 1.3.2,

$$\alpha(-\mathbf{x}) = -\alpha(\mathbf{x}), \qquad \alpha(\mathbf{x}_R) = -\alpha(\mathbf{x}). \qquad (1.5.68, 69)$$

Equations (5.68, 69) supplement assumptions such as (3.23) about the symmetry of $\mathbf{V}$, in such a way that statements (5.24–27) remain applicable to $\overline{\mathbf{B}}$: purely dipole and purely quadrupole solutions exist to the mean field equations, but such pure solutions may be unstable; mixed-mode solutions may be preferred; see for example Rädler & Wiedemann (1989) and Glatzmaier & Roberts (1993). In this context, it may be noted that the solar magnetic field has an appreciable quadrupolar component even though the two hemispheres of the Sun are surely structurally identical.

1.5.4 Mean Field Models; Equilibration

The majority of studies of mean field dynamo models have assumed pure α- and β-effects. On restoring $\mathbf{V}$, the resulting induction equation is (see (5.49))

$$\partial_t \mathbf{B} = \nabla \times (\alpha\mathbf{B} + \mathbf{V} \times \mathbf{B}) + \eta\nabla^2\mathbf{B}, \qquad (1.5.70)$$

where we have omitted the ^{T} from η^T and have temporarily discarded the overbars on mean field quantities. If, as before, we seek axisymmetric solutions, we may write $\mathbf{B}$ and $\mathbf{V}$ in the forms (3.45–50), and obtain from (5.70) the two equations

$$\frac{1}{s}\frac{D_M}{Dt}(sA) = \Delta A + R_\alpha \alpha B, \qquad (1.5.71)$$

$$s\frac{D_M}{Dt}\left(\frac{B}{s}\right) = \Delta B - R_\alpha\left[\alpha\Delta A + \frac{1}{s}\nabla\alpha\cdot\nabla(sA)\right] + R_\omega s\mathbf{B}_M\cdot\nabla\zeta, \quad (1.5.72)$$

where D_M/Dt denotes the motional derivative following $\mathbf{V}_M$:

$$\frac{D_M}{Dt} = \partial_t + \mathbf{V}_M\cdot\nabla. \qquad (1.5.73)$$

We have again reverted to dimensioned quantities.

Equations (5.71, 72) define what is sometimes called an $\alpha^2\omega$-dynamo, the two extreme members of which are:

(a) *the $\alpha\omega$-dynamo.* This case, which arises when $R_\alpha \ll R_\omega$ was introduced in section 1.4.3 and will be considered no further here, except to remark that, since only A can be inferred from observations outside a naturally-occurring dynamo (B remaining unseen within it — see (3.57)), the total field strength in V greatly exceeds that of $\mathbf{B}_M$ and, for this reason, these solutions are often called 'strong field dynamos';

(b) *the α^2-dynamo.* This case, defining what are often called 'weak field dynamos', arises when $R_\alpha \gg R_\omega$. It is governed by (5.71) and by (5.72) with the last term on its right-hand side omitted:

$$s\frac{D_M}{Dt}\left(\frac{B}{s}\right) = \Delta B - R_\alpha \left[\alpha\Delta A + \frac{1}{s}\nabla\alpha\cdot\nabla(sA)\right]. \qquad (1.5.74)$$

As before, (3.56–3.61) must also be applied.

Assuming $\mathbf{V}$ and α are independent of t, these equations pose an eigenvalue problem for the growth rate λ of field very similar to that defined in section 1.3.2. The marginal α^2-dynamos are characterised by

$$R_{\alpha c} = O(1), \qquad \mathbf{B}_M = O(\mathbf{B}_\phi). \qquad (1.5.75, 76)$$

As in section 1.3.2, the eigenvalue problem is generally non-self-adjoint, but the dynamos are nevertheless generally either DC, or they are AC with a very small frequency $\Im(\lambda)$ (Shukurov, Sokoloff & Ruzmaikin 1985). The MEGA method described in section 1.4.3 has been applied to α^2-dynamos (see Sokoloff, Shukurov & Ruzmaikin 1983). Many α^2-dynamos have been integrated numerically, for they provide a basis for understanding dynamos that are near the threshold of excitation, $R = R_c$.

Figure 1.5 shows an amusing α^2-model: the double doughnut dynamo or 'DDD' (Steenbeck *et al.* 1966). The two hollow circular tori are filled with turbulence of negative helicity. the planes of the two tori are perpendicular, each passing through the centre of the other. The mean field following round each loop is associated with a parallel mean current (α is positive) that, through Ampère's law, produces a field in the correct direction in the other loop. Thus, if R_α is sufficiently large, the device will sustain field indefinitely.

Finally, another important dynamical modification of α_{ij} and β_{ijk} should be touched on: *quenching*, also known as *saturation*, i.e., the effect of the mean magnetic field itself on these tensors through the Lorentz force. This force may be expected to reduce the vigour of the turbulence, and hence diminish $|\alpha|$ and β. Presumably $|\alpha|$ will be reduced more than β, since one intuitively expects that their ratio (which is proportional to R_α when $\beta \gg \eta$)

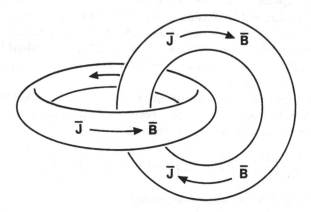

Figure 1.5. The double doughnut dynamo.

should decrease with increasing $|\overline{\mathbf{B}}|$, and prevent the α^2-models from runaway field growth when R_α exceeds $R_{\alpha c}$, and so allow the dynamo to equilibrate at finite field strength.[9] It is easy to picture α-quenching, in terms of the process illustrated in figure 1.4: as $\overline{\mathbf{B}}$ intensifies, the helical motions have greater difficulty in creating and twisting the giant Ω. This also leads to the idea that $|\alpha|$ will be reduced by a term proportional to $\overline{\mathbf{B}} \cdot \nabla \times \overline{\mathbf{B}}$ (Roberts 1971). Other arguments suggest that $O(\overline{B}^2)$ reductions, which are $O(\mathcal{L}/l)$ times greater, will exist and be more significant. The $\alpha\omega$-dynamo is subject to both α-quenching and ω-quenching, the latter being a reduction in $|\nabla\zeta|$ arising from the mean magnetic stresses of the fluctuating field, $\overline{B_i'B_j'}$, which is a quadratic functional of $\overline{\mathbf{B}}$. (In principle, there is also a quenching of $\overline{\mathbf{V}}_M$, and all components of $\overline{\mathbf{V}}$ are in addition acted upon by the Lorentz force, $\overline{\mathbf{J}} \times \overline{\mathbf{B}}$, of the mean field.) Quenching, which is properly a topic of 'mean field MHD', is strictly beyond the scope of this chapter.

---

[9] It may be noted, however, that such runaway growth is in fact observed in the weak field regime of the Childress–Soward dynamo, in which it operates in a roughly α^2-way; see Soward (1974).

REFERENCES

Alfvén, H. 1950 Discussion of the origin of the terrestrial and solar magnetic fields. *Tellus* **2**, 74–82.

Backus, G. 1958 A class of self sustaining dissipative spherical dynamos. *Ann. Phys. N.Y.* **4**, 372–447.

Beck, R., Kronberg, P.P. & Wielebinski, R. (eds.) 1990 *Galactic and Intergalactic Magnetic Fields.* IAU Symposium 140. Kluwer, Dordrecht.

Berthomieu, G. & Cribier, M. (eds.) 1990 *Inside the Sun.* Kluwer, Dordrecht.

Bevir, M.K. 1973 Possibility of electromagnetic self-excitation in liquid metal flows in fast reactors. *J. Brit. Nucl. Engy Soc.* **12**, 455–458.

Bisnovatyi-Kogan, G.S., Ruzmaikin, A.A. & Sunyaev, R.A. 1973 Star contraction and magnetic field generation in protogalaxies. *Sov. Astron.* **17**, 137–139.

Braginsky, S.I. 1964a Self-excitation of a magnetic field during the motion of a highly conducting fluid. *Zh. Exp. Teor. Fiz. SSSR* **47**, 1084–1098. (English transl.: *Sov. Phys. JETP* **20**, 726–735 (1965).)

Braginsky, S.I. 1964b Theory of the hydromagnetic dynamo. *Zh. Exp. Teor. Fiz. SSSR* **47**, 2178–2193. (English transl.: *Sov. Phys. JETP* **20**, 1462–1471 (1965).)

Braginsky, S.I. 1964c Kinematic models of the Earth's hydromagnetic dynamo. *Geomag. Aeron.* **4**, 732–747. (English transl.: **64**, 572–583 (1964).)

Braginsky, S.I. 1978 Nearly axially symmetric model of the hydromagnetic dynamo of the Earth. *Geomag. Aeron.* **18**, 340–351. (English transl.: **78**, 225–231 (1978).)

Braginsky, S.I. 1991 Towards a realistic theory of the geodynamo. *Geophys. Astrophys. Fluid Dynam.* **60**, 89–134.

Braginsky, S.I. & Meytlis, V.P. 1990 Local turbulence in the Earth's core. *Geophys. Astrophys. Fluid Dynam.* **55**, 71–87.

Brandenburg, A. & Tuominen, I. 1991 The solar dynamo. In *The Sun and Cool Stars: Activity, Magnetism, Dynamos* (ed. I. Tuominen, D. Moss, & G. Rüdiger), pp. 223–233. IAU Symposium, No. 130. Springer, Berlin.

Brandenburg, A., Tuominen, I. & Krause, F. 1990 Dynamos with a flat α-effect distribution. *Geophys. Astrophys. Fluid Dynam.* **50**, 95–112.

Bullard, E.C. & Gellman, H. 1954 Homogeneous dynamos and terrestrial magnetism. *Phil. Trans. R. Soc. Lond.* A **247**, 213–255.

Busse, F.H. 1975 A necessary condition for the geodynamo. *J. Geophys. Res.* **80**, 278–280.

Cattaneo, F., Hughes, D.W. & Weiss, N.O. 1991 What is a stellar dynamo? *Mon. Not. R. Astron. Soc.* **253**, 479–484.

Chandrasekhar, S. 1961 Adjoint differential systems in the theory of hydrodynamic stability. *J. Math. Mech.* **10**, 683–690.

Childress, S. 1969 *Théorie magnetohydrodynamique de l'effet dynamo*. Report from Departement Mécanique de la Faculté des Sciences, Univ. Paris..

Cowling, T.G. 1934 The magnetic field of sunspots. *Mon. Not. R. Astron. Soc.* **94**, 39–48.

Cowling, T.G. 1957 The dynamo maintenance of steady magnetic fields. *Quart. J. Mech. Appl. Math.* **10**, 129–136.

Cox A.N., Livingston, W.C. & Matthews, M.S. (eds.) 1991 *Inside the Sun*. Univ. of Arizona Press, Tucson, Arizona.

Donner, K.J. & Brandenburg, A. 1990 Magnetic field structure in differentially rotating discs. *Geophys. Astrophys. Fluid Dynam.* **50**, 121–129.

Dudley, M.L. 1988 *A numerical study of kinematic dynamos with stationary flow*. PhD Thesis, University of Sydney.

Dudley, M.L. & James, R.W. 1989 Time-dependent kinematic dynamos with stationary flows. *Proc R. Soc. Lond. A* **425**, 407–429.

Elstner, D. Meinel, R. & Beck, R. 1992 Galactic dynamos and their radio signatures. *Astron. Astrophys. Suppl.* **94**, 587–600.

Eltayeb, I.A. & Loper, D.E. 1988 On steady kinematic helical dynamos. *Geophys. Astrophys. Fluid Dynam.* **44**, 259–269.

Fearn, D.R. & Proctor, M.R.E. 1987 On the computation of steady, self-consistent spherical dynamos. *Geophys. Astrophys. Fluid Dynam.* **38**, 293–325.

Fearn, D.R., Roberts, P.H. & Soward, A.M. 1988 Convection, stability and the dynamo. In *Energy, Stability and Convection* (ed. G.P. Galdi & B. Straughan), pp. 60–324. Longmans, New York.

Gailitis, A.K. 1970 Self-excitation of a magnetic field by a pair of annular vortices. *Magnit. Gidro.* **6**(1), 19–22. (English transl.: *Magnetohydrodynamics*, 14–17.)

Gailitis, A.K. 1974 Generation of magnetic field by mirror-symmetric turbulence. *Magnit. Gidro.* **10**(2), 31–35. (English transl.: *Magnetohydrodynamics*, 131–134.)

Gailitis, A.K. & Freiberg, Ya. 1974 Self-excitation of a magnetic field by a pair of annular eddies. *Magnit. Gidro.* **10**(1), 37–42. (English transl.: *Magnetohydrodynamics*, 26–30.)

Gailitis, A.K. & Freiberg, Ya. 1976 Theory of a helical MHD dynamo. *Magnit. Gidro.* **12**(2), 3–6. (English transl.: *Magnetohydrodynamics*, 127–130.)

Gailitis, A.K. & Freiberg, Ya. 1980 Nonuniform model of a helical dynamo. *Magnit. Gidro.* **16**(1), 15–19. (English transl.: *Magnetohydrodynamics*, 171–15.)

Gailitis, A.K., Karasev, B.G., Kirillov, I.R., Lielausis, O.A., Luzhanskii, S.M., Ogorodnikov, A.P. & Preslitskii, G.V. 1987 Experiment with a liquid metal model of an MHD dynamo. *Magnit. Gidro.* **23**(4), 3–7. (English transl.: *Magnetohydrodynamics*, 349–352.)

Gailitis, A., Lielausis, O., Karasev, B.G., Kirillov, I.R. & Ogorodnikov, A.P. 1989 The helical MHD dynamo. In *Liquid Metal Magnetohydrodynamics* (ed. J. Lielpeteris & R. Moreau), pp. 413–418. Kluwer, Dordrecht.

Gilbert, A.D. 1988 Fast dynamo action in the Ponomarenko dynamo. *Geophys. Astrophys. Fluid Dynam.* **44**, 241–258.

Glatzmaier, G.A. & Roberts, P.H. 1993 Intermediate geodynamo models. *J. Geomag. Geoelec.* **45**, 1605–1616.

Herzenberg, A. 1958 Geomagnetic dynamos. *Phil. Trans. R. Soc. Lond. A* **250**, 543–583.

Hughes, D.W. 1993 Testing for dynamo action. In *Solar and Planetary dynamos* (ed. M.R.E. Proctor, P.C. Matthews & A.M. Rucklidge), pp. 153–159. Cambridge University Press.

Ivanova, T.S. & Ruzmaikin, A.A. 1977 A nonlinear magnetohydrodynamic model of the solar dynamo. *Sov. Astron.* **21**, 479–485.

Ivers, D.J. & James, R.W. 1986 Extension of the Namikawa–Matsushita antidynamo theorem to toroidal fields. *Geophys. Astrophys. Fluid Dynam.* **36**, 317–324.

Ivers, D.J. & James, R.W. 1988 An antidynamo theorem for partly symmetric flows. *Geophys. Astrophys. Fluid Dynam.* **44**, 271–278.

Jacobs, J.A. (ed.) 1987 *Geomagnetism*, vol. 2. Academic, New York.

Kono, M. & Roberts, P.H. 1991 Small amplitude solutions of the dynamo problem 1. The adjoint system and its solutions. *J. Geomag. Geoelec.* **43**, 839–862.

Kono, M. & Roberts, P.H. 1992 Small amplitude solutions of the dynamo problem 2. The case of α^2-models. *Geophys. Astrophys. Fluid Dynam.* **67**, 65–85.

Krause, F. & Meinel, R. 1988 Stability of simple nonlinear α^2-dynamos. *Geophys. Astrophys. Fluid Dynam.* **43**, 95–117.

Krause, F. & Roberts, P.H. 1973 Some problems of mean-field electrodynamics. *Astrophys. J.* **181**, 977–992.

Kumar, S. & Roberts, P.H. 1975 A three-dimensional kinematic dynamo. *Proc R. Soc. Lond. A* **314**, 235–258.

Larmor, J. 1919 How could a rotating body such as the Sun become a magnet? *Rep. Brit. Assoc. Adv. Sci.* 159–160.

Lortz, D. 1968 Exact solutions of the hydromagnetic dynamo problem. *Plasma Phys.* **10**, 967–972.

Lortz, D. 1990 Mathematical problems in dynamo theory. In *Physical Processes in Hot Cosmic Plasmas* (ed. W. Brinkmann), pp. 221–234. Kluwer, Dordrecht.

Moffatt, H.K. 1969 The degree of knottedness of tangled vortex lines. *J. Fluid Mech.* **35**, 117–129.

Nakajima, T. & Kono, M. 1991 Kinematic dynamos associated with large scale fluid motions. *Geophys. Astrophys. Fluid Dynam.* **60**, 177–209.

Nakajima, T. & Roberts, P.H. 1994 A mapping method for solving dynamo equations. (In preparation).

Namikawa, T. & Matsushita, S. 1970 Kinematic dynamo problem. *Geophys. J. R. Astr. Soc.* **19**, 395–415.

Parker, E.N. 1955 Hydromagnetic dynamo models. *Astrophys. J.* **121**, 293–314.

Parker E.N. 1957 The solar hydromagnetic dynamo. *Proc. Nat. Acad. Sci. Wash.* **43**, 8–14.

Parker, E.N. 1971 The generation of magnetic fields in astrophysical bodies III. The galactic field. *Astrophys. J.* **163**, 255–285.

Pekeris, C.L., Accad, Y. & Shkoller, B. 1973 Kinematic dynamos and the Earth's magnetic field. *Phil. Trans. R. Soc. Lond.* A **275**, 425–461.

Ponomarenko, Yu.B. 1973 On the theory of the hydromagnetic dynamo. *Zh. Prik. Mech. Tech. Fiz. SSSR* (6), 47–51. (English transl.: *J. Appl. Mech. Tech. Phys.* **14**, 775–778 (1973).)

Proctor, M.R.E. 1977 On Backus' necessary condition for dynamo action in a conducting sphere. *Geophys. Astrophys. Fluid Dynam.* **9**, 89–93.

Rädler, K.-H. & Wiedemann, E. 1989 Numerical experiments with a simple nonlinear mean-field dynamo model. *Geophys. Astrophys. Fluid Dynam.* **49**, 71–79.

Roberts, G.O. 1972 Dynamo action of fluid motions with two-dimensional periodicity. *Phil. Trans. R. Soc. Lond.* A **271**, 411–454.

Roberts, P.H. 1960 Characteristic value problems posed by differential equations arising in hydrodynamics and hydromagnetics. *J. Math. Anal. & Applic.* **1**, 195–214.

Roberts, P.H. 1967 *An Introduction to Magnetohydrodynamics*. Longmans, London.

Roberts, P.H. 1971 Dynamo theory. In *Mathematical Problems in the Geophysical Sciences* (ed. W. H. Reid), pp. 129–206. Lectures in Applied Mathematics, vol. 14. Amer. Math. Soc., Providence, R.I..

Roberts, P.H. 1972 Kinematic dynamo models. *Phil. Trans. R. Soc. Lond.* A **272**, 663–703.

Roberts, P.H. 1987 Dynamo theory. In *Irreversible Phenomena and Dynamical Systems Analysis in Geosciences* (ed. C. & G. Nicolis), pp. 73–133. Reidel, Dordrecht.

Roberts, P.H. 1988 Future of geodynamo theory. *Geophys. Astrophys. Fluid Dynam.* **42**, 3–31.

Roberts, P.H. 1989 From Taylor state to model-Z? *Geophys. Astrophys. Fluid Dynam.* **49**, 143–160.

Roberts, P.H. 1992 A mapping method for solving dynamo equations. In unpublished Proc. 3rd SEDI Symposium, *Lateral heterogeneity of the transition between core and mantle*, Mizusawa, Japan, 6–10 July 1992.

Roberts, P.H. & Soward, A.M. 1992 Dynamo theory. *Annu. Rev. Fluid Mech.* **24**, 459–512.

Rosner, R. & Weiss, N.O. 1992 The origin of the solar cycle. In *The Solar Cycle* (ed. K.L. Harvey), pp. 511–531. Astronomical Society of the Pacific, Conference Series vol. 27.

Rüdiger, G. 1990 The alpha-effect in galaxies is highly anisotropic. *Geophys. Astrophys. Fluid Dynam.* **50**, 53–66.

Russell, C.T. 1987 Planetary magnetism. In *Geomagnetism,* vol. 2 (ed. J.A. Jacobs), pp. 457–560. Academic, New York.

Ruzmaikin, A.A. & Starchenko, S.V. 1991 On the origin of the Uranus and Neptune magnetic fields. *Icarus* **93**, 82–87.

Ruzmaikin, A.A., Shukurov, A.M., & Sokoloff, D.D. 1988a *The Magnetic Fields of Galaxies*. Kluwer, Dordrecht.

Ruzmaikin, A.A., Sokoloff, D.D. & Shukurov, A.M. 1988b Magnetism of spiral galaxies. *Nature* **336**, 341–347.

Ruzmaikin, A.A., Sokoloff, D.D. & Shukurov, A.M. 1988c Hydromagnetic screw dynamo. *J. Fluid Mech.* **197**, 37–56.

Ruzmaikin, A.A., Sokoloff, D.D., Solovev, A.A. & Shukurov, A.M. 1989 Couette-Poiseuille flow as a screw dynamo. *Magnit. Gidro.* **24**(1), 9–14. (English transl.: *Magnetohydrodynamics*, 6–11.)

Ruzmaikin, A.A., Shukurov, A.M., Sokoloff, D.D. & Starchenko, S.V. 1990 Maximally-efficient-generation approach in dynamo theory. *Geophys. Astrophys. Fluid Dynam.* **52**, 125–140.

Schmitt, D. 1990 A torus-driven dynamo for magnetic fields in galaxies and accretion disks. *Rev. Mod. Astron.* **3**, 86–97.

Shukurov, A.M., Sokoloff, D.D. & Ruzmaikin, A.A. 1985 Oscillating α^2-dynamo. *Magnit. Gidro.* **21**(1), 9–13. (English transl.: *Magnetohydrodynamics*, 6–10.)

Sokoloff, D.D., Shukurov, A.M. & Ruzmaikin, A.A. 1983 Asymptotic solution of the α^2-dynamo problem. *Geophys. Astrophys. Fluid Dynam.* **25**, 293–307.

Sokoloff, D.D., Shukurov, A.M. & Shumkina, T.S. 1989 The second asymptotic approximation in the screw dynamo problem. *Magnit. Gidro.* **24**(1), 3–8. (English transl.: *Magnetohydrodynamics*, 1–6.)

Soward, A.M. 1972 A kinematic theory of large magnetic Reynolds number dynamos. *Phil. Trans. R. Soc. Lond. A* **272**, 431–462.

Soward, A.M. 1974 A convection-driven dynamo 1. The weak field case. *Phil. Trans. R. Soc. Lond. A* **275**, 611–651.

Soward, A.M. 1990 A unified approach to a class of slow dynamos. *Geophys. Astrophys. Fluid Dynam.* **53**, 81–107.

Soward, A.M. 1991 The Earth's dynamo. *Geophys. Astrophys. Fluid Dynam.* **62**, 191–209.

Soward, A.M. 1992a Thin disk kinematic $\alpha\omega$-dynamo models I. Long length scale modes. *Geophys. Astrophys. Fluid Dynam.* **64**, 163–199.

Soward, A.M. 1992b Thin disk kinematic $\alpha\omega$-dynamo models II. Short length scale modes. *Geophys. Astrophys. Fluid Dynam.* **64**, 201–225.

St. Pierre 1993 The strong field branch of the Childress–Soward dynamo. In *Solar and Planetary dynamos* (ed. M.R.E. Proctor, P.C. Matthews & A.M. Rucklidge), pp. 295–302. Cambridge University Press.

Steenbeck, M. & Krause, F. 1966 Erklärung stellarer und planetarer Magnetfelder durch einen turbulenzbedingten Dynamomechanismus. *Zeit. Naturforsch.* **21a**, 1285–1296.

Steenbeck, M. & Krause, F. 1969 Zur Dynamotheorie stellarer und planetarer Magnetfelder I. Berechnung sonnenähnlicher Wechselfeldgeneratoren. *Astron. Nachr.* **291**, 49–84.

Steenbeck, M., Krause, F. & Rädler, K.-H. 1966 Berechnung der mittleren Lorentz-Feldstärke $\overline{v \times B}$ für ein elektrisch leitendes Medium in turbulenter, durch Coriolis-Kräfte beeinflußter Bewegung. *Z. Naturforsch.* **21a**, 369–376.

Stepinski, T.F. & Levy, E.H. 1990 Generation of dynamo magnetic fields in thin Keplerian disks. *Astrophys. J.* **362**, 318–332.

Tough, J.G. 1967 Nearly symmetric dynamos. *Geophys. J. R. Astr. Soc.* **13**, 393–396. Corrigendum **15**, 343 (1969).

Tuominen, I., Moss, D. & Rüdiger, G. (eds.) 1991 *The Sun and Cool Stars: Activity, Magnetism, Dynamos.* IAU Symposium 140. Springer, Berlin.

Vainshtein, S. I. & Zel'dovich, Ya. B. 1972 Origin of magnetic fields in astrophysics. *Usp. Fiz. SSSR* **106**, 431–457. (English transl.: *Sov. Phys. Usp.* **15**, 159–172 (1972).)

Vishniac, E.T. & Diamond, P. 1992 Local MHD instabilities and the wave-driven dynamo in accretion disks. *Astrophys. J.* **398**, 561–568.

Weiss, N.O. 1989 Dynamo processes in stars. In *Accretion disks and magnetic fields in astrophysics* (ed. G. Belvedere), pp. 11–29. Kluwer, Dordrecht.

Zel'dovich, Ya.B. 1956 The magnetic field in the two-dimensional motion of a conducting turbulent fluid. *Zh. Exp. Teor. Fiz. SSSR* **31**, 154–155. (English transl.: *Sov. Phys. JETP* **4**, 460–462 (1957).)

Zel'dovich, Ya.B. & Ruzmaikin, A.A. 1980 Magnetic field in a conducting fluid in two dimensional motion. *Zh. Exp. Teor. Fiz. SSSR* **78**, 980–986. (English transl.: *Sov. Phys. JETP* **51**, 493–497 (1980).)

CHAPTER 2

Solar and Stellar Dynamos

N. O. WEISS

Department of Applied Mathematics and Theoretical Physics
University of Cambridge
Cambridge CB3 9EW, UK

2.1 INTRODUCTION

The aim of this chapter is to provide a link between observations of magnetic fields in the Sun and other active stars, and the theory that is presented elsewhere in this volume. I shall begin therefore by considering the observational background and the phenomenological picture that emerges from it. Then I shall go on to discuss a hierarchy of idealized dynamo models that help to explain different aspects of these observations. This material has been the subject of several recent reviews (Weiss 1989, Belvedere 1990, Brandenburg & Tuominen 1991, Stix 1991, DeLuca & Gilman 1991, Rosner & Weiss 1992, Schmitt 1993).

This treatment relies heavily on what has become known as the solar-stellar connection. Figure 2.1 shows a Hertzsprung–Russell diagram, with the relative luminosity of the stars plotted as a function of their effective surface temperature (or, equivalently, their spectral type). The stars on the main sequence form a one-parameter family, with their positions determined by their masses. Some of the hot stars to the left of the vertical line have strong magnetic fields, which vary only as a consequence of the star's rotation. The fields in these magnetic stars (the Ap stars) are apparently fossil relics and will not be considered here. Stars to the right of the vertical line are sufficiently cool that hydrogen only becomes ionized beneath their visible surfaces; as a result, they have deep convective envelopes. The combination of convection and rotation is associated with magnetic activity in these cool stars. Their behaviour is similar to that found in the Sun, and it is with them that we are concerned. Solar magnetic fields have been observed in enormous detail over many years but the Sun is a single star with unique properties; by observing many stars with different properties we can discover how magnetic activity depends on such parameters as age, rotation rate and stellar mass. Thus we may hope to construct a theory of stellar magnetism that can be

M.R.E. Proctor & A.D. Gilbert (eds.)
Lectures on Solar and Planetary Dynamos, 59–95
©1994 Cambridge University Press.

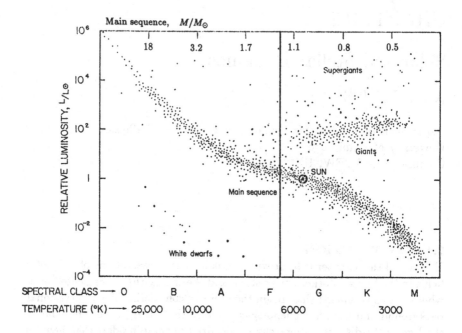

Figure 2.1 Hertzsprung–Russell diagram, showing luminosity relative to that of the Sun as a function of the effective surface temperature and spectral type. Relative masses of stars on the main sequence are indicated at the top of the figure. Late-type stars, to the right of the vertical line, have deep outer convection zones. The lower main-sequence stars exhibit magnetic activity of the type discussed here. (After Noyes 1982.)

applied to all lower main-sequence stars.

This chapter is organized as follows. Observations of stellar magnetic activity are summarized in the next section, while solar activity is described in section 2.3. Two mechanisms have been put forward to explain the solar cycle; the reasons for preferring a dynamo to a magnetic oscillator are assessed in section 2.4. Section 2.5 reviews the case for locating the dynamo just below the base of the convection zone. The emphasis then shifts to calculations, beginning with self-consistent nonlinear dynamos in section 2.6. Mean field dynamo models are discussed in section 2.7 and results obtained from severely truncated toy models are presented in section 2.8. Finally, the chapter concludes with some speculations on the physical properties of dynamos in stars that are much more active than the Sun.

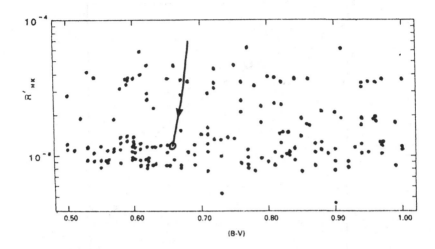

Figure 2.2 Magnetic activity as a function of spectral type for local field stars on the lower main sequence. The index R'_{HK}, a normalized measure of Ca^+ emission, is plotted against the colour index $B - V$. The Sun's evolutionary track is indicated. (After Soderblom & Baliunas 1988.)

2.2 STELLAR MAGNETIC ACTIVITY

Magnetic fields can be measured directly through broadening of spectral lines caused by the Zeeman effect (Saar 1991 a, b). Stars that are much more active than the Sun have field strengths of around 2000 G over 50% of the unspotted fraction of their surfaces. Such stars also show substantial photometric variations, which are interpreted as being due to the appearance of starspots as the star rotates. The luminosity of an active star may change by up to 30%, implying that spots cover up to 60% of its surface area (Byrne 1992). By contrast, the solar luminosity only varies by amounts of order 0.1%. X-ray emission is also associated with magnetic activity: the solar corona is magnetically heated and the Einstein and Rosat satellites have detected thermal X-ray emission from stellar coronae. Further evidence comes from stellar flares, observed at radio as well as X-ray frequencies.

The most significant results have been obtained by measuring Ca^+, H and K emission from stars (Baliunas & Vaughan 1985, Hartmann & Noyes 1987, Soderblom & Baliunas 1988, Noyes, Baliunas & Guinan 1991). It has been known for many years that Ca^+ emission is closely correlated with magnetic fields on the Sun. In 1966 Wilson (1978) initiated a systematic programme at Mt. Wilson of monitoring H and K emission from a group of nearby stars. Figure 2.2 shows a measure of the Ca^+ emission from nearby field stars, plotted against the colour index $B - V$, which increases with decreasing surface temperature and defines the spectral type. For stars of a fixed spectral type

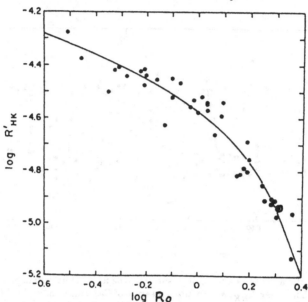

Figure 2.3 Dependence of stellar magnetic activity on rotation. Logarithmic plot of the time-averaged index R'_{HK} against the Rossby number $Ro = 2\pi/\sigma$. The Sun's position is marked and the curve shows the suggested functional relationship. (After Noyes *et al.* 1984*a*.)

there is a range of activity. If, however, we select stars from a cluster such as the Hyades, which are all of the same age, then they lie close to a roughly horizontal line in the figure. Moreover, stars with high velocities perpendicular to the galactic plane, which are considered to be old, are only feebly active. Thus the scatter is apparently a function of age: as stars grow older they become less active. The Sun, at 4.7×10^9 yr, is a weakly active middle-aged star; its evolutionary track is indicated on the figure.

What is the physical mechanism that controls magnetic activity? The rotation periods of active stars can be estimated from Doppler broadening of spectral lines, or determined more precisely from rotational modulation of the Ca^+ emission. Results for stars with different structures can be brought together by introducing the inverse Rossby number $\sigma = \Omega\tau_c$, where Ω is the angular velocity and τ_c is the computed convective turnover time at the base of the convection zone, as a parameter (Noyes *et al.* 1984*a*). Figure 2.3 shows that magnetic activity, as measured by Ca^+ emission, is a function of the normalized rotation rate. The most active stars, with rotation periods of a day or less, are either very young or else close binaries, with tidal coupling between spin and orbital momentum. The angular momentum of a single star on the main sequence decreases owing to the torque exerted by a stellar

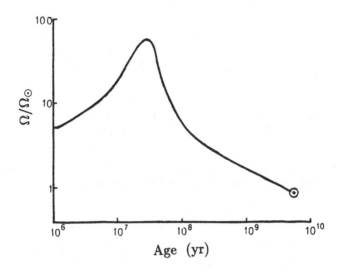

Figure 2.4 Rotational history of the Sun. Ratio of surface angular velocity Ω to the rotation rate $\Omega_\odot$ of the present Sun, as a function of age, for a G-star. Rapid acceleration as the star approaches the main sequence is followed by gradual spin-down through magnetic braking. (After Rosner & Weiss 1985.)

wind (Schatzman 1962, Mestel 1968, 1990), which itself depends on the magnetic field and therefore on Ω (Durney & Latour 1978, Mestel & Spruit 1987, Kawaler 1988). Figure 2.4 shows Ω as a function of time for a G-star like the Sun: the star spins up as it collapses onto the main sequence, conserving its angular momentum while its moment of inertia is decreased. Stars in a very young cluster, like α Persei, rotate 50 to 100 times more rapidly than the Sun but the outer regions spin down very rapidly, so that stars in the Pleiades (which are 2×10^7 yr older) already appear to be rotating much more slowly (Stauffer & Soderblom 1991). Thereafter, the rate of spindown through magnetic braking is much more gradual and can be represented by a power law of the form $\Omega \propto t^{-1/2}$ (Skumanich 1972). The coupling between rotation and magnetic fields is apparently such that activity never fades out completely during the star's main sequence lifetime (Baliunas & Jastrow 1990).

If the Sun were observed as a nearby star its cyclic activity could be detected from variations in Ca^+ emission (White *et al.* 1992). Similar patterns of cyclic activity have been found in a number of other slowly rotating stars. For instance, the G2 star HD 81809 has a cycle with a period of about 8 yr, as shown in figure 2.5. Unfortunately, there are only a dozen clear examples of stars with cycles like the Sun's and not all of their rotation periods have

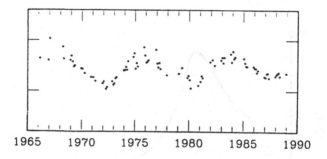

1965   1970   1975   1980   1985   1990

Figure 2.5 A stellar analogue of the solar cycle. Chromospheric Ca$^+$ emission from 1966 to 1989 for the G2 star HD81809, showing cyclic activity with a period of 8.3 yr. (After Baliunas & Jastrow 1990.)

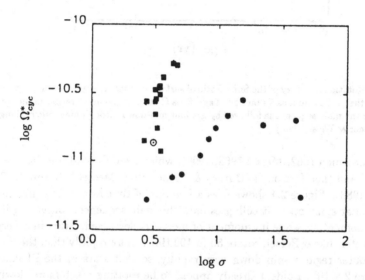

Figure 2.6 Dependence of cycle frequency on angular velocity. Logarithmic plot of a scaled cycle frequency Ω_{cyc} as a function of the inverse Rossby number σ. The Sun's position is marked and squares indicate stars with solar-like cycles; circles indicate other stars. Note the two distinct families. (After Saar & Baliunas 1992.)

been precisely determined. Although this sample is small, it is of interest to discover how the cycle frequency depends on the rotation rate (Noyes, Weiss & Vaughan 1984b). Figure 2.6 shows the cycle frequency Ω_{cyc} as a function of the inverse Rossby number σ for the relevant stars (Saar & Baliunas 1992). It is clear that the cycle period increases with increasing rotation period: for a star of fixed mass the results are consistent with a power law of the form

$\Omega_{\rm cyc} \propto \sigma^3$, though the exponent is very poorly determined. Note, however, the presence of a second group of stars with periods that obey a different law. They serve as a warning that a model of the solar cycle cannot automatically be extended to cover stars with very different properties. Nevertheless, we infer that any slowly rotating G-star should share the magnetic properties of the Sun, which is the only star whose activity can be observed in detail.

2.3 SOLAR ACTIVITY

The sun has been observed through telescopes for almost four centuries and it is now possible to resolve structures on scales of a few hundred km on its surface. Soft X-rays provide the striking representation of solar activity and long time-sequences are available from Skylab and from the current Yohkoh mission. These images show that coronal emission is dominated by active regions and compact bright points. The exceptionally high-resolution picture reproduced in figure 2.7(a) was obtained with the Normal Incidence X-ray Telescope (Sams, Golub & Weiss 1992): the X-ray image is superimposed on a white light image (owing to a fortunate accident) so that the relation between coronal loops and sunspots is apparent. The corresponding magnetic fields are displayed in figure 2.7(b). The magnetogram shows that photospheric magnetic fields are highly intermittent: active regions contain sunspots and pores; within faculae magnetic flux is confined to intergranular lanes and the weaker photospheric network is made up of isolated flux tubes with intense magnetic fields. These magnetic structures persist down to the smallest scales that can be resolved.

Despite this fine structure the large-scale features follow a systematic pattern of cyclic activity (Stix 1989). The area covered by sunspots varies aperiodically with a mean period of about 11 yr, as shown in figure 2.8. At sunspot minimum new spots appear at latitudes around $\pm 30°$; the zones of activity swell until they approach the equator, where they die out at the next minimum, so producing the familiar butterfly diagram in figure 2.8. Sunspot polarities follow Hale's laws: the magnetic field emerging in an active region is predominantly toroidal, with the axis inclined by about 10° so that preceding spots are slightly closer to the equator. In each hemisphere preceding and following spots have opposite polarities; these polarities are different in the two hemispheres and reverse from one cycle to the next. Thus the *magnetic* cycle has a 22-year period. Solar activity is conventionally measured by the relative sunspot number R, which is rather arbitrarily defined (Stix 1989) but provides a qualitatively reliable indicator. Figure 2.9 shows the variation of magnetic activity over the past 300 yr, constructed from the record of sunspot numbers. The cycle appears to be chaotic with an aperiodically modulated amplitude. It has been suggested that this record shows a characteristic timescale of around 90 yr (the Gleissberg cycle) though the time-sequence is

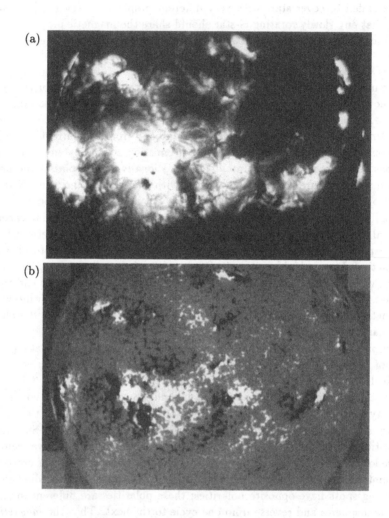

Figure 2.7 (a) NIXT X-ray image superposed on a white-light image of the Sun. Note the prominent sunspot group. (b) Corresponding KPNO magnetogram; black and white represent fields of opposite polarities. X-ray loops indicate the structure of the field in the corona. (From Sams et al. 1992.)

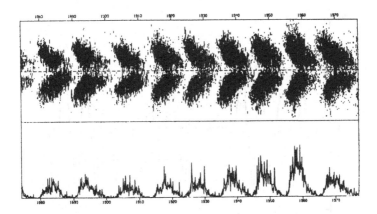

Figure 2.8 The cyclic pattern of solar activity. Butterfly diagram, showing the distribution of sunspots in latitude and time (upper panel), with the corresponding area covered by sunspots (lower panel). Note the irregular 11-yr cycle and the equatorward drift of the sunspot zones. (From the Greenwich photoheliographic observations, 1874–1976, after Yallop & Hohenkerk 1980.)

rather short.

Figure 2.10 shows sunspot numbers from Galileo's earliest observations in 1610. Observations were sporadic in the first half of the seventeenth century but for an interval of 70 years there was a prolonged dearth of sunspots, coinciding with the reign of Louis XIV, the Roi Soleil, and indeed with most of Isaac Newton's own lifetime (Eddy 1976). There is no doubt that this effect was real. Contemporary astronomers commented on the shortage of sunspots and reported excitedly on occasional spots when they appeared. It was even referred to by the poet Andrew Marvell in 1667 (Weiss & Weiss 1979). In his satirical poem *The Last Instructions to a Painter* he addresses Charles II:

To the King

So his bold Tube, Man to the Sun apply'd
And Spots unknown to the bright Star descry'd;
Show'd they obscure him, while too near they please,
And seem his courtiers, are but his disease.
Through Optick Trunk the Planet seem'd to hear,
And hurls them off, e'er since, in his Career.

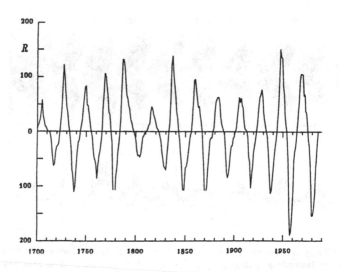

Figure 2.9 The 22-yr magnetic cycle. A record of alternating magnetic activity, constructed by de-rectifying the record of sunspot numbers, R, from 1700 to 1985. (After Bracewell 1988.)

> And you, *Great Sir*, that with him Empire share,
> Sun of our World, as he the *Charles* is there,
> Blame not the *Muse* that brought those spots to sight,
> Which, in your Splendor hid, Corrode your Light....

This reference shows that the lack of sunspots had become common knowledge by this date. (It remained familiar to Lalande and William Herschel but was forgotten after Schwabe identified the 11-year activity cycle in 1843, until it was rediscovered by Spörer in the 1880s. Maunder drew attention to Spörer's work and the interruption is now known as the Maunder Minimum.)

Was the Maunder Minimum a unique event or are grand minima a recurrent feature of solar and stellar magnetic activity? Fortunately, it is possible to extend the record by making use of proxy data. Galactic cosmic rays impinging on the earth's atmosphere lead to the production of radioisotopes including ^{10}Be and ^{14}C. Since the cosmic ray flux is modulated by magnetic fields in the solar wind, there are fluctuations (anticorrelated with solar activity) in the rates at which these cosmogenic isotopes are produced. Figure 2.11(a) shows variations in the abundance of ^{10}Be, which is preserved in polar icecaps, over the last millennium: the Maunder Minimum is apparent,

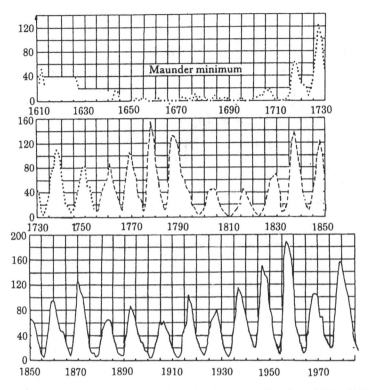

Figure 2.10 The Maunder minimum. Estimated sunspot numbers from 1610 to 1850 and monthly sunspot numbers from 1850 to 1986. (After Ribes 1990.)

together with several previous minima, spaced irregularly at intervals of 200–300 yr (Raisbeck *et al.* 1990, Beer, Raisbeck & Yiou 1991). This pattern is confirmed by the variations in ^{14}C production, which are shown in the same figure. The curve for ^{14}C lags behind that for [10]Be, owing to the fact that ^{14}C spends 30–40 yr in the atmosphere before being absorbed into trees, while [10]Be is deposited within 2 yr. Thus the 11 yr activity cycle can only be followed in the [10]Be record. The ^{14}C abundance anomalies affect radiocarbon dating and have therefore been investigated in considerable detail (Stuiver & Braziunas 1988, Damon & Sonett 1991). Figure 2.11(*b*) shows fluctuations in production of ^{14}C as a function of time for the last 9000 yr. The long-term trend can be ascribed to variations in the geomagnetic field and the shortlived positive anomalies correspond to grand minima in solar activity. This record suggests that such minima occur aperiodically, with a characteristic timescale of around 200 yr; frequency analysis yields spectra with peaks at a period of 208 yr (Damon & Sonett 1991). There is also a tendency for different minima to have the same form (Stuiver & Braziunas 1988, 1989).

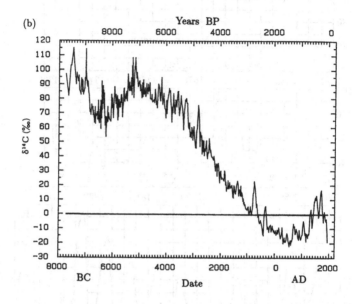

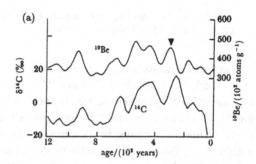

Figure 2.11 (*a*) Variations in ^{10}Be and ^{14}C production over the past 1200 yr. The triangle marks the position of the Maunder minimum. (After Raisbeck *et al.* 1990.) (*b*) Fluctuations in ^{14}C abundance over the last 9500 years. (After Stuiver & Braziunas 1988.)

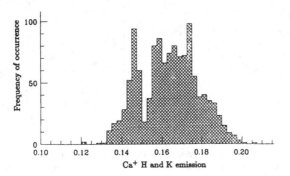

Figure 2.12 The bimodal distribution of Ca^+ emission in a group of weakly active stars. The two humps may correspond to epochs of cyclic activity and grand minima. (After Baliunas & Jastrow 1990.)

This pattern of recurrent grand minima is consistent with the observed behaviour of solar-type stars. Figure 2.12 shows the distribution of Ca^+ emission from a group of weakly active stars (Baliunas & Jastrow 1990). The distribution appears to be bimodal. The larger hump corresponds to stars in active phases, with a range of activity similar to that found in the Sun. The smaller hump is explained by assuming that stars spend about 30% of the time in a quiescent state, with emission at the level found locally in field-free regions of the Sun at sunspot minimum (White *et al.* 1992). Thus we have to explain not only the generation of cyclic magnetic fields but also their aperiodic modulation.

2.4 THE ORIGIN OF THE SOLAR CYCLE

The ohmic decay time for the magnetic field **B** in a star or planet of radius R is given by $\tau_\eta = R^2/\eta$ where η is the ohmic diffusivity. For a star like the Sun, $\tau_\eta \approx 10^{10}$ yr, so a fossil magnetic field could survive for the star's lifetime on the main sequence. Planets are quite different: $\tau_\eta \approx 10^4$ yr for the earth, so a dynamo is required to explain the persistence of the geomagnetic field. The problem with the Sun is to explain why the magnetic field reverses on a timescale so much shorter than τ_η. Some inductive process is needed and two mechanisms have been advanced. The first, more obvious mechanism is a magnetic oscillator: given a stationary poloidal field in a highly conducting medium, an oscillatory angular velocity will generate an alternating toroidal field. It follows that Ω should vary with a period of 22 yr in the Sun.

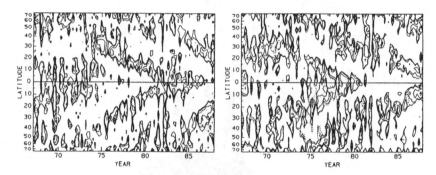

Figure 2.13 Fluctuations in solar differential rotation ('torsional oscillations'). Regions with anomalously rapid (left) and anomalously slow (right) rotation rates, as functions of latitude and time. At a fixed latitude Ω varies with an 11-yr period. (After Ulrich *et al.* 1988.)

The second mechanism is a self-excited hydromagnetic dynamo. Two processes are involved (Parker 1955, 1979). The first is easy: differential rotation creates toroidal flux (antisymmetric about the equator) from a dipolar poloidal field. The second process is more subtle: cyclonic convection (with a net helicity that is antisymmetric about the equator) generates a reversed poloidal field from the toroidal field. Dynamo action is limited by the nonlinear action of the Lorentz force, which is quadratic in **B**. Hence any fluctuations in angular velocity should depend on $|\mathbf{B}|^2$ and should therefore vary with an 11 year period.

Detailed observations show that the Sun's poloidal field reverses at sunspot maximum (90° out of phase with the toroidal field). Sheeley (1992) and his collaborators have argued that the polar fields are produced by dispersal of the inclined fields in active regions (e.g., Wang & Sheeley 1991, Wang, Sheeley & Nash 1991); if so, the reversal of the polar fields might just be a by-product of the mechanisms that maintain the field. Studies of geomagnetic storms have, however, established that sunspot activity is correlated with activity at higher latitudes during the *preceding* rather than the *following* half-cycle (Legrand & Simon 1991); this provides firm support for a dynamo. Furthermore, the observed variations in surface differential rotation have a period of 11 rather than 22 years. Figure 2.13 shows complementary images of positive and negative anomalies in the Sun's differential rotation, the so-called torsional oscillations (Howard & LaBonte 1980, Ulrich *et al.* 1988). Although the zones of abnormally rapid rotation migrate from pole to equator in about 22 yr, the angular velocity at a fixed latitude varies with a period of 11 yr. At low latitudes the rapidly rotating regions coincide with the sunspot zones. That is precisely what would be expected from a nonlinear dynamo (Schüssler

1981, Yoshimura 1981) and this result provides the strongest argument for preferring dynamos to oscillators.

Of course, it is always possible to assert that velocities and fields observed at the solar surface are merely secondary effects — epiphenomena, to use Cowling's term. Then one can only fall back on Einstein's dictum: *raffiniert ist der Herrgott, aber boshaft ist er nicht* (subtle is the Lord, but he is not malicious). In what follows it will therefore be assumed that cyclic activity in late-type stars is maintained by a dynamo operating in or near the convection zone.

2.5 WHERE IS THE SOLAR DYNAMO?

The first essential is to establish a phenomenological picture of the solar cycle. Where is the dynamo located? How does it operate? How is it related to magnetic activity observed at the surface of the Sun? The scale of active regions and the systematic behaviour of sunspots suggest that they are formed by flux tubes emerging from depths of order 100 Mm (Thomas & Weiss 1992). For a long while it was supposed that the dynamo was in the lower part of the convection zone but that picture ran into several difficulties.

Observations suggest that magnetic fields within the convection zone have a highly intermittent structure. Flux tubes that emerge in sunspots have fluxes of 10^{22} mx (100 TWb) and fields greater than 3000 G. Such an isolated flux tube cannot remain in static equilibrium. Consider a horizontal flux tube with a magnetic field **B** and let P_i, ρ_i and P_e, ρ_e be the pressure and density inside and outside the tube, respectively. Then magnetohydrostatic equilibrium requires that $P_i + B^2/2\mu_0 = P_e$, so that $P_i < P_e$. For a tube in thermodynamic equilibrium with its surroundings it follows that $\rho_i < \rho_e$. The tube will be buoyant and float upwards. Moreover the escape time is about a month (comparable with the turnover time τ_c). Thus magnetic buoyancy prevents the accumulation of isolated magnetic flux tubes within the bulk of the convection zone (Parker 1979).

A consequence is that the toroidal flux generated within the convection zone by shearing a weak poloidal field would escape long before the field strength observed at the surface could be attained. To cope with this problem it was suggested that magnetic flux is expelled from the convection zone itself and held down in a thin layer of weak convective overshoot at the interface between the convective and radiative zones, at a depth of 200 Mm (Spiegel & Weiss 1980, Golub *et al.* 1981, Galloway & Weiss 1981, van Ballegooijen 1982). The case for such a shell dynamo has been considerably strengthened by recent measurements of differential rotation with the Sun.

The variation of angular velocity with latitude was discovered 130 years ago. Observations of sunspots show a marked equatorial acceleration: the rotation rate decreases from 14.5° per day at the equator to 13.9° per day at latitudes of ±30°. Doppler measurements reveal that the photospheric plasma rotates less rapidly; its rotation rate falls from 14.2° per day at the equator to 10.8° per day at a latitude of 70° (Howard 1984). Autocorrelation analyses indicate that small scale magnetic features follow the photospheric plasma, as one would expect, while longer-lived, larger-scale features show less differential rotation, with a rate of 13.2° per day for latitudes greater than 50° (Stenflo 1989). The interpretation of these results is still controversial (e.g., Snodgrass 1992).

Theoretical considerations, supported by numerical experiments, suggest that in a rapidly rotating star the Proudman–Taylor constraint should lead to convection cells that are elongated parallel to the rotation axis, giving rise to a pattern that has been likened to a bunch of bananas. It should then follow that the angular velocity tends to be constant on cylindrical surfaces. This picture received strong support from experiments, carried out in a zero-gravity environment, on convection driven by electrostatic forces in a rotating system (Hart, Glatzmaier & Toomre 1986a, Hart et al. 1986b). These experiments in space confirmed that an irregularly tesselated pattern of convection gives way to banana cells as the angular velocity is increased. So it came to be believed that surfaces of constant Ω in the outer part of the sun should have the form shown in figure 2.14(a), with $\Omega \approx \Omega(s)$ and $\partial\Omega/\partial s > 0$, referred to cylindrical polar co-ordinates (s, ϕ, z). In this picture the gradients of Ω could generate a toroidal field B_ϕ from the radial component B_s of the poloidal field throughout the convection zone or the upper part of the radiative zone.

This consensus was shattered by helioseismological measurements. Five-minute oscillations at the solar surface correspond to acoustic modes of high degree. The frequencies of these p-modes show rotational splitting which can be used to determine the variation of Ω with latitude and depth in the interior of the Sun. These results reveal the very different isorotational surfaces in figure 2.14(b) (Dziembowski, Goode & Libbrecht 1989, Rhodes et al. 1990). The surface differential rotation persists throughout the convection zone, where $\Omega \approx \Omega(\theta)$, referred to spherical polar co-ordinates (r, θ, ϕ), but there is an apparent transition to uniform rotation in a layer (the tachocline) of thickness ≤ 50 Mm at the interface between the convective and radiative zones. Within this layer, which is too thin to be resolved, $|\partial\Omega/\partial r|$ is large; at the equator $\partial\Omega/\partial r > 0$ but $\partial\Omega/\partial r < 0$ at the poles. This pattern of differential rotation has to be maintained somehow by the effects of turbulent motion in the convection zone. The structure of the tachocline is determined by a combination of shear-generated turbulence on spherical surfaces and meridional circulation in the radiative zone (Spiegel & Zahn 1992).

A new consensus has now developed (e.g., Schmitt 1993). In this picture there is a magnetic layer in the overshoot region, where strong toroidal fields can be generated from the component B_r of the poloidal field. The filling fac-

(a) (b)

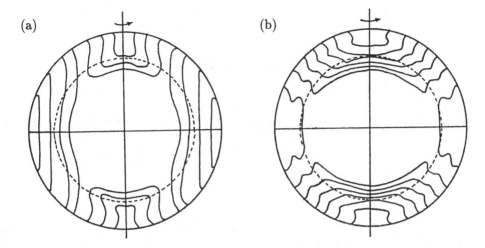

Figure 2.14 Internal rotation of the Sun. (*a*) The old picture, with Ω constant on cylindrical surfaces, relevant if convection is dominated by rotation. (After Rosner & Weiss 1985). (*b*) The rotation rate determined by helioseismology, with Ω dependent on latitude only in the convection zone, and uniform rotation in the deep interior. (After Libbrecht 1988.)

tor in this layer cannot be established (Schüssler 1993); if the magnetic field is smoothly distributed the toroidal field could be in magnetohydrostatic equilibrium but such an equilibrium is liable to instabilities driven by magnetic buoyancy (Hughes & Proctor 1988, Hughes 1992). These instabilities lead to the formation of isolated flux tubes which enter the convection zone and rise through it to emerge as active regions. The flux that emerges could be supplied by a layer 10 Mm thick with a field of 10^4 G; the magnetic energy density would then be comparable with the kinetic energy of convection in the lower part of the convective zone. Recent estimates suggest, however, that flux tubes that escape to form sunspots should have fields of 10^5 G in the overshoot region (Moreno-Insertis 1992). The magnetic energy density is still much less than the thermal energy of the gas but buoyancy forces can overcome Coriolis effects, while an adiabatically rising flux tube remains relatively compact, with a strong magnetic field, until it approaches the surface.

2.6 SELF-CONSISTENT NONLINEAR DYNAMO MODELS

Given this observational picture, how should one try to model dynamo action in a star? Any attempt to simulate the whole process on a computer is clearly doomed; all that is feasible is to construct idealized model problems that can be solved with some precision. The natural starting point is to consider Boussinesq convection in a rapidly rotating spherical annulus that is heated at its inner boundary. Numerical solutions demonstrate the presence

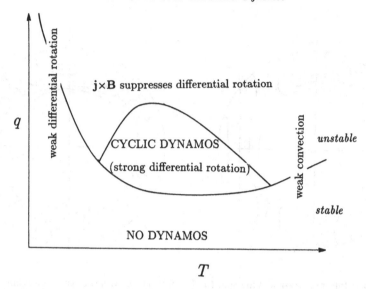

Figure 2.15 Behaviour of self-consistent nonlinear dynamos. Schematic representation of different parameter regimes in the Tq-plane. (After Gilman 1983.)

of banana cells which generate a pattern of differential rotation like that illustrated in figure 2.14(a) (Gilman 1979). Dynamo action is confirmed by introducing a seed magnetic field which grows and develops to give waves of activity — though they propagate towards the poles and not towards the equator (Gilman 1983). The process works but the model fails to reproduce an essential feature of the solar cycle. Moreover, anelastic calculations yield essentially similar results (Glatzmaier 1985).

Gilman (1983) investigated behaviour as the magnetic diffusivity and the angular velocity were varied; similar results have been obtained for geo-dynamo models by Zhang & Busse (1988, 1989). Gilman's results are summarized in figure 2.15, which shows the Tq-parameter plane, where the Taylor number $T \propto \Omega^2$ is a dimensionless measure of the rotation rate and q is the ratio of the thermal to the magnetic diffusivity; the thermal forcing, measured by a Rayleigh number, was held fixed. As q is increased it passes through a critical value, depending on the magnitude of T, where dynamo action sets in as a magnetic instability. The bifurcation value is a minimum for moderate values of T where there is a subcritical Hopf bifurcation leading to a cyclic dynamo in the nonlinear regime. If the rotation rate is too low, differential rotation is weak and the dynamo is hard to excite; if T is very large, convection is inhibited by rotation and the critical value of q rises accordingly. Nonlinear cyclic dynamos appear only in a restricted parameter range, where there is strong differential rotation. As q is further increased the Lorentz force becomes more powerful, differential rotation is suppressed and regular cyclic behaviour disappears. These computations clearly demonstrate that a combination of convection and rotation can produce cyclic magnetic activity, qualitatively similar to that observed in the Sun. They also show that this pattern only appears in a limited region of parameter space. In order to carry out an honest calculation it is necessary to adopt relatively large laminar diffusivities, which have the effect of suppressing small-scale motion. The results can only be applied to a star by assuming that turbulent diffusion is adequately represented by such a straightforward formulation. So it comes as no surprise that such a simplified model cannot reproduce the detailed pattern of solar activity.

Others have attempted to represent turbulent processes more accurately by restricting the calculation to a box at some intermediate latitude. Thus Nordlund *et al.* (1992) obtained local dynamo action in a compressible layer rotating about an axis inclined to the vertical, with periodic lateral boundary conditions, as described by Brandenburg in chapter 4. These models confirm that convection can generate a disordered field in a rotating system but it is not clear how far they are relevant to the large-scale ordered pattern of the solar cycle.

Computers are rapidly growing more powerful, making it feasible to tackle increasingly ambitious simulations. It would, however, be premature to start a detailed computation until the phenomenological picture has been rendered more precise. For the moment, therefore, it seems preferable to rely on simpler models.

2.7 MEAN FIELD DYNAMOS

Mean field dynamo theory has already been reviewed by Roberts in the opening chapter of this book. For turbulent dynamo action the effect of differential rotation (the ω-effect) is measured by a magnetic Reynolds number $R_\omega = \Omega' d^3/\eta_T$, where Ω' is the relevant angular velocity gradient, η_T is the turbulent diffusivity and d is an appropriate lengthscale. The action of helical or cyclonic eddies in regenerating the poloidal field from the toroidal field (the α-effect) is measured by a parameter α with the dimensions of a velocity, so the appropriate magnetic Reynolds number is $R_\alpha = \alpha d/\eta_T$. Then the critical parameter that determines dynamo action is the dynamo number

$$D = R_\alpha R_\omega = \frac{\alpha \Omega' d^4}{\eta_T^2}.$$

For turbulent eddies with velocity v, length-scale l and turnover time $\tau_c = l/v$ we can estimate the effect of the Coriolis force to obtain $\alpha \approx \Omega l$, while

$\Omega' \approx \Omega/l$ (say), $\eta_T \approx l^2/\tau_c$ and hence

$$D \approx \Omega^2 \tau_c^2 \left(\frac{d}{l}\right)^4.$$

Finally, if we assume that there is a unique length-scale, so that $l = d$, then $D \approx \sigma^2$, where the inverse Rossby number $\sigma = \Omega\tau_c$ (Durney & Latour 1978). We have already seen in section 2.2 that stellar magnetic activity does indeed depend on σ.

Since the pioneering paper of Steenbeck & Krause (1969) many kinematic $\alpha\omega$-dynamo models have been put forward to describe the solar cycle (Parker 1979, Stix 1989). The most promising current version is the $\alpha^2\omega$-dynamo developed by Prautzsch (1993) and Schmitt (1993) at Göttingen. In this model the isorotational surfaces are derived from helioseismology, as in figure 2.14(b), and the ω-effect is concentrated near the base of the convection zone, while $\partial\Omega/\partial r$ has different signs at high and low latitudes. Strong toroidal fields accumulate in the region of convective overshoot; in a rapidly rotating system they support magnetostrophic waves, which are excited by instabilities powered by magnetic buoyancy (e.g., Acheson 1979, Hughes 1985, Schmitt 1987). These waves have a net helicity, which provides an α-effect that is antisymmetric about the equator and concentrated near the base of the convection zone; α is negative at low latitudes in the northern hemisphere but changes sign at higher latitudes, where it drops in magnitude. This combination guarantees the production of dynamo waves that migrate towards the equator. Introducing a radial dependence of the diffusivity η_T, such that η_T falls to zero at the base of the overshoot layer, ensures that strong fields are concentrated there. The results are summarized by the butterfly diagrams in figures 2.16(a, b), which show the toroidal field near the bottom and top of the convection zone for a marginally stable solution (Prautzsch 1993, Schmitt 1993). This linear model is consistent both with observations and with the phenomenological picture outlined in the previous section.

The next stage is to include nonlinear effects in order to limit the growth of the magnetic instability. To be consistent, this requires an extension of mean field electrodynamics to cover the equation of motion. This has been attempted by Rüdiger (1989), who introduced the Λ-effect to generate differential rotation, and a nonlinear dynamo model has been developed by Brandenburg *et al.* (1991). Most treatments have, however, been content to introduce parametrized effects, in the spirit of the mean field approximation. The α-effect is delicate and therefore sensitive to the Lorentz force (cf. Tao, Cattaneo & Vainshtein 1993), so α-quenching — for example, by setting $\alpha \propto (1 + |\mathbf{B}|^2/B_0^2)^{-1}$ for some B_0 (Jepps 1975) — is a popular procedure; for consistency, η_T ought to be similarly reduced. Gilman's calculations, as well as observations, suggest that differential rotation is affected by a strong

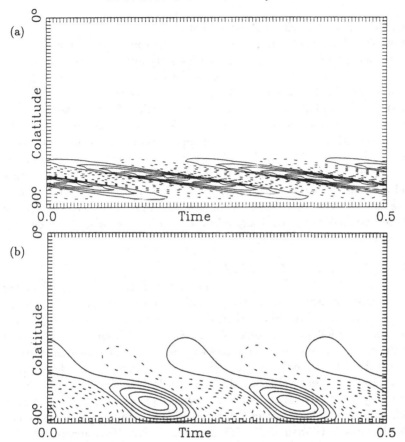

Figure 2.16 Butterfly diagrams for a kinematic mean field dynamo. Contours of the toroidal field (*a*) near the base of the convection zone and (*b*) near the surface, as functions of colatitude and time, for a model constructed to match our current understanding of the solar cycle. (After Schmitt 1993.)

magnetic field, so ω-quenching can also be introduced. Finally, there is the possibility of enhanced losses through magnetic buoyancy, which can be represented by introducing a loss rate proportional to $|B_T|B_T$ or B_T^3 in the equation for the toroidal field B_T. It should be remarked that the formation of active regions by rising Ω-shaped flux tubes does not in itself cause any net loss of toroidal flux; rather, one should consider entire flux tubes that are lost to the dynamo as they escape into the convection zone, where they are mangled and destroyed by turbulent motion. There are many papers describing non-linear mean field models in the literature (e.g., Parker 1979, Stix 1989, Moss, Tuominen & Brandenburg 1990). Some have been designed to give butterfly diagrams like that in figure 2.8; others have generated more exotic patterns (e.g., Belvedere, Pidatella & Proctor 1990).

How should we regard mean field theory in the context of stellar dynamos? The theory is well-founded and can be rigorously justified in certain circumstances. Unfortunately, there is no separation of spatial scales in stars like the sun. Nor is there a separation of scales in time that would justify the first-order smoothing (or quasilinear) approximation, for the magnetic Reynolds number is large and the lifetime of convective eddies is comparable with the turnover time τ_c. So the mean field approximation is not valid (Cowling 1981, Weiss 1983). Mean field dynamos must be regarded as conveniently parametrized models, based on a theory that captures the essential physics (and relies implicitly on the existence of fast dynamos). Unless one retains a naive faith in models of turbulent behaviour, one must, however, consider any detailed calculation as essentially arbitrary. So it seems wiser to to develop models that are less elaborate.

2.8 ILLUSTRATIVE MODELS

Mean field dynamos are usually axisymmetric, with $\mathbf{B} = \mathbf{B}(r, \theta, t)$. The most obvious simplification is to average radially through the magnetic layer, so as to obtain a one-dimensional shell dynamo, with $\mathbf{B} = \mathbf{B}(\theta, t)$ only. This procedure was followed, for example, by Schmitt & Schüssler (1989): the nonlinear solutions shown in figure 2.17 were all obtained for the same value of D, with the field limited by α-quenching. The existence of multiple solutions for the same parameter values is a characteristic feature of nonlinear problems.

An easy alternative is to reduce the system to planar geometry (Parker 1955, 1979, Stix 1972, Kleeorin & Ruzmaikin 1981, Weiss, Cattaneo & Jones 1984, DeLuca & Gilman 1986, 1988, Schmalz & Stix 1991, Jennings 1991). Such models have been used to compare the effects of different nonlinear saturation mechanisms, for instance by comparing the dependence of the cycle period on the parameter σ with that found observationally (Noyes *et al.* 1984*b*, Jennings & Weiss 1991). Parker (1993) has recently developed an interesting kinematic model in which the α-effect and ω-effect are confined to the half spaces $z > 0$ and $z < 0$, respectively, where the z-axis points upwards, and the magnetic diffusivity is much greater in the upper half-space. This ingenious configuration provides a tractable means of analysing dynamo action in a thin magnetic layer at the base of the convection zone, with a strong shear in the region of weak convective overshoot. The Hopf bifurcation leads to surface waves which are localised near the interface, with the poloidal structure illustrated in figure 2.18. Further properties of this model are currently being explored by S. Tobias.

Much can be learned by studying severely truncated toy models. Although they cannot be used to predict the behaviour of a realistic stellar dynamo, such models do provide a means of demonstrating generic patterns of be-

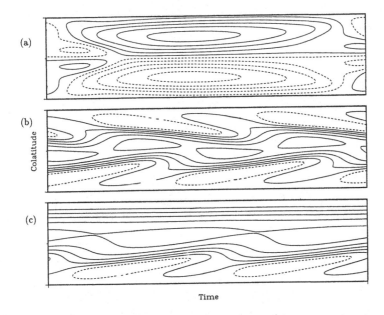

Figure 2.17 Butterfly diagrams for nonlinear mean field dynamos. Contours of the toroidal field for a spherical dynamo model, showing (*a*) a periodic dipole solution, with the field antisymmetric about the equator, (*b*) a periodic solution with spatiotemporal symmetry of type *mq* and (*c*) an asymmetric periodic solution of type *e*. (After Schmitt & Schüssler 1989.)

haviour that are shared by mean field dynamo models and by fully self-consistent dynamos (Weiss 1993). The toy systems can be analysed in detail, taking advantage of recent developments in nonlinear dynamics. Spiegel describes this approach as 'astromathematics' in chapter 8.

The most obvious problem to consider is the apparently chaotic behaviour of the solar cycle, which was described in section 2.3. The sunspot cycle is a classic example of an aperiodic oscillator (e.g., Yule 1927, Spiegel 1985, Tong 1983, 1990, Weiss 1988) and the available evidence suggests that the magnetic activity shown in figure 2.9 is intrinsically irregular, rather than a result of noise superposed upon a periodic oscillation (Gough 1990). Unfortunately, the sunspot record is too short to distinguish conclusively between stochastic and deterministic fluctuations (Weiss 1990, Morfill, Scheingraber & Sonett 1991). It has been suggested that the solar dynamo is so close to being marginally stable that the amplitude of the magnetic cycle is controlled by turbulent fluctuations (e.g., Hoyng 1988, Choudhuri 1992). That would place the Sun in a special category, where it cannot be compared with other stars; such an approach does not contribute to the general problem. On gen-

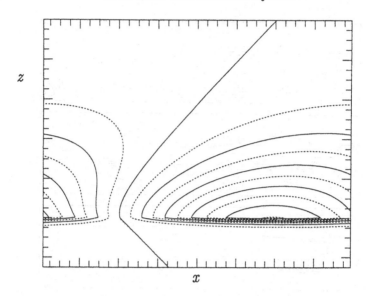

Figure 2.18 Surface wave dynamo. Lines of force of the leftward-propagating poloidal field for a two-dimensional kinematic model with $\alpha = 0$ for $z < 0$ and $\omega = 0$ for $z > 0$. The x and z co-ordinates are scaled differently. (After Parker 1993.)

eral physical grounds we might indeed expect nonlinear dynamos to become chaotic, and turbulent noise should only be a minor perturbation.

The irregular behaviour of the solar cycle since 1700 is typical of many chaotic oscillators but the systematic modulation associated with grand minima indicates the presence of some deeper structure. This can be modelled by low-order systems of coupled nonlinear ordinary differential equations. All that is needed is to take a second-order system with a Hopf bifurcation (to describe the basic cycle) and to couple it to the output of some third-order system with slow chaotic oscillations. Spiegel (chapter 8) and Platt (1993) choose the Lorenz system for this purpose and obtain a fifth-order model that generates intermittent cyclic behaviour. Weiss *et al.* (1984) followed a somewhat different approach: they extended Parker's (1955, 1979) linear dynamo waves — already described by Roberts in chapter 1 — to include nonlinear interactions with spatially fluctuating differential rotation (the observed 'torsional oscillations') and constructed a sixth-order system. The solutions revealed three successive Hopf bifurcations leading to quasiperiodic behaviour, with trajectories lying on a three-torus in phase space; as the dynamo number increased the torus was actually destroyed, giving rise to aperiodically modulated solutions. In this model system the amplitude of the basic cycle is modulated on two different time-scales (Jones, Weiss & Cattaneo 1985),

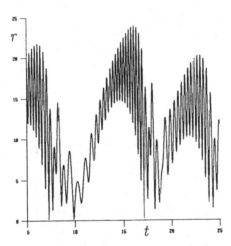

Figure 2.19 Chaotic modulation in a sixth-order model of nonlinear dynamo waves. Variation of the envelope of the oscillatory poloidal field, which is proportional to $r(t)\exp i\theta(t)$, for a dynamo number $D = 8$ (cf. Jones *et al.* 1985).

as shown in figure 2.19. Similar results have been obtained by Schmalz & Stix (1991) and by Feudel, Jansen & Kurths (1993). In fact, aperiodic modulation only requires a third-order system (Langford 1983, Weiss 1993). The appropriate bifurcation structure has recently been analysed by Kirk (1991) and the behaviour of a simple model is currently being investigated (Tobias & Weiss 1994, Tobias, Weiss & Kirk, in preparation).

The function of these models needs to be carefully explained, as it is frequently misunderstood. It is well known that arbitrarily truncated systems often display chaotic behaviour that is an artifact of the truncation: the chaos disappears as the order of the truncated system is increased. In certain problems, such as thermosolutal convection (Proctor & Weiss 1990, Knobloch, Proctor & Weiss 1992), it is possible to derive low-order models that exhibit chaotic behaviour in a parameter regime where they are asymptotically valid approximations to the nonlinear partial differential equations. This is not possible for stellar dynamos. The linear mean field equations themselves have no proper justification and there is no accepted procedure for extending them into the nonlinear regime. Nevertheless, we can recognize behaviour in a real system and describe it by constructing appropriate low-order models. So

(a)   (b)

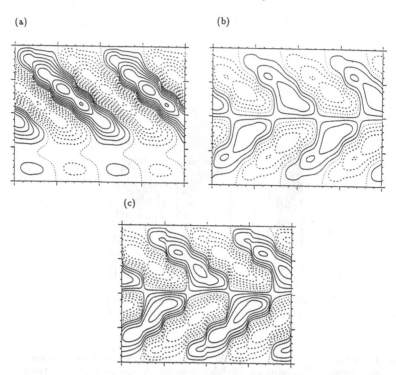

(c)

Figure 2.20 Symmetry-breaking in nonlinear dynamos. Butterfly diagrams for a one-dimensional dynamo model, showing periodic solutions with spatiotemporal symmetries (a) of type mi and (b) of type mq. (c) An asymmetric periodic solution of type e. (After Jennings & Weiss 1991.)

we turn to the dynamo in the sky: the observed modulation of solar activity has a characteristic time-scale and exhibits patterns that repeat. This suggests that intermittency is associated with mechanisms of the type described above, which can be investigated systematically by studying simple models. For the moment, we are unable to proceed much further.

Spatial structure raises other issues. The azimuthally averaged solar magnetic field has approximate dipole symmetry, with a toroidal field that is antisymmetric about the equator, but there are significant deviations which persist for several cycles. Mean field dynamos exhibit similar behaviour (e.g., Brandenburg *et al.* 1989, Schmitt & Schüssler 1989, Moss *et al.* 1990, Rädler *et al.* 1990). For example, the toroidal field in figure 2.17(a) is antisymmetric about the equator, while those in figures 2.17(b, c) are asymmetric. In order to investigate symmetry-breaking systematically it is necessary to follow branches of stable and unstable oscillatory solutions. That can only be done for a highly simplified model. Jennings (1991) and Jennings & Weiss

(1991) considered a one-dimensional cartesian dynamo (cf. Stix 1972) with parametrized nonlinearities and represented the solution by a total of 14 Fourier modes. The first bifurcation from the field-free trivial solution led to steady quadrupole (qs) solutions, with a toroidal field symmetric about the equator; it was followed by Hopf bifurcations giving rise to branches of pure dipole (d) and quadrupole (q) oscillatory solutions. Symmetry-breaking bifurcations led to mixed-mode oscillations of the types shown in figure 2.20. The solution in figure 2.20(a) (of type mi) has a symmetry with respect to displacement by half a period in time at the same latitude. On the other hand, the solution in figure 2.20(b) (of type mq) has a symmetry corresponding to reflection about the equator and advancing half a period in time; the solution for a spherical dynamo in figure 2.17(b) displays the same symmetry. A tertiary bifurcation leads to solutions (of type e) that are periodic but have no other symmetry, like those in figures 2.20(c) and 2.17(c). The full range of symmetries can be classified by constructing the appropriate symmetry group and the advantage of the simplified treatment is that the bifurcation structure can be established (Jennings & Weiss 1991). This is shown in figure 2.21, where the different branches are labelled according to the types of solution on them. Although the details are all model-dependent, the general pattern of primary, secondary and tertiary bifurcations, leading to a plethora of stable and unstable branches, must be a feature of any nonlinear dynamo. Further complications arise if we consider non-axisymmetric systems (Gubbins & Zhang 1993), which will be discussed by Knobloch in chapter 11.

2.9 STELLAR DYNAMOS
The theory of stellar dynamos has been increasingly influenced by the solar-stellar connection, although most models have been devised in order to explain the sunspot cycle. Theory predicts that stellar activity depends on the dynamo number $D \propto \sigma^2$ and observations confirm that magnetic activity is indeed a function of the inverse Rossby number σ. As a star evolves on the main sequence it spins down and becomes less active. For theoreticians it is, however, more appropriate to start with a slowly rotating star and to consider the effect of increasing the dynamo number. In this concluding section I present some physical speculations on magnetic activity in stars of fixed mass as the angular velocity is increased: this is equivalent to following the Sun's evolution in reverse.

Slow rotators, with σ of order unity, apparently behave like the Sun. The Coriolis force is relatively weak and convection leads to an angular velocity distribution with a steep radial gradient at the base of the convection zone. As a result, there is a shell dynamo (of $\alpha\omega$ or $\alpha^2\omega$ type) in the overshoot region, which generates the relatively ordered field that is responsible for the solar cycle. This large-scale dynamo is coupled to a small-scale turbulent

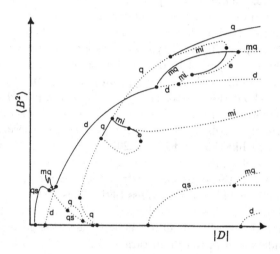

Figure 2.21 Bifurcations in a nonlinear dynamo. Schematic bifurcation diagram for a one-dimensional model, showing the averaged magnetic energy $\langle B^2 \rangle$ as a function of the dynamo number D. Full and broken lines denote stable and unstable solutions, respectively, and the different solution branches are labelled according to their spatiotemporal symmetries. (After Jennings & Weiss 1991.)

dynamo (of α^2 type) within the convection zone (cf. Zel'dovich, Ruzmaikin & Sokoloff 1983). Left on its own, the latter would generate a disordered fibril field with a highly intermittent structure; in addition, it is fed by flux that enters the convection zone from below (Golub *et al.* 1981).

The level of magnetic activity in the Sun implies that the dynamo number $D_\odot$ is comfortably supercritical. The observational evidence suggests, first, that D does not drop below the critical value D_c during a G-star's lifetime on the main sequence (Baliunas & Jastrow 1990) and, secondly, that cyclic activity — interrupted by grand minima — persists over a moderate parameter range with $D_c < D \leq 2D_\odot$ (Noyes *et al.* 1984b, Saar & Baliunas 1992). Within this range we should expect spatiotemporal behaviour to grow more and more complicated as D increases.

There is no reason why this cyclic regime should persist for large values of D. We saw from Gilman's (1983) model in figure 2.15 that cyclic dynamos occurred in a limited region of parameter space. In rapidly rotating stars the pattern of convection is dominated by the Coriolis force, leading to the formation of banana cells when $D \gg 1$. The changed pattern of convection must lead to a different distribution of angular momentum and we expect the isorotational surfaces to be cylinders. Then a different type of dynamo will operate (Knobloch, Rosner & Weiss 1981). Indeed, it seems likely that cyclic behaviour is itself a feature of slowly rotating stars with a magnetic layer at the base of the convection zone. It is therefore dangerous to extrapolate from solar dynamo models to describe rapid rotators, where the field structure will be very different.

In hyperactive stars, with $\sigma \approx 100$, the magnetic field must interfere substantially with convection. Indeed, magnetic flux is likely to be concentrated into isolated regions so that convection can proceed elsewhere. Such a configuration may be responsible for polar starspots. These strong fields exert a Lorentz force which could cancel the Coriolis force, allowing a different pattern of convection (e.g., Roberts 1991). It is not even obvious that the field will reverse; indeed, a fossil field might be able to survive without requiring a dynamo to maintain it (cf. Cattaneo, Hughes & Weiss 1991b).

Finally, we need to consider fully convective late M stars, where there is no possibility of a shell dynamo. Their observed activity has to be ascribed to a dynamo acting in the convection zone, which may be a fibril dynamo giving rise to a disordered field. Durney, De Young & Roxburgh (1993) suggest that the rate of spindown will consequently be reduced.

Future developments in the theory of solar and stellar dynamos will rely heavily on numerical experiments and simulations. It seems wise to focus on the Sun, where there are strong observational constraints. It is therefore possible to isolate different physical mechanisms and to treat them separately. The first problem is the structure of compressible convection. Here three-dimensional numerical experiments are already yielding significant results (Stein & Nordlund 1989, Cattaneo *et al.* 1991a, Simon & Weiss 1991). The effects of convection on passive fields can be investigated through kinematic simulations (Simon, Title & Weiss 1991, Cattaneo & Vainshtein 1991) but the interaction with rotation has still to be explored. It is clear that turbulent dissipation and the α-effect are both sensitive to the presence of magnetic fields (Cattaneo & Vainshtein 1991, Vainshtein & Cattaneo 1992, Rosner & Weiss 1992, Tao *et al.* 1993, Jones & Galloway 1993, Brandenburg in chapter 4), though the details remain controversial. In the next chapter Proctor describes recent progress in studying two- and three-dimensional magnetoconvection. All these ingredients have to be incorporated in a nonlinear dynamo. So a successful model of the solar cycle seems a long way off.

Acknowledgements
This review has benefited from lively discussions with many participants in
the dynamo theory programme at the Isaac Newton Institute. I am especially
grateful to Michael Stix for suggesting improvements to the text and I thank
SERC for financial support while I held a Senior Fellowship.

REFERENCES

Acheson, D.J. 1979 Instability by magnetic buoyancy. *Solar Phys.* **62**, 23–50.

Baliunas, S.L. & Jastrow, R. 1990 Evidence for long-term brightness changes of solar-type stars. *Nature* **348**, 520–523.

Baliunas, S.L. & Vaughan, A.H. 1985 Stellar activity cycles. *Annu. Rev. Astron. Astrophys.* **23**, 379–412.

Beer, J., Raisbeck, G.M. & Yiou, F. 1991 Time variations of ^{10}Be and solar activity. In *The Sun in Time* (ed. C.P. Sonett, M.S. Giampapa & M.S. Matthews), pp. 343–359. University of Arizona Press.

Belvedere, G. 1990 Solar and stellar cycles. In *Inside the Sun* (ed. M. Cribier & G. Berthomieu), pp. 371–382. Kluwer.

Belvedere, G., Pidatella, R.M. & Proctor, M.R.E. 1990 Nonlinear dynamics of a stellar dynamo in a spherical shell. *Geophys. Astrophys. Fluid Dynam.* **51**, 263–286.

Bracewell, R.N. 1988 Varves and solar physics. *Q. J. R. Astron. Soc.* **29**, 119–128.

Brandenburg, A. & Tuominen, I. 1991 The solar dynamo. In *The Sun and Cool Stars: Activity, Magnetism, Dynamos* (ed. I. Tuominen, D. Moss & G. Rüdiger), pp. 223–233. Springer.

Brandenburg, A., Krause, F., Meinel, R., Moss, D. & Tuominen, I. 1989 The stability of nonlinear dynamos and the limited role of kinematic growth rates. *Astron. Astrophys.* **213**, 411–422.

Brandenburg, A., Moss, D., Rüdiger, G. & Tuominen, I. 1991 Hydromagnetic $\alpha\Omega$-type dynamos with feedback from large-scale motions. *Geophys. Astrophys. Fluid Dynam.* **61**, 179–198.

Byrne, P.B. 1992 Starspots. In *Sunspots: Theory and Observations* (ed. J.H. Thomas & N.O. Weiss), pp. 63–73. Kluwer.

Cattaneo, F. & Vainshtein, S.I. 1991 Suppression of turbulent transport by a weak magnetic field. *Astrophys. J.* **376**, L21–L24.

Cattaneo, F., Brummell, N.H., Toomre, J., Malagoli, A. & Hurlburt, N.E. 1991a Turbulent compressible convection. *Astrophys. J.* **370**, 282–294.

Cattaneo, F., Hughes, D.W. & Weiss, N.O. 1991b What is a stellar dynamo? *Mon. Not. R. Astron. Soc.* **253**, 479–484.

Choudhuri, A.R. 1992 Stochastic fluctuations of the solar dynamo. *Astron. Astrophys.* **253**, 277–285.

Cowling, T.G. 1981 The present status of dynamo theory. *Annu. Rev. Astron. Astrophys.* **19**, 115–135.

Damon, P.E. & Sonett, C.P. 1991 Solar and terrestrial components of the atmospheric ^{14}C variation spectrum. In *The Sun in Time* (ed. C.P. Sonett, M.S. Giampapa & M.S. Matthews), pp. 360–388. University of Arizona Press.

DeLuca, E.E. & Gilman, P.A. 1986 Dynamo theory for the interface between the convection zone and the radiative interior of a star Part I. Model equations and exact solutions. *Geophys. Astrophys. Fluid Dynam.* **37**, 85–127.

DeLuca, E.E. & Gilman, P.A. 1988 Dynamo theory for the interface between the convection zone and the radiative interior of a star Part II. Numerical solutions of the nonlinear equations. *Geophys. Astrophys. Fluid Dynam.* **43**, 119–148.

DeLuca, E.E. & Gilman, P.A. 1991 The solar dynamo. In *Solar Interior and Atmosphere* (ed. A.N. Cox, W.C. Livingston & M.S. Matthews), pp. 275–303. University of Arizona Press.

Durney, B.R. & Latour, J. 1978 On the angular momentum loss of late-type stars. *Geophys. Astrophys. Fluid Dynam.* **9**, 241–255.

Durney, B.R., De Young, D.S. & Roxburgh, I.W. 1993 On the generation of the large-scale and turbulent magnetic fields in solar-type stars. *Solar Phys.* **145**, 207–225.

Dziembowski, W., Goode, P.R. & Libbrecht, K.G. 1989 The radial gradient in the Sun's rotation. *Astrophys. J.* **337**, L53–L57.

Eddy, J.A. 1976 The Maunder minimum. *Science* **192**, 1189–1202.

Feudel, U., Jansen, W. & Kurths, J. 1993 Tori and chaos in a nonlinear model for solar activity. *J. Bifurc. Chaos* (in press).

Galloway, D J. & Weiss, N.O. 1981 Convection and magnetic fields in stars. *Astrophys. J.* **243**, 945–953.

Gilman, P.A. 1979 Model calculations concerning rotation at high solar latitudes and the depth of the solar convection zone. *Astrophys. J.* **231**, 284–292.

Gilman, P.A. 1983 Dynamically consistent nonlinear dynamos driven by convection in a rotating shell. II. Dynamos with cycles and strong feedbacks. *Astrophys. J. Suppl.* **53**, 243–268.

Glatzmaier, G.A. 1985 Numerical simulations of stellar convective dynamos. II. Field propagation in the convection zone. *Astrophys. J.* **291**, 300–307.

Golub, L., Rosner, R., Vaiana, G.S. & Weiss, N.O. 1981 Solar magnetic fields: the generation of emerging flux. *Astrophys. J.* **243**, 309–316.

Gough, D.O. 1990 On possible origins of relatively short-term changes in the solar structure. *Phil. Trans. R. Soc. Lond.* A **330**, 627–640.

Gubbins, D. & Zhang, K. 1993 Symmetry properties of the dynamo equations for palaeomagnetism and geomagnetism. *Phys. Earth Planet. Inter.* **75**,

225–241.

Hart, J.E., Glatzmaier, G.A. & Toomre, J. 1986a Space-laboratory and numerical simulations of thermal convection in a rotating hemispherical shell with radial gravity. *J. Fluid Mech.* **173**, 519–544.

Hart, J.E., Toomre, J., Deane, A.E., Hurlburt, N.E., Glatzmaier, G.A., Fichtl, G.H., Leslie, F., Fowlis, W.W. & Gilman, P.A. 1986b Laboratory experiments on planetary and stellar convection performed on Spacelab 3. *Science* **234**, 61–64.

Hartmann, L.W. & Noyes, R.W. 1987 Rotation and magnetic activity in main-sequence stars. *Annu. Rev. Astron. Astrophys.* **25**, 271–301.

Howard, R.F. 1984 Solar rotation. *Annu. Rev. Astron. Astrophys.* **22**, 131–155.

Howard, R.F. & LaBonte, B.J. 1980 The sun is observed to be a torsional oscillator with a period of 11 years. *Astrophys. J.* **239**, L33–L36.

Hoyng, P. 1988 Turbulent transport of magnetic fields. III. Stochastic excitation of global magnetic modes. *Astrophys. J.* **332**, 857–871.

Hughes, D.W. 1985 Magnetic buoyancy instabilities incorporating rotation. *Geophys. Astrophys. Fluid Dynam.* **34**, 99–142.

Hughes, D.W. 1992 The formation of flux tubes at the base of the convection zone. In *Sunspots: Theory and Observations* (ed. J.H. Thomas & N.O. Weiss), pp. 371–384. Kluwer.

Hughes, D.W. & Proctor, M.R.E. 1988 Magnetic fields in the solar convection zone: magnetoconvection and magnetic buoyancy. *Annu. Rev. Fluid Mech.* **20**, 187-223.

Jennings, R.L. 1991 Symmetry breaking in a nonlinear $\alpha\omega$-dynamo. *Geophys. Astrophys. Fluid Dynam.* **57**, 147–189.

Jennings, R.L. & Weiss, N.O. 1991 Symmetry breaking in stellar dynamos. *Mon. Not. R. Astron. Soc.* **252**, 249–260.

Jepps, S.A. 1975 Numerical models of hydromagnetic dynamos. *J. Fluid Mech.* **67**, 625–646.

Jones, C.A, & Galloway, D.J. 1993 Cylindrical convection in a twisted field. In *Solar and Planetary dynamos* (ed. M.R.E. Proctor, P.C. Matthews & A.M. Rucklidge), pp. 161–170. Cambridge University Press.

Jones, C.A., Weiss, N.O. & Cattaneo, F. 1985 Nonlinear dynamos: a complex generalization of the Lorenz equations. *Physica D* **14**, 161–176.

Kawaler, S.D. 1988 Angular momentum loss in low-mass stars. *Astrophys. J.* **333**, 236–247.

Kirk, V. 1991 Breaking of symmetry in the saddle-node-Hopf bifurcation. *Phys. Lett. A* **154**, 243–248.

Kleeorin, N.I. & Ruzmaikin, A.A. 1981 Properties of simple nonlinear dynamos. *Geophys. Astrophys. Fluid Dynam.* **17**, 281–296.

Knobloch, E., Rosner, R. & Weiss, N.O. 1981 Magnetic fields in late-type stars. *Mon. Not. R. Astron. Soc.* **197**, 45P–49P.

Knobloch, E., Proctor, M.R.E. & Weiss, N.O. 1992 Heteroclinic bifurcations in a simple model of double-diffusive convection. *J. Fluid Mech.* **239**, 273–292.

Langford, W.F. 1983 A review of interactions of Hopf and steady-state bifurcations. In *Nonlinear Dynamics and Turbulence* (ed. G.I. Barenblatt, G. Iooss & D.D. Joseph), pp. 215–237. Pitman.

Legrand, J.P. & Simon, P.A. 1991 A two-component solar cycle. *Solar Phys.* **131**, 187–209.

Libbrecht, K.G. 1988 Solar *p*-mode frequency splittings. In *Seismology of the Sun and Sun-like Stars* (ed. E.J. Rolfe), pp. 131–136. ESA–SP286.

Mestel, L. 1968 Magnetic braking by a stellar wind I. *Mon. Not. R. Astron. Soc.* **138**, 359–391.

Mestel, L. 1990 Magnetic braking. In *Basic Plasma Processes on the Sun* (ed. E.R. Priest & V. Krishan), pp. 67–76. Kluwer.

Mestel, L. & Spruit, H.C. 1987 On magnetic braking of late-type stars. *Mon. Not. R. Astron. Soc.* **226**, 57–66.

Moreno-Insertis, F. 1992 The motion of magnetic flux tubes in the convection zone and the subsurface origin of active regions. In *Sunspots: Theory and Observations* (ed. J.H. Thomas & N.O. Weiss), pp. 385–410. Kluwer.

Morfill, G.E. Scheingraber, H. & Sonett, C.P. 1991 Sunspot number variations: stochastic or chaotic? In *The Sun in Time* (ed. C.P. Sonett, M.S. Giampapa & M.S. Matthews), pp. 30–58. University of Arizona Press.

Moss, D., Tuominen, I. & Brandenburg, A. 1990 Nonlinear dynamos with magnetic buoyancy in spherical geometry. *Astron. Astrophys.* **228**, 284–294.

Nordlund, Å., Brandenburg, A., Jennings, R.L., Rieutord, M., Ruokolainen, J., Stein, R.F. and Tuominen, I. 1992 Dynamo action in stratified convection with overshoot. *Astrophys. J.* **392**, 647–652.

Noyes, R.W. 1982 *The Sun, our Star.* Harvard University Press.

Noyes, R.W., Hartmann, L.W., Baliunas, S.L., Duncan, D.K. & Vaughan, A.H. 1984*a* Rotation, convection, and magnetic activity in lower main-sequence stars. *Astrophys. J.* **279**, 763–777.

Noyes, R.W., Weiss, N.O. & Vaughan, A.H. 1984*b* The relationship between stellar rotation rate and activity cycle periods. *Astrophys. J.* **287**, 769–773.

Noyes, R.W., Baliunas, S.L. & Guinan, E.F. 1991 What can other stars tell us about the Sun? In *Solar Interior and Atmosphere* (ed. A.N. Cox, W.C. Livingston & M.S. Matthews), pp. 1161–1185. University of Arizona Press.

Parker, E.N. 1955 Hydromagnetic dynamo models. *Astrophys. J.* **122**, 293–314.

Parker, E.N. 1979 *Cosmical Magnetic Fields: their origin and activity.* Clarendon Press.

Parker, E.N. 1993 A solar dynamo surface-wave at the interface between convection and nonuniform rotation. *Astrophys. J.* **408**, 707–719.

Platt, N. 1993 On-off intermittency: general description and feedback models. In *Solar and Planetary dynamos* (ed. M.R.E. Proctor, P.C. Matthews & A.M. Rucklidge), pp. 233–240. Cambridge University Press.

Prautzsch, T. 1993 The dynamo mechanism in the deep convection zone of the Sun. In *Solar and Planetary dynamos* (ed. M.R.E. Proctor, P.C. Matthews & A.M. Rucklidge), pp. 249–264. Cambridge University Press.

Proctor, M.R.E. & Weiss, N.O. 1990 Normal forms and chaos in thermosolutal convection. *Nonlinearity* **3**, 619–637.

Rädler, K.-H., Wiedemann, E., Brandenburg, A., Meinel, R. & Tuominen, I. 1990 Nonlinear mean-field dynamo models: stability and evolution of three-dimensional magnetic field configurations. *Astron. Astrophys.* **239**, 413–423.

Raisbeck, G.M., Yiou, F., Jouzel, J. & Petit, J.R. 1990 ^{10}Be and δ^2H in polar ice cores as a probe of the solar variability's influence on climate. *Phil. Trans. R. Soc. Lond. A* **330**, 463–470.

Rhodes, E.J., Cacciani, A., Korzennik, S., Tomczyk, S., Ulrich, R.K. & Woodward, M.F. 1990 Depth and latitude dependence of the solar internal angular velocity. *Astrophys. J.* **351**, 687–700.

Ribes, E. 1990 Astronomical determinations of solar variability. *Phil. Trans. R. Soc. Lond. A* **330**, 487–497.

Roberts, P.H. 1991 Magnetoconvection patterns in rotating convection zones. In *The Sun and Cool Stars: Activity, Magnetism, Dynamos* (ed. I. Tuominen, D. Moss & G. Rüdiger), pp. 37–56. Springer.

Rosner, R. & Weiss, N.O. 1985 Differential rotation and magnetic torques in the interior of the Sun. *Nature* **317**, 790–792.

Rosner, R. & Weiss, N.O. 1992 The origin of the solar cycle. In *The Solar Cycle* (ed. K. Harvey), pp. 511–531. Astronomical Society of the Pacific.

Rüdiger, G. 1989 *Differential Rotation and Stellar Convection.* Akademie-Verlag/Gordon & Breach.

Saar, S.H. 1991a Recent advances in the observation and analysis of stellar magnetic fields. In *The Sun and Cool Stars: Activity, Magnetism, Dynamos* (ed. I. Tuominen, D. Moss & G. Rüdiger), pp. 389–400. Springer.

Saar, S.H. 1991b The time-evolution of magnetic fields on solar-like stars. In *The Sun in Time* (ed. C.P. Sonett, M.S. Giampapa & M.S. Matthews), pp. 832–847. University of Arizona Press.

Saar, S.H. & Baliunas, N.O. 1992 Recent advances in stellar cycle research. In *The Solar Cycle* (ed. K. Harvey), pp. 150–167. Astronomical Society of the Pacific.

Sams, B.J., Golub, L. & Weiss, N.O. 1992 X-ray observations of sunspot penumbral structure. *Astrophys. J.* **399**, 313–317.

Schatzman, E. 1962 A theory of the role of magnetic activity during star formation. *Ann. Astrophys.* **25**, 18–29.

Schmalz, S. & Stix, M. 1991. An $\alpha\Omega$ dynamo with order and chaos. *Astron. Astrophys.* **245**, 654–662.

Schmitt, D. 1987. An $\alpha\omega$-dynamo with an α-effect due to magnetostrophic waves. *Astron. Astrophys.* **174**, 281–287.

Schmitt, D. 1993. The solar dynamo. In *The Cosmic Dynamo* (ed. F. Krause, K.-H. Rädler & G. Rüdiger), pp. 1–12. Kluwer.

Schmitt, D. & Schüssler, M. 1989. Nonlinear dynamos I. One-dimensional model of a thin layer dynamo. *Astron. Astrophys.* **223**, 343–351.

Schüssler, M. 1981 The solar torsional oscillation and dynamo models of the solar cycle. *Astron. Astrophys.* **94**, 755–756.

Schüssler, M. 1993 Flux tubes and dynamos. In *The Cosmic Dynamo* (ed. F. Krause, K.-H. Rädler & G. Rüdiger), pp. 27–39. Kluwer.

Sheeley, N.R. 1992 The flux-transport-model and its implications. In *The Solar Cycle* (ed. K. Harvey), pp. 1–13. Astronomical Society of the Pacific.

Simon, G.W. & Weiss, N.O. 1991 Convective structures in the Sun. *Mon. Not. R. Astron. Soc.* **252**, 1P–5P.

Simon, G.W., Title, A.M. & Weiss, N.O. 1991 Modeling mesogranules and exploders on the solar surface. *Astrophys. J.* **375**, 775–788.

Skumanich, A. 1972 Time scales for CaII emission decay, rotational braking, and lithium depletion. *Astrophys. J.* **171**, 565–567.

Snodgrass, H.B. 1992 Synoptic observations of large scale velocity patterns on the Sun. In *The Solar Cycle* (ed. K. Harvey), pp. 205–240. Astronomical Society of the Pacific.

Soderblom, D.R. & Baliunas, S.L. 1988 The Sun among the stars: what the stars indicate about solar variability. In *Secular Solar and Geomagnetic Variations in the last 10,000 Years* (ed. F.R. Stephenson & A.W. Wolfendale), pp. 25–49. Kluwer.

Spiegel, E.A. 1985 Cosmic arrhythmias. In *Chaos in Astrophysics* (ed. J.R. Buchler, J.M. Perdang & E.A. Spiegel), pp. 91–135. Reidel.

Spiegel, E.A. & Weiss, N.O. 1980 Magnetic activity and variations in solar luminosity. *Nature* **287**, 616–617.

Spiegel, E.A. & Zahn, J.-P. 1992 The solar tachocline. *Astron. Astrophys.* **265**, 106–114.

Stauffer, J.R. & Soderblom, D.R. 1991 The evolution of angular momentum in solar-mass stars. In *The Sun in Time* (ed. C.P. Sonett, M.S. Giampapa & M.S. Matthews), pp. 832–847. University of Arizona Press.

Steenbeck, M. & Krause, F. 1969 Zur Dynamotheorie stellarer und planetarer Magnetfelder I. Berechnung sonnenähnlicher Wechselfeldgeneratoren. *As-*

tron. Nachr. **291**, 49–84.

Stein, R.F. & Nordlund, Å. 1989 Topology of convection beneath the solar surface. *Astrophys. J.* **342**, L95–L98.

Stenflo, J.O. 1989 Differential rotation of the Sun's magnetic field pattern. *Astron. Astrophys.* **210**, 403–409.

Stix, M. 1972 Nonlinear dynamo waves. *Astron. Astrophys.* **20**, 9–12.

Stix, M. 1989 *The Sun.* Springer.

Stix, M. 1991 The solar dynamo. *Geophys. Astrophys. Fluid Dynam.* **62**, 211–228.

Stuiver, M. & Braziunas, T.F. 1988 The solar component of the atmospheric ^{14}C record. In *Secular Solar and Geomagnetic Variations in the last 10,000 Years* (ed. F.R. Stephenson & A.W. Wolfendale), pp. 245–266. Kluwer.

Stuiver, M. & Braziunas, T.F. 1989 Atmospheric ^{14}C and century-scale solar oscillations. *Nature* **338**, 405–408.

Tao, L., Cattaneo, F. & Vainshtein, S.I. 1993 Evidence for the suppression of the α-effect by weak magnetic fields. In *Solar and Planetary dynamos* (ed. M.R.E. Proctor, P.C. Matthews & A.M. Rucklidge), pp. 303–310. Cambridge University Press.

Thomas, J.H. & Weiss, N.O. 1992 The theory of sunspots. In *Sunspots: Theory and Observations* (ed. J.H. Thomas & N.O. Weiss), pp. 3–59. Kluwer.

Tobias, S.M. & Weiss, N.O. 1994 Nonlinear stellar dynamos. In *Cosmical Magnetism: Contributed Papers* (ed. D. Lynden-Bell), Institute of Astronomy, Cambridge, UK.

Tong, H. 1983 *Threshold models in non-linear time series analysis.* Springer.

Tong, H. 1990 *Nonlinear time series: a dynamical systems approach.* Clarendon Press.

Ulrich, R.K., Boyden, J.E., Webster, L., Snodgrass, H.B., Podilla, S.P., Gilman, P. & Shieber, T. 1988 Solar rotation measurements at Mount Wilson V. Reanalysis of 21 years of data. *Solar Phys.* **117**, 291–328.

Vainshtein, S.I. & Cattaneo, F. 1992 Nonlinear restrictions on dynamo action. *Astrophys. J.* **393**, 165–171.

van Ballegooijen, A.A. 1982 The overshoot layer at the base of the solar convective zone and the problem of magnetic flux storage. *Astron. Astrophys.* **113**, 99–112.

Wang, Y.-M. & Sheeley, N.R. 1991 Magnetic flux transport and the Sun's dipole moment: new twists to the Babcock–Leighton model. *Astrophys. J.* **375**, 761–770.

Wang, Y.-M., Sheeley, N.R. & Nash, A.G. 1991 A new solar cycle model including meridional circulation. *Astrophys. J.* **383**, 431–442.

Weiss, J.E. & Weiss, N.O. 1979 Andrew Marvell and the Maunder Minimum. *Q. J. R. Astron. Soc.* **20**, 115–118.

Weiss, N.O. 1983 Solar magnetism. In *Stellar and Planetary Magnetism*, vol. 2 (ed. A.M. Soward), pp. 115–131. Gordon & Breach.

Weiss, N.O. 1988 Is the solar cycle an example of deterministic chaos? In *Secular Solar and Geomagnetic Variations in the last 10,000 Years* (ed. F.R. Stephenson & A.W. Wolfendale), pp. 69–78. Kluwer.

Weiss, N.O. 1989 Dynamo processes in stars. In *Accretion Discs and Magnetic Fields in Astrophysics* (ed. G. Belvedere), pp. 11–29. Kluwer.

Weiss, N.O. 1990 Periodicity and aperiodicity in solar magnetic activity. *Phil. Trans. R. Soc. Lond. A* **330**, 617–625.

Weiss, N.O. 1993 Bifurcations and symmetry-breaking in simple models of nonlinear dynamos. In *The Cosmic Dynamo* (ed. F. Krause, K.-H. Rädler & G. Rüdiger), pp. 219–229. Kluwer.

Weiss, N.O., Cattaneo, F. & Jones, C.A. 1984 Periodic and aperiodic dynamo waves. *Geophys. Astrophys. Fluid Dynam.* **30**, 305–341.

White, O.R., Skumanich, A., Lean, J., Livingston, W.C. & Keil, S.L. 1992 The Sun in a non-cycling state. *Publ. Astron. Soc. Pacific.* **104**, 1139–1143.

Wilson, O.C. 1978 Chromospheric variations in main-sequence stars. *Astrophys. J.* **226**, 379–396.

Yallop, B.D. & Hohenkerk, C.Y. 1980 Distribution of sunspots 1874–1976. *Solar Phys.* **68**, 303–305.

Yoshimura, H. 1981 Solar cycle Lorentz force: waves and the torsional oscillations of the sun. *Astrophys. J.* **247**, 1102–1111.

Yule, G.U. 1927 On a method of investigating periodicities in disturbed series, with special reference to Wolf's sunspot numbers. *Phil. Trans. R. Soc. Lond. A* **226**, 267–298.

Zel'dovich, Ya.B., Ruzmaikin, A.A. & Sokoloff, D.D. 1983 *Magnetic Fields in Astrophysics*. Gordon and Breach.

Zhang, K. & Busse, F.H. 1988 Finite amplitude convection and magnetic field generation in a rotating spherical shell. *Geophys. Astrophys. Fluid Dynam.* **44**, 33–53.

Zhang, K. & Busse, F.H. 1989 Magnetohydrodynamic dynamos in rotating spherical shells. *Geophys. Astrophys. Fluid Dynam.* **49**, 97–116.

CHAPTER 3

Convection and Magnetoconvection in a Rapidly Rotating Sphere

M. R. E. PROCTOR

Department of Applied Mathematics and Theoretical Physics
University of Cambridge
Cambridge CB3 9EW, UK

3.1 INTRODUCTION

In this chapter I attempt a review of theories of convection in a spherical geometry in the presence of magnetic fields and rotation. The understanding of such motion is essential to a proper theory of the geodynamo. Even though, as discussed by Malkus and Braginsky (chapters 5 and 9), the nature of the driving mechanism for the convection is not certain, and is likely to be compositional in nature, we shall generally, following other authors, look only at thermal convection, which is the simplest to study. In addition, it will be assumed (incorrectly) that the core fluid has essentially constant viscosity, density, etc., allowing the Boussinesq approximation to be employed.

The excuse for these simplifications is readily to hand: the dynamical complexities induced by the interaction of Coriolis and Lorentz forces are still not fully resolved, and transcend the details of the forcing or of compressibility effects. The effects of this interaction on global fields are discussed by Fearn (chapter 7) but here we shall confine ourselves to a small part of the complete picture: the non-axisymmetric instabilities of an imposed (and prescribed) axisymmetric magnetic field and differential rotation in a rotating sphere. This task is the mirror-image of the 'intermediate' models of Braginsky (chapter 9) and the non-linear 'macrodynamic' dynamos driven by the α-effect, described by Fearn (chapter 7), in that these works parametrize the small, rather than the global fields.

In what follows, we shall begin by defining a geometry and non-dimensionalization for the system. We shall mainly be working in a spherical geometry, but use for illustration simplified (e.g., planar, cylindrical) geometry where appropriate. We shall first treat the onset of convection when the magnetic field is weak, and then describe the changes that occur as the field gets stronger, finally treating recent work on non-axisymmetric 'magnetostrophic' instabilities. We shall throughout remain resolutely linear, leaving any dis-

M.R.E. Proctor & A.D. Gilbert (eds.)
Lectures on Solar and Planetary Dynamos, 97–115
©1994 Cambridge University Press.

cussion of the effects of the dynamical back-reaction of the magnetic field to Fearn's lecture in chapter 7.

3.2 SCALING, GEOMETRY AND GOVERNING EQUATIONS

We recall that the dynamics of the core is certainly characterised by strong rotation and (probably) by strong magnetic fields, with viscosity and inertial forces much smaller in magnitude, though not insignificant. It is desirable to scale all variables to reflect this balance. We shall simplify the core to a sphere of radius r_0, rotating at angular velocity $\Omega \mathbf{1}_z$. (We shall use both cylindrical (s, z, ϕ) and spherical (r, ϑ, ϕ) coordinate systems as seems convenient.) Then the rotation is rapid in the sense that the *Ekman number*

$$E \equiv \nu/2r_0^2, \tag{3.2.1}$$

measuring the ratio of viscous to Coriolis forces, is very small. The actual value of E in the core depends on the (unreliable) determination of ν. Most recent estimates give $\nu \sim 10^{-5}\text{--}10^{-7}$ m^2 s^{-1}, so that $E \sim 10^{-12}\text{--}10^{-14}$ (Poirier 1988). If we further simplify by supposing that the core density is fairly constant and of magnitude ρ_0 then the required balance between Lorentz and Coriolis forces is achieved if the *Elsasser number*

$$\Lambda = B_0^2/2\Omega\mu\rho_0\eta \tag{3.2.2}$$

is of order unity, where B_0 is a typical magnitude of $\mathbf{B}$ and η is the magnetic diffusivity as before. Based on the observed value of the poloidal magnetic field on the core–mantle boundary $\Lambda \sim 10^{-2}$, but if we infer that the (invisible) toroidal field is rather larger than this then a value of $O(1)$ seems reasonable.

To provide the driving force for our convective model, we suppose that the temperature and hence density stratification is superadiabatic. In reality the most plausible source for core convection is the buoyancy produced by the release of light fluid at the inner core boundary (see Malkus, chapter 5), but the theory of such convection is not well developed. Assuming a basic radial temperature gradient of typical magnitude β, then, in keeping with our assumption of small density variation (so that we may use the Boussinesq approximation), we can write the temperature T in the form

$$T = \beta r_0(T_0(r) + \theta), \tag{3.2.3}$$

so that θ is a dimensionless measure of the perturbation temperature. In addition we suppose the density $\rho \approx \rho_0(1 - \alpha(T - \beta r_0 T_0))$. We take the thermal diffusivity to be κ, and complete the basic state with an imposed axisymmetric magnetic field $\mathbf{B} = B_0\hat{B}\mathbf{1}_\phi$, and zonal flow $\mathbf{U} = U_0\hat{U}\mathbf{1}_\phi$ so that $\hat{B}, \hat{U}$ are dimensionless functions of position. In reality these fields are in fact produced by the instabilities that we are going to describe, but here we will suppose them given and time independent. Then scaling the perturbation magnetic field **b** with B_0, the perturbation velocity **v** with κ/r_0 and time with r_0^2/κ, substituting into the equations and dropping hats, we obtain the dimensionless system (after removing the hydrostatic pressure):

$$Ep^{-1}\left[\frac{\partial \mathbf{v}}{\partial t} + Rq^{-1}U\mathbf{1}_\phi \cdot \nabla \mathbf{v}\right] + \mathbf{1}_z \times \mathbf{v} = -\nabla P +$$

$$+\Lambda q^{-1}((\nabla \times B\mathbf{1}_\phi) \times \mathbf{b} + (\nabla \times \mathbf{b}) \times B\mathbf{1}_\phi) + E\nabla^2 \mathbf{v} + Ra\, E\, \theta \mathbf{r}, \quad (3.2.4)$$

$$q\left[\frac{\partial \mathbf{b}}{\partial t} - \nabla \times (\mathbf{v} \times B\mathbf{1}_\phi)\right] = R_m \nabla \times (U\mathbf{1}_\phi \times \mathbf{b}) + \nabla^2 \mathbf{b}, \quad (3.2.5)$$

$$\frac{\partial \theta}{\partial t} + \mathbf{v} \cdot \nabla T + R_m q^{-1} U\mathbf{1}_\phi \cdot \nabla \theta = \nabla^2 \theta, \quad (3.2.6)$$

$$\nabla \cdot \mathbf{v} = \nabla \cdot \mathbf{b} = 0. \quad (3.2.7)$$

Here the new dimensionless parameters are the Prandtl number $p = \nu/\kappa$, the Roberts number $q = \kappa/\nu$ (though $\zeta = q^{-1}$ is usually used in the astrophysical literature), the magnetic Reynolds number $R_m = U_0 r_0/\eta$, and the Rayleigh number $Ra = g\alpha\beta r_0^4/\kappa\nu$. The quantity RaE is of special importance in the theory, and we shall denote it by $\widetilde{Ra}$ when convenient.

We note that the nonlinear terms in (2.4–7) have not been included. This is because we shall concentrate primarily on the linear instability of the basic state, which is more than difficult enough. Nonlinear effects will be discussed only in qualitative terms and without detailed calculation.

We have already seen that for the Earth's core $E \ll 1$. The material parameters p and q are not accurately known but it is likely that $p \leq O(1)$ while q is likely to be small $(O(10^{-6}))$ if molecular values are used. The small-q limit has its very own dynamical peculiarities however and there is some question as to whether such a small value can give an accurate picture of the actual dynamics, dominated as it is by compositional effects. A naive view of the effects of small-scale turbulence might suggest that a 'turbulent' q should be, like p, of order unity. Most published work has concentrated on the two cases $q = 10^{-6}$, $q = 1$, and so we shall too.

3.3 NON-MAGNETIC CONVECTION

The nature of the convection that arises in the core depends crucially on the Elsasser number Λ. In the context of the geodynamo this parameter is of course not at our disposal, but emerges as a result of the dynamo process. It seems clear that Λ cannot be very large, since the magnetic fields would tend to suppress convection, or to induce magnetic instabilities, as we shall discuss later. Let us first begin with the supposition that $\Lambda \ll 1$, so that the magnetic field can be regarded as making a negligible contribution to the dynamics. Then the momentum equation can be written

$$1_z \times \mathbf{v} + \nabla P = E \left[-p^{-1} \frac{\partial \mathbf{v}}{\partial t} + R \theta \mathbf{r} + \nabla^2 \mathbf{v} \right]. \qquad (3.3.1)$$

The nature of the dominant balance for $E \ll 1$ now depends on whether $Ep^{-1}\partial/\partial t = O(1)$, and this in turn seems to depend on the size of p. If we assume first 'slow' time dependence so that all the terms on the right-hand side are small, we are left with the geostrophic equation

$$1_z \times \mathbf{v} + \nabla P \approx 0, \qquad \text{implying} \qquad \frac{\partial \mathbf{v}}{\partial z} \approx 0. \qquad (3.3.2)$$

The leading order flow thus depends on s and ϕ only, and, in the absence of viscosity, for a perfect sphere the only exact solutions are *geostrophic flows* $\mathbf{v} = V_G(s)1_\phi$ which cannot be coupled directly to convection. Instead, convection takes the form of columnar motion. The shortest scale is azimuthal, represented by the inverse of the azimuthal wavenumber m. Simple balances in the governing equation then suggest that for $p = O(1)$

$$Ra\,|\theta| \sim m^2|\mathbf{v}| \sim \left| \frac{\partial \mathbf{v}}{\partial t} \right| \qquad \text{and} \qquad m^2|\theta| \sim |\mathbf{v}|, \qquad (3.3.3)$$

so that Ra, m and the frequency ω are ordered by the relations $Ra \sim \omega^2 \sim m^4$. The optimum size of m is determined by a subtle balance between the requirement that the cells be of small aspect ratio and the effects of tilt of the boundaries at $r = 1$ (the magnitude of which depends on the distance s_c of the dominant rolls from the rotation axis). Busse (1970), following Roberts (1968), was able to show that the optimum $m \sim E^{-1/3}$, so that $Ra \sim E^{-4/3}, \widetilde{Ra} \sim E^{-1/3}$, while the preferred value of s_c is about 0.5 (for $p \to 0$, $s_c \sim 0.52$), and s_c increases with p.

There are serious deficiencies in the linearised theory, however. The radial scale of motion in the columns is not determined properly at leading order, although the appropriate scale would appear to be $E^{2/9}$. Soward (1977) has noted that the leading order balances employed by Busse and Roberts are inconsistent, in that the variation of local phase speed across the layer renders the convection rolls liable to shear and ultimate destruction. Soward was able to find weakly nonlinear solutions close to the critical value Ra_c, by the asymptotic methods above, in which boundary curvature and nonlinear effects are both important in preventing the shear blowing up. He reasoned that if there *were* a solution to the linearised problem obeying the Roberts/Busse scalings, the details would be different and the critical value of Ra would not coincide with the earlier asymptotic calculation. It is only relatively recently that these ideas have been given support by Zhang (1992) who has presented an accurate numerical solution to the linearised equation, at small Ekman numbers. (E was made as small as 10^{-6} using spherical harmonics with degree up to about 100.) Convincing evidence of asymptotic behaviour was found, with the Roberts/Busse scaling but different numerical constants. For example in Zhang's calculation, when $p = 10$ we have $Ra_c \sim R_{\text{LIN}} = 1.4(E/2)^{-4/3}$ while for the asymptotic calculation the constant 1.4 is replaced by $R_{\text{ASYM}} = 0.802 \times (10/11)^{4/3} = 0.706$. For every Prandtl number, the asymptotic theory underestimates the actual stability boundary, and it is also in error in giving the critical wavenumber and frequency (see figure 1 of Zhang 1992). The form of the convection depends strongly on the Prandtl number. In every case the azimuthal derivative $s^{-1}\partial/\partial\phi$ is the largest of the three space derivatives. At small p, the radial derivative $\partial/\partial s$ appears to be comparable with $s^{-1}\partial/\partial\phi$, reflecting the *spiral* form of the convection (see figure 3.1). The radial extent of the cells is very large. As p increases both the radial extent and the spiralling angle decrease, to yield a convective pattern more similar to the original (and famous) 'columns' sketch by Busse (1970, 1975). The spiral regime has been shown by Yano (1992), in a detailed paper using WKB theory, to be part of the same branch of solutions as the columnar modes. His asymptotic theory gives reasonable agreement with Zhang's computations over a wide range of Prandtl numbers.

The role of the mechanical boundary conditions is obscure in these asymptotic theories, as the reduced equations are of second order in the limit and only require one (normal velocity) condition. For finite values of E, however, one may expect the distinction between rigid and stress-free boundaries to become apparent. Zhang & Jones (1993) have carried out the relevant calculations, and find that for rigid boundaries there is some reduction in the critical Rayleigh numbers when the Prandtl number is of order unity or larger. The reverse hold for small Prandtl numbers. However the differences are unlikely to be significant at the very small values of E obtaining in the core.

Zhang (1992) has also solved the full nonlinear problem for values of Ra slightly greater than R_{LIN}. Though nonlinearity is outside the remit of this chapter, it should be noted that he finds stable finite amplitude solutions while Soward's analysis suggests that stable finite amplitude convection should be possible for Ra close to R_{ASYM}. Though the role of mechanical boundary conditions (rigid versus stress-free) is unclear, these results can only be reconciled by a curve of amplitude against Ra of the form shown in figure 3.2.

Before leaving the non-magnetic problem, we should mention a recently-discovered mode of convection of quite a different type. Zhang (1993) has shown that certain inertial waves in an unstratified inviscid fluid (which represent a balance between inertia, Coriolis forces and pressure gradient) can

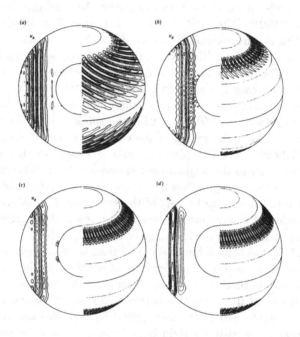

Figure 3.1 Contours of zonal velocity (left) and contours of surface toroidal flow (right) at the onset of convection for $E = 10^{-5.5}$ and $p = 0.1, 1.0, 10$ and 100 respectively. (After Zhang 1992.)

take the form of equatorially trapped modes. In the presence of unstable stratification these waves can be sustained against viscous dissipation. These inertial modes of convection, which are analogous to the 'wall modes' previously known to occur for convection in rotating cylinders (for references, see Knobloch, chapter 11) were earlier noted by Zhang & Busse (1987); they occur for lower values of Ra than the columnar mode when the Prandtl number p is sufficiently small. If the wavenumber m satisfies the inequality $m^{5/2} \gg pE^{-1}$ then the critical value of Ra is of order unity, while if this inequality fails Ra_c depends on p and E through the ratio pE^{-1}. The columnar modes occur at much higher azimuthal wave numbers than the equatorial modes, so

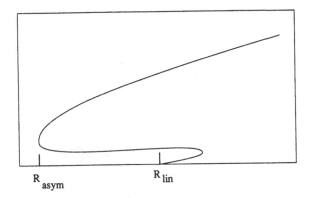

Figure 3.2 Sketch of presumed amplitude–Rayleigh number curve to reconcile the results of Zhang (1992) and Soward (1977).

although one expects the changeover to occur when $p \sim E$, in practice it occurs at larger values of p. For the very small value of E obtaining in the Earth's core it seems unlikely that the equatorial modes will be preferred, since $p > O(10^{-2})$ for the core fluid, but they certainly provide a fascinating alternative form of convection.

3.4 EFFECTS OF A WEAK IMPOSED MAGNETIC FIELD

If we now modify our convection problem by adding a uniform axial current $J_0 \mathbf{1}_z$, this will induce a magnetic field $B_0 \mathbf{1}_\phi$, $B_0 = \mu_0 J_0 s/2$ (in dimensional terms). In the dimensionless equations, the induced Lorentz force due to the perturbation magnetic field will now be scaled with Λ where Λ is a measure of J_0^2. Although this current distribution leads to a Lorentz force even in the absence of perturbation the force is irrotational since J_0 is uniform and so a static equilibrium is still self-consistent (if we neglect centrifugal forces).

The effect of the magnetic field depends not only on Λ but on the diffusivity ratio (Roberts number) q. An inspection of the equations shows that the Lorentz and viscous forces for the columnar type of convection are comparable when (recalling that $\mathbf{B} \cdot \nabla \sim Bm$)

$$\Lambda q^{-1} mb \sim Em^2 v, \tag{3.4.1}$$

while for small q or $q = O(1)$ we have a balance between advection and diffusion, so that

$$mv \sim q^{-1} m^2 b, \tag{3.4.2}$$

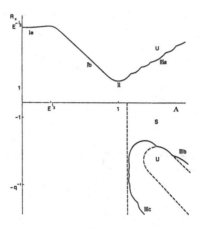

Figure 3.3 Sketch of stability boundary in the Rayleigh number–Λ plane for convection in an imposed uniform axial current. Unstable regions are denoted by U, stable regions by S. For this picture $p = O(1)$, $q \ll 1$ and $\Lambda \ll qE^{-1}$. The scallops in the curve are due to changes in the most unstable wavenumber. (After Fearn 1979.)

with $m \sim E^{-1/3}$. This shows that Lorentz forces become important when $\Lambda \sim E^{1/3}$. So far the convection is still columnar. When Λ becomes of order unity the preferred mode of convection no longer has large zonal wavenumber, and asymptotic methods fail. A careful numerical study by Fearn (1979) has revealed that for $\Lambda \leq E^{1/3}$ the effect of an azimuthal field is stabilizing, due presumably to enhanced total dissipation. At larger Λ ($E^{1/3} \ll \Lambda \leq O(1)$) the Lorentz forces progressively break the constraint of rotation, the optimum m decreases and the critical value of $\widetilde{Ra}$ decreases until $Ra \sim E^{-1}$ ($\widetilde{Ra} = O(1)$) at $\Lambda = O(1)$ (figure 3.3). Since $\widetilde{Ra}$ is independent of ν, we conclude that for $\Lambda = O(1)$ the convection is characterised by a balance that does not involve viscosity at leading order, and in fact involves Coriolis, buoyancy, Lorentz and pressure forces only (a *magnetostrophic* balance). For yet larger values of Λ the magnetic constraints begin to dominate, and the instability begins to be driven primarily by the magnetic field itself, with buoyancy forces only acting as a catalyst. This is evident from the figure, where it can be seen that for large enough Λ instability can occur for either sign of Ra. These magnetic instabilities had previously been noted for convection in a cylindrical geometry (Soward 1979), but are of course absent in a plane geometry, where the initial field is uniform (Eltayeb & Roberts 1970, Eltayeb 1972). In the latter case the positive $\widetilde{Ra}$ branch still exists, complicating the simple characterisation of these solutions as buoyancy-catalysed. Indeed, for general magnetic fields the magnetic instability modes discussed below are more important than the negative-Ra diffusive modes.

Because, on reasonable estimates of field strength, Λ is indeed of order unity in the geophysical context, there has been a great deal of attention paid to the situation in which the basic force balance is magnetostrophic. It is then convenient to adopt a new scaling to bring the important terms into prominence. If we define the ohmic decay timescale τ by $\tau = q^{-1}t$, and set $\mathbf{u} = q\mathbf{v}$, so that the velocity is now scaled with the ohmic diffusion velocity η/r_0, we may write, instead of (2.4–7),

$$\mathbf{1}_z \times \mathbf{u} = -\nabla P + \Lambda(\nabla \times B\mathbf{1}_\phi \times \mathbf{b} + \nabla \times \mathbf{b} \times B\mathbf{1}_\phi) + \widetilde{Ra}\, q\, \theta r\mathbf{1}_r\ [+E\nabla^2\mathbf{u}], \quad (3.4.3)$$

$$\mathbf{b}_\tau - \nabla \times (\mathbf{u} \times B\mathbf{1}_\phi) = R_m(U\mathbf{1}_\phi \times \mathbf{b}) + \nabla^2\mathbf{b}, \quad (3.4.4)$$

$$\theta_\tau + \mathbf{u} \cdot \nabla T + R_m\, U\mathbf{1}_\phi \cdot \nabla\theta = q\nabla^2\theta. \quad (3.4.5)$$

Note that we have now neglected fluid inertia entirely, but retain the (nominally even smaller) viscous term since it may have important dynamical consequences. It is essentially these equations which form the basis for the subsequent discussion.

3.5 INSTABILITY OF GENERAL TOROIDAL FIELDS IN THE MAGNETOSTROPHIC LIMIT

We now concentrate on the magnetostrophic equations above, which are held to be defined, as before, within a conducting sphere surrounded by insulator. Since even the constant-current field is analytically intractable in this geometry, we lose little by choosing more realistic forms of magnetic field (for example, fields whose toroidal part changes sign at the equator), and including the effects of differential rotation. The first calculations along these lines were performed by Fearn & Proctor (1983a) who solved the equations using grid point methods, converting the stability problem into an eigenvalue problem for a (large) matrix, and finding the most rapidly growing solutions. Toroidal magnetic fields were chosen that were odd about the equator $\vartheta = \pi/2$, while the differential rotation profile was taken to be even, in keeping with the usual ideas concerning the parity of the mean fields in the geodynamo. Subsidiary calculations in a cylindrical geometry, with buoyancy perpendicular to the rotation axis, for which the problem can be separated in certain circumstances, were discussed by Fearn (1983). Viscosity is neglected entirely in these calculations: for the linearised problem studied this does not cause any difficulties with satisfying Taylor's condition (discussed by Fearn, chapter 7), as the instability has non-zero azimuthal wavenumber and thus the leading order Lorentz force has zero zonal average. At the quadratic level, however, the Lorentz forces will drive zonal flows, so that the viscosity is necessary for the description of weakly nonlinear convection (Fearn, Proctor & Sellar 1994). These calculations revealed a crucial difference between the stability problem for the general toroidal field and the constant current case, namely that for Λ sufficiently large instability can occur even when $\widetilde{Ra}$ is negative; this implies that for for $\Lambda \gg O(10)$ the instability is driven by the field itself. Unfortunately Fearn & Proctor (1983a) used an incorrect boundary condition for the modes with $m = 1$, leading to incorrect results for that case. In consequence the similarity between the convective stability boundaries for the general toroidal field and the constant current case were somewhat exaggerated. Zhang (personal communication) with the benefit of much greater processing power, has carried out very detailed calculations with an inner core, very small though not actually zero Ekman number and $q = 1$. Figure 3.4 shows the neutral stability curves for different wavenumbers, and the contours of the eigenfunctions in various cases. The columnar nature of the modes is evident, even though Λ is of order unity, so the Proudman-Taylor theorem does not apply. Modes of quadrupole parity have much higher critical values of $\widetilde{Ra}$.

Although there is no evidence of a separate buoyancy catalysed mode in this case, it is possible that one could be identified for very small values of q by the characteristic frequency, which is $O(q)$ for a thermal mode and $O(1)$ for a magnetic mode. More detailed computations will eventually give a complete picture. Meanwhile we note that for larger Λ an instability can occur even when $\widetilde{Ra} < 0$ and so the mechanism must be energy stored in the field itself. A review of recent work on such magnetically driven instabilities has been given by Fearn (1993) (see also section 3.7 below).

Although most work on the stability problem without differential rotation has focussed on the case where the axisymmetric basic field is toroidal, we expect in general that the mean field has a significant poloidal part. Recent work on mean field dynamos, discussed by Fearn (chapter 7), has shown that dynamos of α–ω type may be less favoured than those where the poloidal field is large enough compared with the toroidal for the α-effect in the toroidal equation to be significant. We should, therefore, add a poloidal component to the basic field. This has been investigated recently by Fearn *et al.* (1994), whose preliminary results seem to show that the stability boundaries are sensitive to the addition of a poloidal component, with the form of the instability becoming much more columnar in the direction of rotation. Further work is clearly needed here.

Fearn & Proctor (1983a) also investigated the effects of adding a prescribed differential rotation to the basic state. Once again, a full dynamo model is necessary properly to determine such a flow, which can be driven by Rey-

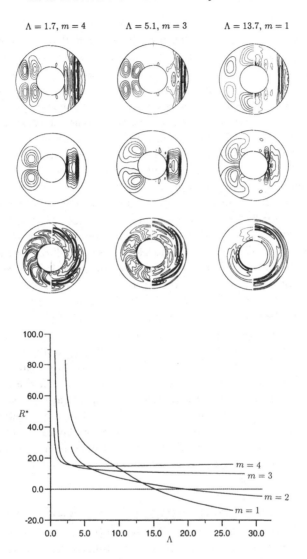

Figure 3.4 (*a*) Eigenfunctions for linear stability of the non-uniform toroidal field $B = 8r^2(1 - r^2)\sin\vartheta\cos\vartheta$, for various values of Λ, with $q = 1$ and dipole parity. Top row shows toroidal field and flow in a meridian plane, second row poloidal field and radial velocity in a meridian plane and third row radial velocity and toroidal flow in the equatorial plane. (*b*) The stability boundaries for this field (cf. figure 3.3). Curves for different wavenumber dipole modes are labelled (Courtesy of K. Zhang.)

nolds stresses and horizontal temperature gradients in the rotating medium (thermal winds). The flow is taken to be even about the equator, which maintains the parities of the solutions. It is found that differential rotation in general exerts a stabilizing influence, which becomes significant when $\max(R_m, R_m/q) > 1$. If we concentrate on the case $q = 1$ we find that the eigenfunctions become concentrated in the neighbourhood of locations where the angular velocity is extremal (so shears are small), and the phase velocity of the modes is close to the local angular velocity. When the angular velocity has no extrema, the location of the critical point is determined by subtle balances, which have only been understood in a simple cylindrical model (Fearn & Proctor 1983*b*). The mean zonal emf $(\mathbf{u} \times \mathbf{b})_\phi$ formed by the convection is also strongly localized. (When $q \ll 1$ the thermal field is the first to become concentrated, while the magnetic field structure is largely unaffected.) Braginsky (chapter 9) has noted the importance of localised α-effects in his Model-Z geodynamo, which features large zonal flows, and it is interesting to speculate on the possibilities for the development of a self-consistent model.

Because of resolution difficulties encountered by Fearn & Proctor (1983*a*), who solved a discretized version of the eigenvalue problem so as to determine all the eigenvalues, Fearn & Proctor (1983*b*) and Fearn (1989) have investigated details of the eigenstructure, effects of boundary conditions, etc., in two different simplified cylindrical geometries. Fearn and Proctor's basic state has a density gradient parallel to the rotation axis, while Fearn's has one in the s direction. While Fearn and Proctor were only able to consider separable solutions with non-uniform shear when $q \to 0$, R_m/q finite, Fearn was able to choose arbitrary q. The main results of the spherical work are qualitatively reproduced in the spherical geometry. Fearn has also investigated the dependence on Λ of the effects of differential rotation; he finds that provided q iss of order unity or smaller the only important parameter determining the strength of the shear is R_m/q. Figure 3.5 shows a representative set of results for $q = 10^{-6}$.

3.6 RESISTIVE AND DIFFUSIONLESS INSTABILITIES

Because the timescales based on the diffusivities $(r_0/\kappa, r_0/\eta)$ are rather long, it might be wondered whether there are new, diffusionless modes of instability that appear at high values of $\overline{Ra}$. It has been known for some time (e.g., Braginsky 1964, 1967) that if diffusion is completely ignored, then the resulting system (essentially (4.3–5) without the diffusive terms, though a rescaling of the variables is required) exhibits instability if the destabilizing temperature gradient is too large. We can investigate this effect with a simple model system in which the fluid is confined to a layer with horizontal boundaries, gravity $\mathbf{g} = -g\mathbf{1}_z$ and the rotation vector $\Omega\mathbf{1}_z$ are vertical, and there is an imposed horizontal uniform field $\mathbf{B}_0$ and vertical temperature gradient β. Then the linearised, dimensional perturbation equations are (where

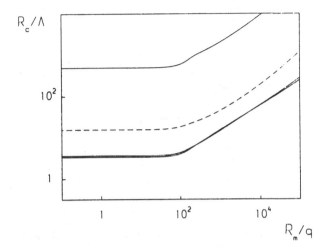

Figure 3.5 Graphs of Ra/Λ versus R_m/q for the cylindrical geometry of Fearn (1989) with $q = 10^{-6}$ and angular velocity $\propto s^2$ for various values of Λ. The top two curves are for $\Lambda = 1$, $\Lambda = 10$ respectively, while the lower, almost coincident, curves are for $\Lambda = 10^2$, 10^3, 10^4 and 10^5. The instability has azimuthal wavenumber 2.

the temperature $T = T_0 - \beta z + \beta\theta)$

$$2\Omega \mathbf{1}_z \times \mathbf{v} = -\nabla P + \mathbf{B}_0 \cdot \nabla \mathbf{b} + g\alpha\beta\theta \mathbf{1}_z, \qquad (3.6.1a)$$

$$\frac{\partial\theta}{\partial t} = \mathbf{v} \cdot \mathbf{1}_z, \qquad (3.6.1b)$$

$$\frac{\partial\mathbf{b}}{\partial t} = \mathbf{B}_0 \cdot \nabla\mathbf{v}, \qquad (3.6.1c)$$

and then if we look for solutions proportional to $\exp(i\mathbf{k}\cdot\mathbf{x} - i\omega t)$, we obtain the dispersion relation

$$\omega^2 = \frac{(\mathbf{V}_0 \cdot \mathbf{k})^4}{4\Omega^2(\mathbf{k}\cdot\mathbf{1}_z)^2} \left[k^2 - \frac{g\alpha\beta(k^2 - (\mathbf{k}\cdot\mathbf{1}_z)^2)}{(\mathbf{V}_0 \cdot \mathbf{k})^2} \right], \qquad (3.6.2)$$

where $\mathbf{V}_0$ is the Alfvén velocity $\mathbf{B}_0/\sqrt{\mu\rho}$. Waves obeying this dispersion relation were termed MAC waves (Magnetic–Archimedean–Coriolis) by Braginsky (1967). The singularity at $\mathbf{k}\cdot\mathbf{1}_z = 0$ is due to the neglect of inertia. We can see that the timescale of the motions is $O(\Omega r_0^2/V_0^2)$, translating in our previous units to $r_0^2/\Lambda\eta$, and that we have diffusionless instability when $g\alpha\beta \geq V_0^2$ or $\widetilde{Ra} \geq R_{\text{MAC}} \sim \Lambda/q$. Note that this is the only possible relation not involving the diffusivities. Thus for *small* q, we see that the convective modes, which exist for $\widetilde{Ra} = O(1) \ll R_{\text{MAC}}$, are more easily destabilized than the MAC waves, but that their growth rates $(O(q)$ on the magnetic diffusion

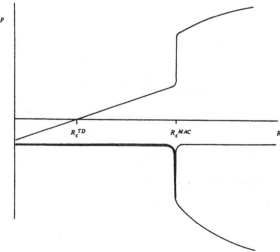

Figure 3.6 Sketch of the growth rates of convective modes (represented by p) as functions of $\widetilde{Ra}$. (After Fearn 1989.)

timescale) are rather small. On the other hand, if Λ, q are $O(1)$ then there is no clear distinction between the resistive and diffusionless timescales or stability criteria.

Figure 3.6, from Fearn (1989) shows how the growth rates of the modes depend upon $\widetilde{Ra}$ when $\Lambda = O(1)$ and q is small. It will be seen that the same mode that becomes diffusively unstable at $\widetilde{Ra} = O(1)$ is transformed into a MAC type mode, with much larger growth rate, near $\widetilde{Ra} = \Lambda/q$, but that there is no new bifurcation. It therefore seems unnecessary, especially as q is not expected to be very small for a turbulent diffusivity model, to distinguish between the two types of solution.

3.7 INSTABILITIES OF THE MAGNETIC FIELD

We now give a slightly fuller, though still brief discussion of a set of instabilities distinct from those driven by density gradients, namely those associated with gradients of the magnetic field itself. We have already seen that thermal modes can be destabilized by the field even when the temperature gradient is stabilizing. Here there is no influence of stratification and the instability is caused directly by the field itself. Since field curvature is important in these modes, the easiest geometry likely to yield results useful to the sphere is that of an annulus rotating about its axis of symmetry, with or without differential rotation. This problem has been extensively studied by Fearn in a series of papers beginning with Fearn (1983); for a full list see Fearn (1988) and Fearn, Roberts & Soward (1988). Fearn & Weiglhofer (1991a, b) have extended the calculations to a spherical geometry.

If, as before, we ignore inertia and viscosity and now regard the fluid as unstratified, we can ignore the heat equation. Then the only parameters remaining, apart from those describing the geometry, are the Elsasser number Λ and the differential rotation parameter R_m. This bald statement is deceptive, however, since the stability or otherwise of any imposed field can depend strongly on its *spatial form* as well as on its typical strength. Certainly the 'ideal instability' described below is typically a locally driven one, depending crucially on field gradients. Four types of instability can be distinguished: (a) the catalysed instabilities previously referred to, (b) a dynamic instability (Malkus 1967); this can occur only when the magnetic energy is comparable with the kinetic energy of the rotating body and so is not relevant to the earth, (c) resistive instability, only existing for finite Λ, and (d) ideal instability, occurring on the nondiffusive timescale $\Omega r_0^2 \mu \rho / B_0^2$ (this is also the timescale for MAC waves), in the limit $\Lambda \to \infty$, and only existing for sufficiently large Λ. Since for the Earth Λ is not too large, and in fact is commonly thought of as $O(1)$, there is no real distinction between the resistive and (damped) 'ideal' modes since the dynamical MAC time and the diffusion time r_0^2/η are then comparable. They may be distinguished by the limiting values of the growth rates as Λ becomes large. Fearn (1988) has also noted that the two types of mode may respond differently to the imposition of differential rotation.

Most of the studies of magnetic instability have been carried out for a toroidal axisymmetric field. Zhang & Fearn (1993) have recently investigated the effect of adding a poloidal field to the basic configuration. They find that, as for the convective instability, the threshold for magnetic instability is considerably reduced, and conjecture that the value of Λ in the core is limited by this mechanism. Only nonlinear studies will be able to resolve the question.

Finally, we remark that while the effects of differential rotation will be variable when R_m is sufficiently small, for large enough differential rotation rates the shear flow will itself be unstable. The mechanism for the instability is different from that obtaining in non-magnetic shear flows, since fluid inertia has been neglected. The question has been addressed by Fearn & Weiglhofer (1992), who find that the optimum conditions for the destabilization of a layer by differential rotation are achieved for $\Lambda \approx 50$–100 with $R_m > O(100)$. For larger Λ the critical value of R_m increases with Λ, while for smaller Λ the critical R_m increases dramatically. For $\Lambda < 10$–30 no instabilities of this form were found. It is clear that these instabilities may be of importance in limiting the growth of the field in the Earth's core.

3.8 CONCLUSIONS

This short review of rotating magnetoconvection has left many questions unanswered. The most important omission is any discussion of nonlinear effects. It is very hard to separate nonlinear properties of convection in a useful way from their interaction with the evolving basic field: indeed a proper discussion requires a complete theory of the geodynamo! Nonetheless there are some broad questions that can be addressed by a weakly nonlinear analysis, such as whether the criteria desired on the basis of linearized theory has any connexion with nonlinear effects, or whether there are strong subcritical effects that render a full nonlinear analysis essential. There has been a small beginning in this direction with the work of Fearn *et al.* (1994), who have found that the first nonlinear effect of Taylor's condition on the differential rotation is mildly destabilizing, but much remains to be done.

The analyses described here have also ignored the presence of the solid inner core. It is possible that the magnetostrophic approximation breaks down when an inner core is present. Hollerbach & Proctor (1993) carry out a simple calculation to show that arbitrary nonaxisymmetric force fields in a spherical annulus lead to a severe singularity in the resulting velocity field at the cylinder circumscribing the inner core. Figure 3.7 shows an example. In addition Jones & Hollerbach (1993) have recently exhibited the unexpectedly important role of the inner core in controlling the evolution of nonlinear mean field dynamo models. Preliminary work by Fearn and collaborators, following the earlier studies of Fearn, Proctor & Weiglhofer, but in a spherical annulus, does not seem however to yield any undue difficulties. Plainly more work is desirable here.

Finally, a proper study of core dynamics should ideally use a proper kind of convection — namely that due to compositional gradients. Because the inner core boundary is widely believed to be a 'mushy zone' (see, e.g., Loper 1989) there are clearly difficulties in formulating proper boundary conditions — indeed it is not even known whether or not the 'chimneys' of buoyant pure fluid seen in terrestrial experiments on ammonium chloride are representative of the core. It is, I believe, in this area that important progress can be made, though it is, as always, sad that the high magnetic diffusivity even of liquid metals makes realistic experiments rather unlikely.

Acknowledgements

I am grateful to my co-editor, Andrew Gilbert, for his forbearance while this work progressed through an over long gestation. Research support from the SERC is gratefully acknowledged.

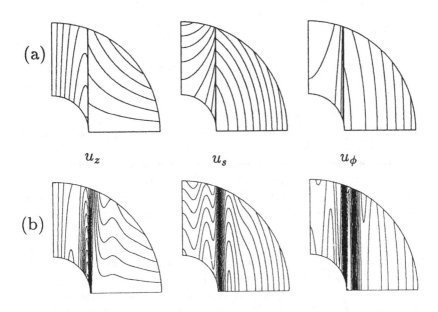

Figure 3.7 Response of the fluid (a) without and (b) with a small amount of viscosity to a continuous forcing (proportional to $e^{i\phi}$ in the magnetostrophic limit in the presence of an inner core. (After Hollerbach & Proctor 1993.)

REFERENCES

Braginsky, S.I. 1964 Magnetohydrodynamics of the Earth's core. *Geomag. Aeron.* **4**, 698–712.

Braginsky, S.I. 1967 Magnetic waves in the Earth's core. *Geomag. Aeron.* **7**, 851–859.

Busse, F.H. 1970 Thermal instabilities in rapidly rotating systems. *J. Fluid Mech.* **44**, 441–460.

Busse, F.H. 1975 A model of the geodynamo. *Geophys. J.* **42**. 437–459.

Eltayeb, I.A. 1972 Hydromagnetic convection in a rapidly rotating fluid layer. *Proc R. Soc. Lond. A* **326**, 227–254.

Eltayeb, I.A. & Roberts, P.H. 1970 On the hydromagnetics of rotating fluids. *Astrophys. J.* **162**, 699–701.

Fearn, D.R. 1979 Thermal and magnetic instabilities in a rapidly rotating fluid sphere. *Geophys. Astrophys. Fluid Dynam.* **14**, 103–126.

Fearn, D.R. 1983 Hydromagnetic waves in a differentially rotating annulus. I. A test of local stability analysis. *Geophys. Astrophys. Fluid Dynam.* **27**, 137–162.

Fearn, D.R. 1988 Instabilities of toroidal magnetic fields. In *Structure and Dynamics of Earth's Deep Interior* (ed. D.E. Smylie & R. Hide), pp. 129–133. American Geophysical Union, Washington, DC.

Fearn, D.R. 1989 Differential rotation and thermal convection in a rapidly rotating magnetic system. *Geophys. Astrophys. Fluid Dynam.* **49**, 173–193.

Fearn, D.R. 1993 Magnetic instabilities in rapidly rotating systems. In *Solar and Planetary dynamos* (ed. M.R.E. Proctor, P.C. Matthews & A.M. Rucklidge), pp. 59–68. Cambridge University Press.

Fearn, D.R. & Proctor, M.R.E. 1983*a* Hydromagnetic waves in a differentially rotating sphere. *J. Fluid Mech.* **128**, 1–20.

Fearn, D.R. & Proctor, M.R.E. 1983*b* The stabilising role of differential rotation on hydromagnetic waves. *J. Fluid Mech.* **128**, 21–36.

Fearn, D.R. & Weiglhofer, W.S. 1991*a* Magnetic instabilities in rapidly rotating spherical geometries. I. From cylinders to spheres. *Geophys. Astrophys. Fluid Dynam.* **56**, 159–181.

Fearn, D.R. & Weiglhofer, W.S. 1991*b* Magnetic instabilities in rapidly rotating spherical geometries. II. More realistic fields and resistive instabilities. *Geophys. Astrophys. Fluid Dynam.* **60**, 275–294.

Fearn, D.R. & Weiglhofer, W.S. 1992 Magnetic instabilities in rapidly rotating spherical geometries. III. The effect of differential rotation. *Geophys. Astrophys. Fluid Dynam.* **67**, 163–184.

Fearn, D.R., Roberts, P.H. & Soward, A.M. 1988 Convection, stability and the dynamo. In *Energy, Stability and Convection* (ed. G.P. Galdi & B. Straughan), pp. 60–324, Longman, Harlow, UK.

Fearn, D.R., Proctor, M.R.E. & Sellar, C.C. 1994 Nonlinear magnetoconvection in a rapidly rotating sphere and Taylor's constraint. *Geophys. Astrophys. Fluid Dynam.* (submitted).

Hollerbach, R. & Proctor, M.R.E. 1993 Non-axisymmetric shear layers in a rotating spherical shell. In *Solar and Planetary dynamos* (ed. M.R.E. Proctor, P.C. Matthews & A.M. Rucklidge), pp. 145–152. Cambridge University Press.

Jones, C.A. & Hollerbach, R. 1993 Influence of the Earth's inner core on geomagnetic fluctuations and reversals. *Nature* **365**, 541–543.

Loper, D.E. 1989 Dynamo energetics and the structure of the core. *Geophys. Astrophys. Fluid Dynam.* **49**, 213–220.

Malkus, W.V.R. 1967 Hydromagnetic planetary waves. *J. Fluid Mech.* **28**, 379–802.

Poirier, J.P. 1988 Transport properties of liquid metals and viscosity of the Earth's core. *Geophys. J.* **92**, 99–105.

Roberts, P.H. 1968 On the thermal instability of a rotating fluid sphere containing heat sources. *Phil. Trans. R. Soc. Lond. A* **263**, 93–117.

Soward, A.M. 1977 On the finite amplitude thermal instability of a rapidly rotating fluid sphere. *Geophys. Astrophys. Fluid Dynam.* **9**, 19–74.

Soward, A.M. 1979 Thermal and magnetically driven convection in a rapidly rotating fluid layer. *J. Fluid Mech.* **90**, 669–684.

Yano, J.-I. 1992 Asymptotic theory of thermal convection in rapidly rotating systems. *J. Fluid Mech.* **243**, 103–131.

Zhang, K. 1992 Spiralling columnar convection in rapidly rotating spherical fluid shells. *J. Fluid Mech.* **236**, 535–556.

Zhang, K. 1993 On equatorially trapped boundary inertial waves. *J. Fluid Mech.* **248**, 203–217.

Zhang, K. & Busse, F.H. 1987 On the onset of convection in rotating spherical shells. *Geophys. Astrophys. Fluid Dynam.* **39**, 119–147.

Zhang, K. & Fearn, D.R. 1993 How strong is the invisible component of the magnetic field in the Earth's core? *Geophys. Res. Lett.* **20**, 2083–2086.

Zhang, K. & Jones, C.A. 1993 The influence of Ekman boundary layers on rotating convection. *Geophys. Astrophys. Fluid Dynam.* **71**, 145–162.

CHAPTER 4

Solar Dynamos: Computational Background

AXEL BRANDENBURG

High Altitude Observatory/National Center for Atmospheric Research*
P.O. Box 3000
Boulder, CO 80307–3000, USA

4.1 INTRODUCTION

The picture of the solar dynamo has evolved continuously over the past twenty years. Both observations and theoretical considerations have introduced new aspects to the problem. Meanwhile, there has been a growing recognition of the importance of the overshoot layer beneath the solar convection zone (CZ) as the place where magnetic flux tubes can be 'stored' over time scales comparable with the solar cycle period (e.g., Spiegel & Weiss 1980). Less clear, however, is the question of whether the field is also generated there, or whether the magnetic field is actually generated in the convection zone, and only then transported into the overshoot layer where it accumulates (Brandenburg et al. 1991, Nordlund et al. 1992).

Numerical simulations of hydromagnetic convection show generation of magnetic fields in the entire CZ, but there is a strong turbulent downward pumping, that causes the magnetic field to build up at the interface between the CZ and the radiative interior. Magnetic buoyancy causes magnetic flux tubes to float upwards, but at the same time convective motions push them down again. In numerical simulations it is seen that under these conditions the magnetic field plays an active role and can still be amplified. It is questionable whether the interface can be considered in isolation. Consequently, throughout this chapter we consider the evolution of the magnetic field in the entire CZ and allow for the interaction between the CZ and the radiative interior in most of the cases.

* The National Center for Atmospheric Research is sponsored by the National Science Foundation.

M.R.E. Proctor & A.D. Gilbert (eds.)
Lectures on Solar and Planetary Dynamos, 117–159
©1994 Cambridge University Press.

The magnetic fields generated by turbulence are so intermittent that it is difficult to understand how the solar magnetic field has such a systematic orientation, as demonstrated by Hale's polarity law (cf. Schüssler 1987). This might be due to the possibility that the field becomes more and more systematically oriented, and perhaps less intermittent, as it is continuously generated and pumped from the CZ into the interface. Once such a systematic large-scale field has built up, it could be the source for the generation of bipolar regions via the Parker–Babcock mechanism (Parker 1955, Babcock 1961).

We begin this chapter by considering the roles of small-scale and large-scale dynamos. In subsequent sections we present direct simulations of small-scale dynamos in Cartesian geometry and, later, mean-field dynamos that explain the large-scale magnetic field. We discuss numerical methods and conclude by summarising crucial issues of the solar dynamo and differential rotation.

4.2 SMALL-SCALE AND LARGE-SCALE MAGNETIC FIELDS
Nonlinear dynamos may be divided into small-scale and large-scale dynamos (Vainshtein & Zel'dovich 1972). Small-scale dynamos typically generate an irregular, random magnetic field, whereas a large-scale dynamo is thought to be responsible for the systematic magnetic field associated with the solar activity cycle.

4.2.1 Small-scale Dynamos
The theory for small-scale dynamo action has been developed by Kazantsev (1968), Novikov, Ruzmaikin & Sokoloff (1983) and Molchanov, Ruzmaikin & Sokoloff (1985, 1988), and numerical simulations have been presented by Meneguzzi, Frisch & Pouquet (1981), Meneguzzi & Pouquet (1989), Kida, Yanase & Mizushima (1991) and Nordlund et al. (1992). According to an argument by Batchelor (1950) a small-scale magnetic field is generated in a turbulent, conducting fluid if the magnetic diffusivity, η, is smaller than the kinematic viscosity ν, i.e., if the magnetic Prandtl number $P_m = \nu/\eta$ is larger than unity. The analysis of Novikov et al. (1983) supports this result. When P_m is small the velocity field is more intermittent than the magnetic field. In this way, ν enters indirectly even in the kinematic problem, because it determines the smallest length scales of the velocity field.

In the sun the magnetic Prandtl number is small ($P_m \lesssim 10^{-4}$), i.e., small-scale dynamo action may be irrelevant. However, there are reasons to expect that a small-scale dynamo may already operate once the magnetic Reynolds number $R_m = u_t d/\eta$ exceeds a certain threshold, regardless of how small the value of P_m is. Molchanov et al. (1988) have shown that dynamo action is possible, even if the velocity is completely 'non-smooth'. Nordlund et al. (1992) argue that the role of viscosity is to provide a viscous drag so that the flow is able to shape the magnetic flux tubes. For small values of P_m there is a turbulent drag force that governs the interaction between the flow and the magnetic flux tubes, replacing the role of the viscous drag force. The solar value, $R_m \approx 10^8$, is well above the critical value of about one hundred. Therefore small-scale dynamo action may work after all.

4.2.2 Large-scale Dynamos

The traditional aim of astrophysical dynamo theory is to explain the origin of a large-scale magnetic field. One usually invokes an inverse cascade process that builds up magnetic energy at larger and larger scales (Frisch *et al.* 1975, Pouquet, Frisch & Léorat 1976). For this process to operate one needs a large enough container of conducting fluid with small-scale helical motions giving rise to an α-effect (Steenbeck, Krause & Rädler 1966); see Roberts' lecture in chapter 1. If the system size is too small, this large-scale dynamo process will not work, although a small-scale dynamo may still operate. The α-effect theory can be understood as a linearised approach to the inverse cascade process. In recent calculations of α, Rüdiger & Kitchatinov (1993) used renormalised values for viscosity and magnetic diffusivity that are independent of the (molecular) magnetic Prandtl number. Therefore the α-effect should operate even for small values of P_m.

4.2.3 Mean-field Approach

The solar magnetic field shows a component with remarkably regular behaviour. This includes the (approximately) 22-year magnetic cycle, the equatorward migration of the sunspot activity belts, and systematic magnetic field orientation of strong bipolar regions and sunspot pairs; see the lecture of Weiss (chapter 2). Such behaviour can be described well by mean-field dynamos, as was demonstrated by Steenbeck & Krause (1969). A mean-field dynamo only describes the evolution of a statistically averaged magnetic field, which is not directly observable. Ideally one would like to adopt combined spatial and temporal averages over a small area and time span, for example by filtering out the high temporal and spatial frequencies in Fourier space. If there is no perfect scale separation, however, this only corresponds approximately to statistical averages (e.g., Krause & Rädler 1980).

It should be noted that in the context of planetary dynamos one usually adopts azimuthal averages. Such an approach is useful for the sun as well, but not for certain late-type stars whose magnetic field seems to show large-scale nonaxisymmetric structures (e.g., Saar, Piskunov & Tuominen 1992). An appropriate average would be a Fourier filter that retains only a few Fourier modes in the azimuthal direction. The sun also shows a weak large-scale azimuthal variation of activity, the so-called active longitudes. Even though this surface manifestation is a distinctive feature of the solar cycle, it is likely not to be important for the dynamo itself.

At the surface of the sun the fluctuating part of the observed magnetic field is much larger than the mean magnetic field (e.g., Krause & Rädler 1980). The actual magnetic field is highly intermittent in both space and time, and the mean magnetic field can therefore not easily be extracted from the actual field. It is sometimes suggested that the magnetic field generated by a mean-field dynamo is smooth and 'diffusive', but this is not necessarily the case. This point has been clarified in review papers by Krause (1976) and Stix (1981). Firstly, in the mean-field dynamo equation one only solves for the average magnetic field, $\langle \mathbf{B} \rangle$, whilst the actual magnetic field, $\langle \mathbf{B} \rangle + \mathbf{B}'$, may well be highly intermittent, consisting of a large number of small magnetic flux tubes pointing in different directions, and which all add up to a small, but nonvanishing, average magnetic field. Secondly, the intermittent nature of the magnetic field leads to additional transport effects (e.g., magnetic buoyancy) that have to be included in the mean-field equations.

4.2.4 Intermittency

In the sun the actual magnetic field is highly intermittent, at least at the surface. The filling factor, i.e., the fractional area covered by magnetic fields above a certain threshold, varies between 0.01 and 0.1, in quiescent and active regions, respectively (Schrijver 1987). In numerical simulations of MHD convection one also finds highly intermittent fields, but there is normally no systematic orientation of bipolar regions. This is partly because of the Cartesian geometry adopted in such models, and partly because of the small system size. The simulations of Gilman (1983) and Glatzmaier (1985) in spherical shells do show a significant mean-field component together with a much stronger fluctuating field. The fact that in their simulations there is a poleward migration of the mean magnetic field belts, and not an equatorward migration as in the sun, is perhaps only a minor detail in this respect, but it reminds us that such models do not closely resemble the sun. The important point here is that in all these simulations there is a fluctuating magnetic field whose strength exceeds that of the mean field.

4.2.5 Bipolar Regions

Most of the magnetic field in sunspots and bipolar regions contributes to the fluctuating magnetic field component, $\mathbf{B}'$. It remains a challenge to understand how a fluctuating field can show such a systematic behaviour. The fact that strong fluctuations occur in a turbulent dynamo is quite natural, and the rather systematic behaviour of bipolar regions and sunspot pairs is therefore very surprising indeed.

The evidence for a large-scale magnetic field in the sun comes mainly from the remarkably systematic orientation of bipolar regions. The stronger the

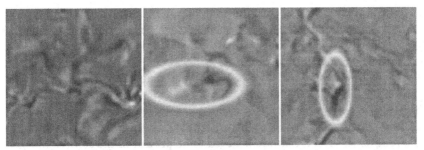

Figure 4.1 Grey-scale representation of the vertical component of the magnetic field, B_z, in a horizontal plane in the upper part of the box ($z = 0$), obtained from three different numerical simulations of MHD convection. *Left*: small-scale dynamo without systematic field in the presence of rotation; *middle*: imposed magnetic field in the x-direction (left–right on the paper), but no dynamo; *right*: small-scale dynamo with imposed shear velocity in the y-direction (down–up on the paper) at the interface ($z = 1$) in the presence of rotation. ('Bipolar regions' are encircled.)

magnetic field in such regions, the better are these bipolar regions oriented in accordance with Hale's polarity law. In analysing the magnetic fields from direct simulations, it is useful to present them in a similar way as do observers, by taking magnetograms of the magnetic field component perpendicular to the surface plane. As an example we show grey-scale plots of the vertical magnetic field B_z, from three different numerical simulations of Nordlund *et al.* (1992) and Brandenburg *et al.* (1993*a*, *b*); see figure 4.1. Some details of these simulations will be given in section 4.5, but for the moment it is important to know that they represent the interaction of an irregular (turbulent) convective flow with magnetic fields in a box. In the first panel of figure 4.1 there is a small-scale dynamo effect operating in the presence of rotation, but there is no imposed magnetic field. In the second case there is a horizontal magnetic field imposed in the x-direction, but the magnetic Prandtl number is smaller ($P_m = 0.2$) and there is no dynamo (this case was presented in Brandenburg *et al.* 1993*b*). The third case is similar to the first one, but there is now a shear velocity imposed in the y-direction (corresponding to longitude). This shear tends to orient bipolar regions, but the evidence is marginal (see the right panel of figure 4.1, where the field is oriented upwards on the paper). In this case the flow structure has also changed considerably in that there are now extended rolls in the y-direction, corresponding to longitude.

Numerical simulations show quite convincingly that the magnetic field occurs in the form of magnetic flux tubes (Nordlund *et al.* 1992). This is an essentially three-dimensional process. Two-dimensional simulations, on the other hand, do not lead to the formation of concentrated flux tubes. The two-dimensional simulations of Cattaneo & Hughes (1988) and Jennings *et al.* (1992) show that buoyancy and advection rapidly destroy initial flux concentrations.

The systematic orientation of strong bipolar regions can be explained if the corresponding flux tubes originate from a strong systematic magnetic field at a depth where fluctuations are less vigorous. It is plausible that such a systematic field can only be located at the bottom of the CZ (e.g., Spiegel & Weiss 1980). In order to understand Hale's polarity law we must expect that the field is either smooth or, if it is intermittent, the majority of flux tubes must have the same orientation. It is possible that this process is facilitated by strong shear layers, that are believed to be present in the sun (cf. Spiegel & Zahn 1992).

4.2.6 Accumulation of Flux in the Overshoot Layer

The systematic properties of strong magnetic fields in bipolar regions lead to the picture that the solar dynamo works in the overshoot layer at the bottom of the CZ; see the lecture of Weiss (chapter 2) and recent reviews by Gilman (1992) and Schmitt (1993). The average field strength in the overshoot layer has often been estimated from the magnetic flux that emerges at the solar surface at any given time. The total flux observed at the solar surface is around $\Phi = 10^{24}$ Mx. Assuming that this flux is distributed over a narrow equatorial belt of 20 Mm thickness and 500 Mm width then leads to an average field strength in the overshoot layer of about 10 kG. This is somewhat above the equipartition value, as given by the local kinetic energy density of the convective motions. With density $\rho = 0.2$ g cm^{-3} and turbulent velocity $u_t =$20–50 m/s (from mixing length models) we have $B_{eq} = \sqrt{\mu_0 \rho}\, u_t =$3–8 kG. Schüssler (1983, 1987) pointed out that flux tubes rising from the bottom of the CZ will be distorted ('brainwashed'). This suggests that the field strength in the overshoot region must be of the order of 10 kG. D'Silva (1993) argues in favour of even larger peak magnetic field strengths of the order of 100 kG in order to explain the observed tilt of bipolar regions. Such strong magnetic fields are not easy to generate by a dynamo alone (Durney, De Young & Passot 1990).

The above-quoted values for the turbulent velocity are obtained from standard solar mixing length models. It is quite possible that the velocity field is sufficiently inhomogeneous to cause a substantially different value for the effective r.m.s. velocity. It is therefore important to produce more realistic models for the deep layers of the solar CZ (cf. Chan 1992).

The question remains: where is the field generated? Whilst there are good arguments indicating that the magnetic field is located in the lower overshoot layer, it does not necessarily mean that the field is also generated there. It is possible that the field is generated in the entire CZ. Video animations of MHD convection show that there is a strong tendency for the magnetic field to be advected downwards by the flow. This effect leads to a maximum field strength in the overshoot layer (Nordlund *et al.* 1992). In order to explain the systematic alignment of the magnetic field in bipolar regions it is necessary to require that the intermittent and randomly oriented magnetic field in the CZ becomes progressively more uniformly oriented towards the lower overshoot layer. This is the missing link that unfortunately has not yet been seen in direct simulations.

The other alternative would be that the field is generated exclusively in the overshoot layer. This would require strong dynamo action producing fields above the equipartition value, but faces the difficulty that beyond B_{eq} the horizontal diagonal components of the α tensor, derived from mean-field theory, decrease rapidly like $|\langle \mathbf{B} \rangle|^{-3}$ (Moffatt 1972, Rüdiger 1974). Whilst these results are only valid for small R_m, Kraichnan (1979) argues that even in the strong field limit and at large R_m dynamo action is still possible, and that the α-effect from interacting Alfvén waves only decreases as $u_t^2/v_a \sim |\langle \mathbf{B} \rangle|^{-1}$.

In the remainder of this chapter we discuss various numerical approaches using both direct simulations and mean-field models. Experience with various numerical schemes is reported. In applying such methods to the sun, the basic problem with direct simulations is to cover a large range of time and length scales. One of the problems with the mean-field approach concerns its applicability. The crucial assumptions underlying the validity of the mean-field approach is scale separation both in space and time. In other words, the correlation length and time of the turbulence have to be small compared to the length and time scales of mean-field quantities (thickness of the CZ and cycle period). This causes difficulties in applying mean-field dynamos to the relatively thin overshoot layer at the bottom of the CZ. We begin by discussing general properties of convective and mean-field dynamos.

4.3 CONVECTIVE VERSUS MEAN-FIELD DYNAMOS

Mean-field dynamos and convective dynamos have quite different mathematical properties: the mean magnetic field generated by a mean-field dynamo is spatially smooth, whilst the field generated by convective dynamos is highly intermittent. In other words, in turbulent flows there is a transport of energy to smaller and smaller scales, whilst mean-field dynamos are governed exclusively by an inverse cascade with energies being transferred to larger scales. This is a peculiar property of the α-effect.

4.3.1 Inverse Magnetic Cascade
Consider the dynamo equation with constant α and η_t in the absence of a mean velocity

$$\partial_t \langle \mathbf{B} \rangle = \alpha \nabla \times \langle \mathbf{B} \rangle + \eta_t \nabla^2 \langle \mathbf{B} \rangle . \qquad (4.3.1)$$

Following the argument of Frisch, She & Sulem (1987), the $\alpha \nabla \times \langle \mathbf{B} \rangle$ term dominates the $\eta \nabla^2 \langle \mathbf{B} \rangle$ term at large scales and tends to lead to a force-free field, i.e., $\langle \mathbf{B} \rangle \propto \nabla \times \langle \mathbf{B} \rangle$. One-dimensional mean-field models of Gilbert & Sulem (1990) and Galanti, Sulem & Gilbert (1991) support this idea.

The property of the mean-field equations to develop large-scale fields has important consequences. In mean-field electrodynamics one expands the electromotive force, $\langle \mathcal{E} \rangle = \langle \mathbf{u}' \times \mathbf{B}' \rangle$, in powers of derivatives of $\langle \mathbf{B} \rangle$, i.e.,

$$\langle \mathcal{E}_i \rangle = \alpha_{ij} \langle B_j \rangle - \eta_{ijk} \partial_j \langle B_k \rangle + \cdots, \qquad (4.3.2)$$

where, in the isotropic case, $\alpha_{ij} = \alpha \delta_{ij}$ and $\eta_{ijk} = \eta_t \epsilon_{ijk}$. The expansion (3.2) would break down if $\langle \mathbf{B} \rangle$ were to develop intense small-scale structures. Fortunately, the mean-field equations tend to oppose this. Nevertheless, Nicklaus & Stix (1988) demonstrated that higher order terms can indeed become important. Of particular interest is the fourth term that contributes to hyperdiffusion, i.e., to a term of the form $-\nabla^4 \langle \mathbf{B} \rangle$ on the right-hand side of (3.1). This term becomes especially important if the ordinary turbulent magnetic diffusion coefficient η_t formally becomes negative. In such a case the evolution of $\langle \mathbf{B} \rangle$ would be ill-defined, unless some hyperdiffusive effects were present. In the one-dimensional case the resulting dynamo equation with negative η_t would resemble the Kuramoto–Sivashinski equation. This equation allows self-excited solutions even in the absence of an α-effect. Unlike a hyperdiffusion that is sometimes used in turbulence simulations to reduce artificially the extent of the dissipation range (e.g., Passot & Pouquet 1987), it would here be a physical effect. A related effect in the hydrodynamical problem of channel flows was studied by Rüdiger (1982). He found that a hyperdiffusive term can become important if boundary effects are crucial.

4.3.2 Consequences for Numerical Models

The different mathematical properties of mean-field and convective dynamos have important consequences for their numerical treatment. In a mean-field dynamo there is a balance between the α-effect and diffusion. It is therefore preferable to use a numerical representation of the Laplace operator that has good stability properties. If mesh point methods are used, the DuFort–Frankel scheme is a possible choice (Jepps 1975, Proctor 1977). On the other hand, in direct simulations diffusion does not play such an important role as it does in mean-field computations. Of prime importance are then the nonlinear terms, including $\mathbf{u} \cdot \nabla \mathbf{u}$, $\mathbf{B} \cdot \nabla \mathbf{B}$ and $\nabla \times (\mathbf{u} \times \mathbf{B})$. These terms are responsible for advection of momentum and magnetic field and, in particular, for the cascade process that transfers energy both to larger and smaller scales. Of course, diffusion is not unimportant, but its purpose is mainly to provide an energy cutoff at large wavenumbers. However, the formation of magnetic flux tubes and vortex tubes is also a diffusive process. In this sense, diffusion can indirectly be important for determining the large-scale evolution of magnetic and velocity fields.

From the considerations above it is evident that there are two quite distinct approaches to the solar dynamo problem. One is to focus on basic processes and solve the three-dimensional MHD problem numerically, and the other is to improve the mean-field concept and apply detailed knowledge to realistic models of the sun. Below we describe recent attempts in these two directions. We begin with a discussion of some numerical methods.

4.4 NUMERICAL METHODS

The properties of the numerical schemes discussed below are particular to mesh point methods with centred differences. Whilst with spectral methods many problems can be avoided, the advantage of mesh point methods is that various physical effects are rather straightforward to implement.

4.4.1 Mesh Point and Spectral Methods

There are many ways to solve the MHD equations numerically. We may distinguish between mesh point and spectral methods or a combination of the two of them for computing spatial derivatives. The most popular spectral method in spherical geometry uses an expansion of all functions in terms of spherical harmonics in the θ- and ϕ-directions and in Tshebyshev polynomials in the r-direction (e.g., Glatzmaier 1984, Zhang & Busse 1989, Valdettaro & Meneguzzi 1991). In the radial direction a representation in terms of spherical Bessel functions is also possible (Rädler & Wiedemann 1989, Rädler *et al.* 1990), although computationally less effective than the Tshebyshev expansion for which fast transformations are available. Another successful approach is to use finite elements (e.g., Charbonneau & MacGregor 1992).

For the axisymmetric mean-field dynamo problem a uniform (r, θ) mesh is frequently used (Jepps 1975, Proctor 1977; see also Brandenburg, Moss & Tuominen 1992). In the nonaxisymmetric case the singularity at the pole can lead to problems, because there the mesh width in the ϕ-direction goes to zero. This problem can be suppressed by using a coarser ϕ mesh in the polar regions (Gilman & Miller 1981). In the special case where the mean-field equations are solved it is possible to use a combined method employing an (r, θ) mesh together with a low number of Fourier modes in the ϕ-direction (Moss, Tuominen & Brandenburg 1991, Barker 1993, Barker & Moss 1993). Only a small number of spherical harmonic orders m is needed, because the solutions often do not vary strongly in the ϕ-direction and the energy falls off rapidly with increasing m (see, e.g., table 2 in Rädler *et al.* 1990).

4.4.2 The Discretisation Error

We now discuss some elementary properties of various discretisation schemes which are useful to understand the origin of numerical instabilities and numerical diffusion. For a comprehensive presentation of this topic we refer to the book by Richtmyer & Morton (1967). Consider for example the discretisation of the diffusion operator, $\partial^2 f/\partial x^2$, in one dimension. Using a Taylor expansion one can see that a simple three-point formula

$$\frac{f_{i-1} - 2f_i + f_{i+1}}{\delta x^2} = \frac{\partial^2 f}{\partial x^2} + \frac{\delta x^2}{12}\frac{\partial^4 f}{\partial x^4} + \cdots \qquad (4.4.1)$$

is accurate to second order in δx, but the error at this order corresponds to a *negative* hyperdiffusion term. In Fourier space the terms on the right-hand side of (4.1) correspond to $[-k^2 + (\delta x^2/12)k^4 + \cdots]\hat{f}(k)$, i.e., the discretisation error acts so as to reduce the effective diffusion. Consider now another example

$$\frac{-f_{i-2} + 10f_{i-1} - 18f_i + 10f_{i+1} - f_{i+2}}{6\delta x^2} = \frac{\partial^2 f}{\partial x^2} - \frac{\delta x^2}{12}\frac{\partial^4 f}{\partial x^4} + \cdots . \qquad (4.4.2)$$

The two schemes are equally accurate to second order in δx, but the error in equation (4.2) corresponds to an additional *positive* hyperdiffusion. This has consequences for the stabilisation of small scales. For a 'zig-zag' mode, where the function value alternates between plus and minus, $(\partial^2 f/\partial x^2)_i$ becomes either $-4f_i/\delta x^2$ or $-6.67f_i/\delta x^2$, depending on which of the two schemes is used. The difference is here not very large, but with other schemes the damping of a zig-zag mode can be much larger. Using Fourier methods, for example, the derivatives are accurate to all orders in δx, and for a zig-zag mode, $(\partial^2 f/\partial x^2)_i = -\pi^2 f_i/\delta x^2$ for the second derivative, i.e., the damping is about twice as big as before. Spline methods (Nordlund & Stein 1990) have the same accuracy properties as the scheme in equation (4.2), but now the damping for a zig-zag mode is $-12f_i/\delta x^2$. This is an advantage of spline methods, because the stabilisation of zig-zag modes is important for high Reynolds number flows with strong gradients. (In practice the accuracy properties of a given scheme can easily be determined numerically by comparing the discretisation error for a given function with its higher derivatives.)

The diffusion terms (4.1, 2) are only accurate to second order in δx. Note, however, that not only the order of the scheme is important, but also the *sign* of the error. On the other hand, higher order schemes are quite crucial for the advection terms in order to minimise unphysical deformations of patterns propagating through the fluid. In fact, using sixth order finite differences on a staggered mesh, Nordlund (personal communication) finds clear advantages over the corresponding fourth order scheme.

4.4.3 Mean-field Computations
We mentioned in section 4.3.2 that the mean-field dynamo equations are of diffusive nature. The maximal time step that satisfies the stability requirements of various schemes may be rather short. The simple DuFort–Frankel

(DF) scheme has the advantage of being unconditionally stable and it is therefore often used. In the DF scheme the diffusion operator is written as

$$\eta[f_{i+1}^n - (f_i^{n+1} + f_i^{n-1}) + f_{i-1}^n]/\delta x^2, \qquad (4.4.3)$$

where n labels the time level and i the mesh point (in one dimension). If the time step is too long the DF scheme no longer represents the original equations. It can easily be shown that a long time step effectively leads to an extra term $\eta(\delta t/\delta x)^2 \partial^2 f/\partial t^2$ (e.g., Peyret & Taylor 1986). This second-order 'semi-implicit' scheme does not involve any expensive matrix inversions, because only the diagonal terms are treated implicitly. The nonlinear terms are usually written explicitly, and this generally requires shorter time steps which in turn improves the accuracy of the time stepping.

A nonuniform diffusivity sometimes leads to problems, if it is straightforwardly implemented. However, it is possible to write

$$\eta(x)\frac{\partial^2 f}{\partial x^2} = \eta_{\max}\left(\frac{\partial^2 f}{\partial x^2}\right)_{DF} + [\eta(x) - \eta_{\max}]\left(\frac{\partial^2 f}{\partial x^2}\right)_{\text{explicit}}, \qquad (4.4.4)$$

so that the maximum value of the diffusivity is treated in the DF (or any other implicit) scheme, and the nonuniform part is solved explicitly. Further experience concerning the discretisation error of the diffusion operator has been reported by Panesar & Nelson (1992).

4.4.4 The Induction Equation in Direct Simulations

In this subsection we discuss methods to solve the induction equation using centred differences. Again, spectral methods as well as the use of a staggered mesh (e.g., Chan & Sofia 1986, Hurlburt & Toomre 1988, Fox, Theobald & Sofia 1991) can avoid certain problems. It is nevertheless useful to explain the properties of centred mesh schemes because they are widely used. In order to preserve $\nabla \cdot \mathbf{B} = 0$ for all times it would be possible to discretise the terms on the right-hand side such that all terms occur under the curl operator. This formulation leads to the problem that the diffusion term, which takes the form $-\nabla \times \nabla \times \mathbf{B}$, is not well behaved at small scales. Here, a first derivative is taken twice in the same direction and, since the centred difference of a zig-zag mode vanishes, there is no diffusion associated with such a zig-zag mode. Consequently, it is not possible to attain large magnetic Reynolds number in such a formulation.

This problem can be avoided by writing $-\nabla \times \nabla \times \mathbf{B} = \nabla^2 \mathbf{B}$. (Here and below we adopt Cartesian coordinates.) In practice, due to discretisation errors, such a formulation can introduce magnetic monopoles, in particular close to the upper boundary. It is possible to adopt a correction procedure (e.g., Schmidt-Voigt 1989), but this involves solving a Poisson equation and can therefore be quite time consuming. Alternatively, we may solve the equations for the vector potential $\mathbf{A}$. Writing $\mathbf{B} = \nabla \times \mathbf{A}$, the induction equation can be uncurled to give

$$\frac{\partial \mathbf{A}}{\partial t} = \mathbf{u} \times \mathbf{B} - \eta \nabla \times \nabla \times \mathbf{A} + \nabla \varphi, \qquad (4.4.5)$$

where $\varphi = 0$ is a convenient gauge. It should be noted that solving for $\mathbf{A}$ instead for $\mathbf{B}$ has generally the disadvantage that the physically relevant field $\mathbf{B}$ is only obtained after differentiation, which involves additional inaccuracies. On the other hand, solving for $\mathbf{A}$ has the advantage that $\nabla \cdot \mathbf{B} = 0$ is guaranteed. Furthermore, the magnetic helicity density $\mathbf{A} \cdot \mathbf{B}$ is readily available and can be used for diagnostic purposes.

In the following we discuss another problem that arises when solving for $\mathbf{A}$. The vector $-\nabla \times \nabla \times \mathbf{A}$ reads explicitly

$$\begin{pmatrix} A_{x,yy} + A_{x,zz} - A_{y,yx} - A_{z,zx} \\ A_{y,xx} + A_{y,zz} - A_{x,xy} - A_{z,zy} \\ A_{z,xx} + A_{z,yy} - A_{x,xz} - A_{y,yz} \end{pmatrix}. \qquad (4.4.6)$$

Note that there are no 'diagonal terms', such as $A_{x,xx}$, and hence no stabilisation of small-scale wiggles in the direction of $\mathbf{A}$. However, by use of the identity $-\nabla \times \nabla \times \mathbf{A} = \nabla^2 \mathbf{A} - \nabla(\nabla \cdot \mathbf{A})$ we may write

$$\begin{pmatrix} A_{x,xx} + A_{x,yy} + A_{x,zz} - (A_{x,x})_{,x} - A_{y,yx} - A_{z,zx} \\ A_{y,xx} + A_{y,yy} + A_{y,zz} - A_{x,xy} - (A_{y,y})_{,y} - A_{z,zy} \\ A_{z,xx} + A_{z,yy} + A_{z,zz} - A_{x,xz} - A_{y,yz} - (A_{z,z})_{,z} \end{pmatrix}, \qquad (4.4.7)$$

which is equivalent to (4.6) and has the advantage that small-scale wiggles can now be damped. It should be noted however that this corresponds to adding a positive hyperviscosity: the usual second order discretisation gives

$$f_{x,xx} - (f_{x,x})_{,x} = 0 - (\delta x^2/4)(\partial^4 f/\partial x^4). \qquad (4.4.8)$$

These considerations show that details of the discretisation can be essential. In the following sections we review some results of numerical simulations and discuss them in the light of the solar dynamo.

4.5 DIRECT SIMULATION OF A DYNAMO
In this section we consider direct simulations of convective MHD turbulence and their relevance for understanding convective dynamos. We focus on the possibility of directly simulating dynamo action, and discuss then their application to estimating turbulent transport coefficients.

4.5.1 Setup of the Model

We consider a particular model setup used by Brandenburg et al. (1993b), which is a slight modification of that described in Brandenburg et al. (1990a) and Nordlund et al. (1992). The main point here is the inclusion of upper and lower overshoot layers. At the boundaries of the computational domain the usual stress-free, perfectly conducting boundaries are used, and a certain radiative flux is imposed at the bottom. The idea is that details of the boundary conditions do not directly influence the general flow behaviour between the upper and lower overshoot layers.

A lower overshoot layer can be modelled by making the heat conductivity, K, depth dependent. In reality, of course, K is a function of temperature and density. In downdrafts, high density and low temperature material will be advected downwards, and will locally lower the border between stably and unstably stratified regions. As a consequence, downdrafts extend further down if this effect is included (Kim & Fox 1993). However, it is difficult to simulate realistically the almost perfect adiabatic stratification in the lower part of the CZ. Consequently, the typical flow velocities in a simulation with less perfect adiabatic stratification are too large compared with the sun. Thus, in such a simulation the downflows are too fast and penetrate deeper into the stably stratified interior than in reality. Therefore, the prescription $K = K(z)$ may be a good compromise after all — especially if the main intention is to remove the restrictions arising from an impenetrable boundary condition. A lower overshoot layer is also physically important in a simulation, because this is the location where a major part of the mean magnetic field is formed and from where the magnetic flux tubes originate that emerge as bipolar regions on the solar surface.

In our model we also adopt a 'cooling layer' at the top by adding ad hoc an additional term in the energy equation

$$\frac{\partial e}{\partial t} + \mathbf{u} \cdot \nabla e = -\frac{p}{\rho} \nabla \cdot \mathbf{u} + \frac{1}{\rho} \nabla[K(z)\nabla T] - f(z)\frac{e - e_0}{\tau_{\text{cool}}}. \qquad (4.5.1)$$

Here, $e = c_v T$ is the internal energy, T temperature, p pressure, ρ density, and $f(z)$ is a profile function that is nonvanishing only in the upper overshoot layer. The cooling is characterised by the cooling time τ_{cool} and the reference 'temperature' e_0. This is a crude simplification of the real processes in the solar atmosphere, but it is a 'cheap' way of introducing a stably stratified upper layer, and it also resembles Newton's cooling law. Alternatively, this could also be achieved by making $K(z)$ larger in the upper part but, because of the low density there, the diffusive time step, $\sim \delta x^2 \rho/K$, would become rather short. An implicit scheme for e (e.g., Hurlburt et al. 1989) could be used, but is perhaps not worth the effort, because in the sun the photospheric layers are optically thin, and the radiative diffusion approximation is invalid anyway. In this sense, a cooling layer seems to be a good compromise.

Many convection simulations presented in the literature are for a polytropic index $m = 1$, that describes the associated hydrostatic equilibrium solution (e.g., Hurlburt, Toomre & Massaguer 1984). This corresponds to a 'radiative nabla' $\nabla_{rad} = 1/(m + 1) = 0.5$. This value is rather small compared to that expected in the upper layers of the sun with $\nabla_{rad} = 10^5$ ($m \to -1$). The related hydrostatic reference solution is of little interest, and certainly not suitable as initial condition in a convection simulation. In practice, one can start with $m \approx 0$ and, once convection is established, m can be reduced to approach the value -1. It would be more physical, of course, if the location of the surface was allowed to adjust itself. In the sun, for example, the related hydrostatic solution would have its surface at a much smaller radius.

4.5.2 Convective Dynamos

Numerical simulations of convective dynamo action in Cartesian geometry have been investigated both in the Boussinesq approximation (Meneguzzi & Pouquet 1989) as well as in a fully compressible flow (Nordlund et al. 1992). Meneguzzi & Pouquet found dynamo action both with and without rotation, if the magnetic Reynolds number, based on the Taylor microscale (the ratio between r.m.s. velocity and r.m.s. vorticity), exceeds a certain threshold around 30. The same critical value is also obtained for compressible MHD convection (Brandenburg et al. 1993a). In these simulations $P_m \equiv \nu/\eta > 1$, and the magnetic energy therefore exceeds the kinetic energy at small scales. This is not representative for the sun, because there $P_m \ll 1$; see section 4.2.1.

The range of length scales covered in such simulations is too short to determine convincingly the possibility of a power law behaviour of spectral energies, but at large scales the spectra are roughly compatible with $k^{-5/3}$ and k^{-1} for kinetic and magnetic energies, respectively. The onset of dynamo action and the saturation level depend somewhat on the numerical resolution, because the majority of magnetic flux tubes are resolved by just a few mesh points. Ideally, a larger number of mesh points is desirable. Also, in order to be sure that there is a true dynamo, as opposed to a transient amplification, the computation has to run for long enough. Cattaneo, Hughes & Weiss (1991) estimate a characteristic diffusion time which, in the case of convection with a continuous spectrum of scales, is a significant fraction of the global diffusion time, $\tau_\eta = d^2/\eta$. In figure 4.2 we show that even though the dynamo is quickly saturated, the magnetic energy can indeed show long term transients, which are associated with slow changes in the flow pattern.

In the following we discuss some properties of convective dynamo action once saturation has set in. The curvature force plays a particular role for the saturation of the dynamo (Nordlund et al. 1992). One can separate the Lorentz force into three components: pressure gradient force, tension force

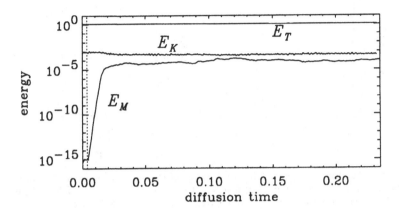

Figure 4.2 Exponential growth of magnetic energy, E_M, over twelve orders of magnitude, followed by saturation, for a model similar to that of Nordlund *et al.* (1992). After saturation, the kinetic energy, E_K, is decreased, but the thermal energy, E_T, remains almost unchanged.

and curvature force (Priest 1982). The separation between tension force and curvature force, i.e., into components of $\mathbf{B} \cdot \nabla \mathbf{B}$ parallel and perpendicular to $\mathbf{B}$, appears somewhat artificial, because $\mathbf{J} \times \mathbf{B}$ obviously has no component parallel to $\mathbf{B}$. However, the magnetic pressure gradient is approximately balanced by the gas pressure gradient, and therefore $\mathbf{B} \cdot \nabla \mathbf{B}$ plays an independent role. The tension force can be regarded as analogous to the force that arises by pulling a rubber band. It is mainly the curvature force that is responsible for saturating the magnetic field, whilst the tension force is relatively unimportant (Nordlund *et al.* 1992). The curvature force prevents the flow from bending and twisting the magnetic flux tubes beyond a certain point, which therefore limits the ability of the flow to amplify the magnetic field further. Even though the magnetic pressure gradient force is much larger than the curvature force, it is unable to cause saturation.

The magnetic field is intermittent, and strong flux tubes occupy only a relatively small fraction of the volume. This can be quantified by using the probability density function for the magnetic field. It turns out that in 40% of the volume, the field strength exceeds its r.m.s. value, and in only 5% of the volume does it exceed 1/4 of its maximum value. The regions where $|\mathbf{B}| \geq B_{rms} = 0.13\,B_{max}$ contribute 80% to the total flux, and those where $|\mathbf{B}| > B_{max}/4$ contribute as much as 25% to the total flux. For further details see Brandenburg *et al.* (1993*a*).

$Pr_M = 0.2$

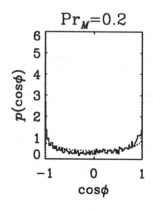

$Pr_M = 1.0$

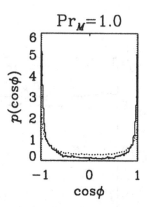

Figure 4.3 Probability density function of the cosine of the angles between u and B for simulations with an imposed horizontal magnetic field and different values of P_m. Solid lines are for regions with $|B| > B_{max}/4$ and dotted lines are for the full space.

4.5.3 Statistical Properties

A variety of statistical properties of the flow can be evaluated from numerical simulations. An interesting investigation of various spectral properties and transfer functions in turbulent dynamos has been presented by Kida *et al.* (1991) for an incompressible fluid driven by forcing instead of convection. The statistics of the angles between various vectors in a compressible simulation of a dynamo has been studied by Brandenburg *et al.* (1993*a*). It turns out that there is alignment between B and ω (vorticity) in some regions and between u and B in other regions. In figure 4.3 the probability density function of the cosine of the angle between u and B is shown for a simulation with an imposed horizontal magnetic field and different values of P_m. For $P_m = 1$ there is a small-scale dynamo effect and there is an enhanced probability for u and B to be aligned. In the absence of a dynamo effect ($P_m = 0.2$) this distribution is more uniform. This has important implications for the α-effect: in the presence of a small-scale dynamo, Alfvén waves cause u and B to be more nearly aligned and thus, since $\langle \mathcal{E} \rangle = \langle u' \times B' \rangle$, α is decreased. This Alfvén effect (Pouquet *et al.* 1976) may be responsible for reducing the α-effect in the solar CZ. In section 4.6.1 we return to the possibility of evaluating α from simulations.

4.5.4 Mean-field Dynamos in a Box

It is interesting to evaluate the average magnetic field obtained in a dynamo simulation. Before doing that let us first discuss what kind of average fields can be expected if this field is to be explained by an α^2-dynamo.

If we adopt (x, y)-averages then the resulting equations become one-dimensional. An α^2-dynamo in Cartesian geometry has been considered by

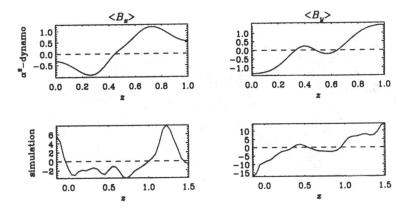

Figure 4.4 Vertical profiles of the horizontal components of the average magnetic field both for the α^2-dynamo of (4.2) (upper row) and the direct simulation (lower row). In the lower row, the convectively unstable region is $0 < z < 1$.

Krause & Meinel (1988) and Meinel & Brandenburg (1990), although not for perfectly conducting boundary conditions. Using the complex variable $\mathcal{B} = \langle B_x \rangle + i \langle B_y \rangle$ the dynamo equation can be written in the form

$$\frac{\partial \mathcal{B}}{\partial t} = \frac{\partial}{\partial z}\left(i\alpha\mathcal{B} + \eta\frac{\partial \mathcal{B}}{\partial z}\right). \qquad (4.5.2)$$

In figure 4.4 we present the marginal solution of $\langle B_x \rangle$ and $\langle B_y \rangle$ as a function of z in the interval $0 \leq z \leq 1$. In this figure we also present the result for the x- and y-components of the average magnetic field obtained from the simulation, where $-0.15 \leq z \leq 1.5$. In this case the marginal value for dynamo action with uniform α is $\alpha d/\eta = 2\pi d/L_z = 3.8$, where $L_z = 1.65d$ is the vertical extension of the box and d the unit length, corresponding to the depth of the unstably stratified layer in the full MHD simulation. Note that in equation (4.2) there is no term corresponding to differential rotation, i.e., $\partial \langle u_y \rangle / \partial z$ does not enter in these equations. This is because the (x, y) average of B_z vanishes (Brandenburg et al. 1990c). If we adopt only y-averages then the problem is two-dimensional and the corresponding $\alpha^2\Omega$-dynamo in the same geometry is described by

$$\left(\frac{\partial}{\partial t} - \nabla^2\right)a = \alpha b, \qquad \left(\frac{\partial}{\partial t} - \nabla^2\right)b = \alpha\nabla^2 a - C_\Omega\frac{\partial a}{\partial x}, \qquad (4.5.3)$$

where $\mathbf{B} = (-\partial a/\partial z, b, \partial a/\partial x)$ and $C_\Omega = \partial \langle u_y \rangle / \partial z$. In figure 4.5 we show the solution of this equation for the marginal value $\alpha d/\eta = 2\pi d/L_x = 2.09$, where $L_x = 3$ is the adopted width of the box. The two-dimensional solution is easier to excite than the one-dimensional solution, and one should thus be able to find it in two-dimensional averages of actual three-dimensional

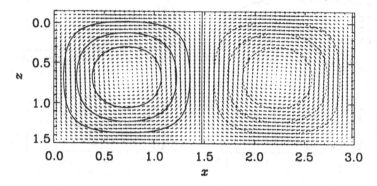

Figure 4.5 Cross-section of the average magnetic field vectors in the (x, z)-plane and contours of the $\langle B_y \rangle$-component, for a kinematic α^2-dynamo in a box with perfectly conducting upper and lower boundaries.

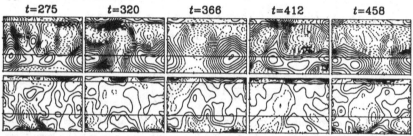

Figure 4.6 Sequence of contours of (x, z) cross-sections of the y-components of the average velocity (upper panel) and magnetic field (lower panel). The two horizontal lines mark the location of the upper and lower overshoot layers.

simulations. In the presence of a shear flow, i.e., $C_\Omega \neq 0$, the magnetic field pattern migrates in the x-direction.

In figure 4.6 we show a sequence of contours of combined (y, t)-averages of u_y and B_y for data from a simulation with $63 \times 63 \times 37$ mesh points. The time average is taken over approximately 45 time units (related to free fall times), corresponding to about 6–7 turnover times. The velocity plotted at different times shows substantial variability, but it is mainly positive in the lower part and negative in the upper part. In our geometry the y-direction corresponds to the toroidal direction, and negative $\partial \langle u_y \rangle / \partial z$ corresponds to $\partial \Omega / \partial r < 0$ in spherical geometry. Note also that $\langle B_y \rangle$ shows concentrations at the top and bottom boundaries of the computational domain. This suggests that the extent of the overshoot layers was not large enough to remove completely the influence of the boundary conditions on the magnetic field.

In summary, our three-dimensional simulations do not show significant mean magnetic fields. It is possible that we need to decrease the value of

P_m to suppress small-scale dynamo action (cf. Gilman 1983) and to increase the box size (at least in the toroidal direction) to favour large-scale dynamo action. Also, in order to see the development of large-scale fields, the simulations have to be continued long after saturation is reached.

4.6 TURBULENT TRANSPORT COEFFICIENTS

The simulation of large-scale magnetic fields in box geometry is a difficult task, and it is not clear how important global effects are. Another application of such simulations is to determine turbulent transport coefficients, such as the α-effect and the turbulent diffusivity. For this purpose a large-scale magnetic field must be present. Since the mean magnetic field generated in such a simulation is weak and quite variable in time it is preferable to apply an external large-scale field. Depending on the direction of this field different components of the α-tensor can be determined.

The computation of turbulent transport coefficients from direct simulations is an important application, because those derived in mean-field theory in the 'first order smoothing approximation' (FOSA) can only strictly be justified if either the magnetic Reynolds number is small or the correlation time of the turbulence is short compared to the turnover time (e.g., Roberts & Soward 1975). In the sun this is not the case and so the FOSA results for the turbulent transport coefficients will be in error. Moreover, there is no clear scale separation in the sun, because the correlation length of the turbulence is comparable with the typical scales of mean-field magnetic structures. In contrast, direct simulations and large eddy simulations are not limited by these constraints. However, the range of time and length scales that can be covered in a simulation is still quite narrow. The mean-field approach on the one hand and direct simulations on the other therefore play *complementary* roles.

4.6.1 The α-effect

In order to determine the α-effect from a simulation we apply a uniform external magnetic field in one of the three directions. A net helicity is generated in the flow when the effects of stratification and Coriolis force are included. Under these conditions there is, in general, a nonvanishing e.m.f. $\langle u' \times B' \rangle$ and three components of the α-tensor can then be determined. In order to obtain the remaining six components of the α-tensor we need to repeat this experiment for the two remaining directions of the imposed magnetic field. To obtain the latitudinal dependence of α we need to perform simulations at different latitudes. Therefore a large number of runs have to be carried out, and all these runs have to be long enough to give good statistics.

It turns out that the horizontal components of the α-tensor, α_{xx} and α_{yy} have the opposite sign to the helicity. This is in agreement with standard theories for α (Moffatt 1978, Krause & Rädler 1980). However, the vertical component α_{zz} can have the opposite sign to the horizontal components (Brandenburg et al. 1990c). This came somewhat as a surprise, but similar results have meanwhile been obtained by Ferrière (1993) and Kaisig, Rüdiger & Yorke (1993). Rüdiger & Kitchatinov (1993) found a negative value for α_{zz} for stratified turbulence and intermediate values of the inverse Rossby number. Thus, there is a clear warning here that the assumption of an isotropic α-tensor, made so often in practice, is not at all justified.

4.6.2 The Λ-effect

An important turbulent transport effect in solar mean-field hydrodynamics is the Λ-effect (Rüdiger 1980b, 1989). It describes the dependence of nondiffusive contributions to the Reynolds stress tensor $Q_{ij} = \langle u_i' u_j' \rangle$ on the angular velocity $\Omega = 1_z \langle u_\phi \rangle / (r \sin \theta)$ via

$$Q_{ij} = \Lambda_{ijk}\Omega_k - N_{ijkl}\partial_l \langle u_k \rangle + \cdots . \qquad (4.6.1)$$

Equation (6.1) resembles (3.2). The Λ-effect and its latitudinal dependence have been determined using simulations of rotating convection in a box located at different latitudes; see Pulkkinen et al. (1991, 1993a, b).

Of particular importance is the horizontal component of the Reynolds stress tensor $Q_{\theta\phi}$. Following Rüdiger (1989) this component takes the form

$$Q_{\theta\phi} = \Lambda_H \cos\theta\Omega - \nu_t \sin\theta\partial\Omega/\partial\theta. \qquad (4.6.2)$$

Ward (1965) found from sunspot proper motions that $Q_{\theta\phi} > 0$ on the northern hemisphere. Since $\partial\Omega/\partial\theta > 0$ for the solar northern hemisphere, and $\nu_t > 0$, this observation would require the presence of Λ_H with $\Lambda_H > 0$. The simulations of Pulkkinen et al. (1991, 1993a, b) confirm this, and the latitudinal dependence of Λ_H is consistent with the observed one (cf. Virtanen 1989).

4.6.3 Turbulent Diffusivities

In computing α it is sufficient to apply a uniform magnetic field. This has then the additional advantage that the eddy diffusivity term $\eta_t \nabla \times \langle \mathbf{B} \rangle$ does not enter. On the other hand, if we want to determine η_t we have to apply a nonuniform external magnetic field. It is then quite typical that the flow develops an electric field that tends to reduce the imposed gradient of the magnetic field. As a result, significant gradients of the magnetic field only remain close to the boundaries of the box. This is a somewhat dangerous situation, because we are not interested in investigating any boundary effects. This prevented us from determining any sensible values for η_t. Another method to determine η_t is to measure the decay time of large-scale magnetic structures in the flow (Brandenburg et al. 1993b).

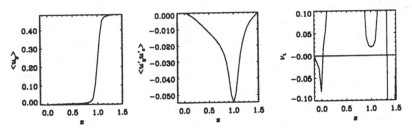

Figure 4.7 Vertical profiles of $\langle u_y \rangle$, $\langle u'_y u'_z \rangle$ and ν_t. At the location of the strongest shear $\nu_t \approx 0.02$.

The simulation with an imposed vertical shear in the y-velocity, mentioned in section 4.2.4, allows us to obtain some information about the turbulent viscosity ν_t. In figure 4.7 we show vertical profiles of $\langle u_y \rangle$, $\langle u'_y u'_z \rangle$, and $\nu_t = -\langle u'_y u'_z \rangle / \partial_z \langle u_y \rangle$. Only in the immediate neighbourhood of the shear layer are the results physically meaningful. Assuming that the effective correlation length is approximately equal to the width of the shear layer, i.e., $\ell = 0.2$, we find $\nu_t \approx 0.2 u_t \ell$. This result is not implausible. It should be noted, however, that we have ignored here the tensor structure of turbulent diffusion.

4.6.4 α-quenching
Measurements of α can be carried out for different values of the imposed magnetic field strength. Nonlinear effects are expected to modify (quench) the various components of the α-tensor. For the horizontal diagonal components of the α-tensor Moffatt (1970, 1972) and Rüdiger (1974) find $\alpha \sim |\langle \mathbf{B} \rangle|^{-3}$. Recent evaluations of α from convection simulations indicate that α-quenching sets in much earlier (Jones & Galloway 1993), although the dependence on $|\langle \mathbf{B} \rangle|$ seems to be weaker (like $|\langle \mathbf{B} \rangle|^{-1}$); see Tao, Cattaneo & Vainshtein (1993) and Brandenburg et al. (1993b). It is interesting to note that a quenching proportional to $|\langle \mathbf{B} \rangle|^{-1}$ has also been predicted by Kraichnan (1979) for interacting Alfvén waves in the strong field limit.

4.6.5 The Strength of Magnetic Fluctuations
It has been suggested that the quenching of α and other turbulent transport coefficients is governed by the magnetic energy density of the actual magnetic field rather than by the energy density of the mean magnetic field (Vainshtein & Cattaneo 1992). It is therefore interesting to determine the strength of magnetic fluctuations from direct simulations. Vainshtein & Cattaneo (1992) argue that the magnetic fluctuations in the sun may actually be so large that there will be an extraordinarily strong quenching of α and η, with the result that a traditional $\alpha\Omega$-type dynamo becomes impossible. According to the formula $\langle \mathbf{B}'^2 \rangle = R_m \langle \mathbf{B} \rangle^2$ (e.g., Krause & Rädler 1980) the fluctuations

can indeed become very large if we use solar values for the magnetic Reynolds number, R_m. However, Gilbert, Otani & Childress (1993) found the scaling to be not R_m but R_m^n with a model-dependent exponent $n = 0.25$–0.35. Furthermore, as the applied mean magnetic field $\langle \mathbf{B} \rangle$ increases, $\langle \mathbf{B}'^2 \rangle$ approaches the equipartition value and nonlinear effects become important that limit the strength of magnetic fluctuations. Thus, even though $\langle \mathbf{B} \rangle$ is increased, $\langle \mathbf{B}'^2 \rangle$ saturates and therefore the ratio $q = \langle \mathbf{B}'^2 \rangle / \langle \mathbf{B} \rangle^2$ decreases. For $|\langle \mathbf{B} \rangle| \approx B_{eq}$, q becomes of order unity (Brandenburg et al. 1993b). In summary, $q = R_m$ does not hold in the nonlinear regime. Unfortunately, direct three-dimensional simulations are at present limited to magnetic Reynolds numbers of the order of one thousand but, based on the analytical model of Kleeorin, Rogachevskii & Ruzmaikin (1990), we may expect that for asymptotically large values of R_m the magnetic fluctuations grow only logarithmically with R_m. This requires that the magnetic power spectrum decreases like k^{-1}, which is the value expected in the presence of an inverse cascade (Pouquet et al. 1976).

Observational evidence for the value of q and its dependence on $\langle \mathbf{B} \rangle$ comes from measurements of the (magnetic) filling factor f at the solar surface. Schrijver (1987) finds $f \approx 0.1$ in active regions and $f \approx 0.01$ in quiescent regions. For a uniformly oriented bunch of flux tubes f is related to q by $q = 1/f$ (Brandenburg et al. 1993b). With this interpretation we estimate $q = 10$ in active regions and $q = 100$ in quiescent regions on the solar surface.

A related aspect should be mentioned here. It is known that during solar minimum the number of X-ray bright points is maximal, but the number is minimal during solar maximum (Golub, Davis, & Krieger 1979). If we assume that the number of X-ray bright points is a measure of the strength of the fluctuating magnetic field, then this observed anticorrelation could be related to the fact that the fluctuating magnetic field is quenched more heavily during sunspot maximum than during sunspot minimum.

4.7 MEAN-FIELD DYNAMOS

The solar magnetic field shows systematic variations over length and time scales large compared to the convective length and time scales. This suggests that some kind of mean-field treatment should be able to describe this. Nevertheless, it is not an easy task to determine a suitable set of turbulent transport coefficients for the sun. In the following we discuss several important properties of dynamos in general before presenting models that are more directly relevant to the sun.

Through the remaining sections we adopt spherical coordinates (r, θ, ϕ). In addition, cylindrical coordinates (ϖ, ϕ, z) are used, with $\mathbf{1}_z$ being the unit vector in the direction of the angular velocity vector $\mathbf{\Omega}$.

4.7.1 The Dynamo Equation

The mean-field dynamo equation is similar to the original induction equation, but all quantities are now averages, and the mean electromotive force, $\langle \mathcal{E} \rangle$ appears on the right-hand side

$$\frac{\partial \langle \mathbf{B} \rangle}{\partial t} = \nabla \times (\langle \mathbf{u} \rangle \times \langle \mathbf{B} \rangle + \langle \mathcal{E} \rangle - \langle \mathbf{J} \rangle / \sigma), \qquad (4.7.1)$$

where $\langle \mathcal{E} \rangle$ has contributions from a number of different terms due to the anisotropies necessarily present in rotating, stratified turbulence. Some of the important terms are

$$\langle \mathcal{E} \rangle = \alpha \langle \mathbf{B} \rangle + \alpha_z \mathbf{1}_z \langle B_z \rangle + \alpha_r \mathbf{1}_r \langle B_r \rangle + \gamma \mathbf{1}_r \times \langle \mathbf{B} \rangle - \eta_t \nabla \times \langle \mathbf{B} \rangle + \cdots . \quad (4.7.2)$$

Here, α and η_t are the isotropic α-effect and turbulent diffusivity, α_z and α_r quantify anisotropies of α in the directions of rotation and gravity, respectively, and the γ term corresponds to a turbulent diamagnetism with $\gamma = -(1/2)\partial \eta_t / \partial r$ (Vainshtein & Zel'dovich 1972). In this connection it should be mentioned that Kitchatinov (1991) found another interesting example of magnetic field advection such that poloidal and toroidal fields are advected in different ways. This effect may have interesting consequences for understanding the magnetic fields in the sun, where it appears that the surface manifestation of the toroidal magnetic field migrates to lower latitudes and the poloidal field to higher latitudes.

There are many more terms contributing to (7.2); see, e.g., Krause & Rädler (1980). A unique theory is needed to derive all these coefficients from the same turbulence model and, although substantial progress has been made in this direction (see Rüdiger & Kitchatinov 1993), much work remains.

Before proceeding further with the discussion of more detailed models let us here comment on some general aspects and properties of the various anisotropies. Weisshaar (1982) found oscillatory α^2-dynamo solutions with the α_r term, and was able to explain the solar butterfly diagram even without any differential rotation in r and θ. However, the degree of anisotropy needed was rather large (three to four orders of magnitude!). However, the dynamo can become oscillatory for much smaller values of α_r. For example, for $\alpha_r = -\alpha$ there is an oscillatory solution, but it lacks any field migration.

The α_z term has different properties from the α_r term, and it does not easily produce an oscillatory solution. However, there is another interesting property of the α_z term if nonaxisymmetric magnetic fields are included. Rüdiger (1980a) found that nonaxisymmetric solutions are more easily excited than axisymmetric ones if the α_z term becomes comparable to the α term in (7.2). Rüdiger & Elstner (1994) obtained similar results even in the presence of solar-like differential rotation. Thus, if mean-field models are to explain the large-scale magnetic field of the sun, it remains to be shown why the effect of nonaxisymmetries (such as the active longitudes) are so weak.

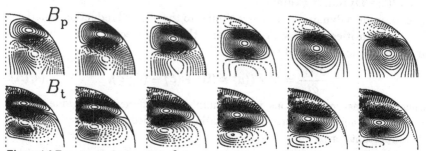

Figure 4.8 Equatorward migrating α^2-dynamo wave for a perfectly conducting outer boundary condition. Poloidal field lines (upper row) and contours of the toroidal field (lower row) are shown at equidistant time intervals covering half a magnetic cycle of 1.1 diffusion times.

4.7.2 Boundary Conditions

In the pre-Maxwell approximation, where the Faraday displacement current is neglected, the magnetic field can instantaneously penetrate the entire space, because radio waves have been filtered out. This is similar to the incompressible and anelastic approximations, where sound waves have been filtered out and pressure perturbations can instantaneously penetrate the entire space. Such approximations transform the originally local problem into a nonlocal one. In the context of the dynamo problem the vacuum boundary condition is nonlocal. Only in the special case where the fluid is embedded into a perfectly conducting medium do the boundary conditions become local.

It is usually assumed that the star is embedded in a vacuum where the magnetic field is a potential field. However, the electrical conductivity in the corona is actually quite large. From Spitzer's formula

$$\eta = 5 \times 10^3 \,\mathrm{cm^2 s^{-1}} \left(\frac{\ln \Lambda}{10}\right) \left(\frac{T}{10^6 \mathrm{K}}\right)^{-3/2}, \qquad (4.7.3)$$

(where $\ln \Lambda = 5$ for the CZ, 10 for the chromosphere, and 20 for the corona; see Priest 1982), we obtain the values $\eta \approx 10^7 \,\mathrm{cm^2 s^{-1}}$ for the photosphere and $\eta \approx 5 \times 10^3 \,\mathrm{cm^2 s^{-1}}$ for the corona. Both values are much smaller than the turbulent value η_t in the CZ. Therefore, a perfectly conducting boundary may be a more plausible outer boundary condition (Parker 1984). In such models, the onset of dynamo action is somewhat delayed compared to dynamos with a vacuum boundary condition, and the solutions can be oscillatory even without differential rotation: a sequence of field plots is given in figure 4.8 (full sphere with isotropic $\alpha = \alpha_0 \cos\theta$ and $C_\alpha = \alpha_0 R/\eta_t = 20$). Another plausible (although not physically rigorous) boundary condition is to assume that the field is radial, i.e., $B_\theta = B_\phi = 0$ (Yoshimura 1975).

4.7.3 Stability of Solutions

The growth rates of stellar dynamos are so large that transient states are irrelevant. (This is not true for galaxies!) Thus, only stable solutions can survive and are of interest for explaining stellar magnetic fields. Therefore, the investigation of the stability of solutions is obligatory.

Various types of solutions can be distinguished according to their symmetries. If, as usual, the α-effect is taken to be perfectly antisymmetric about the equatorial plane and the mean velocity field symmetric, then the magnetic field can be symmetric (S) or antisymmetric (A) with respect to the equatorial plane. If, in addition, the induction effects are axisymmetric then the solutions can show a $2\pi/m$ symmetry in the ϕ direction ($m > 0$) or they can be axisymmetric ($m = 0$). We refer to these solutions as Sm and Am (for further details see Rädler et al. 1990). Despite the symmetry properties of the induction effects there can well be (stable!) solutions with mixed parity. There are several examples of dynamos where, in a certain parameter range, neither pure dipole nor pure quadrupole solutions are stable (Brandenburg, Tuominen & Moss 1989c). Similar studies have been carried out by Jennings (1991) using a one-dimensional model.

As an instructive example we show in figure 4.9 the bifurcation diagram for a dynamo in a sphere with a perfectly conducting outer boundary condition and constant radial differential rotation. The result is rather similar to the case with a vacuum boundary condition outside (cf. Brandenburg et al. 1989c, figure 11): in the two cases, the $A0$-solution becomes excited first, loses stability for a particular value of C_α, goes over into a mixed parity solution, and finally becomes an $S0$-solution.

There is a north–south asymmetry of the sunspot number that may be explained in terms of a solution with a weakly mixed parity. For the sun, a useful quantity characterising the degree of north–south asymmetry is

$$Q = \left[N^{(+)} - N^{(-)} \right] \Big/ \left[N^{(+)} + N^{(-)} \right], \qquad (4.7.4)$$

where $N^{(\pm)}$ are the sunspot numbers in the northern and southern hemispheres, respectively. This quantity has been measured both for the sun and for simple dynamo models (Brandenburg, Krause & Tuominen 1989b). For the sun Q has varied between about -0.1, at the end of last century, and about $+0.15$, at the present time. Mixed parity modes have also been discussed in connection with the Maunder minimum (Brandenburg et al. 1990a, Sokoloff & Nesme-Ribes 1993).

4.7.4 Chaotic Behaviour

The mean solar magnetic field behaves irregularly in time. This may either be described by a noise term included in the mean-field equations (Hoyng 1988, Choudhuri 1992, Moss et al. 1992) or it could be dynamical chaos due to the

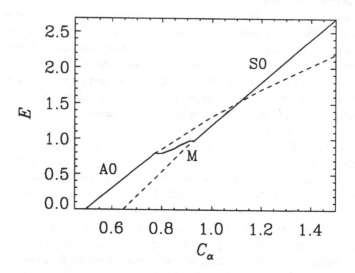

Figure 4.9 Bifurcation diagram for an $\alpha^2\Omega$-dynamo in a full sphere with perfectly conducting boundary condition on $r = R$. Unstable pure parity solutions are shown by dotted lines. The stable mixed parity solution is marked as M.

nonlinear nature of the governing equations (Tavakol 1978, Ruzmaikin 1981, Jones, Weiss & Cattaneo 1985). Chaotic behaviour is found in truncated dynamo models (see the lectures of Weiss in chapter 2 and Spiegel in chapter 8), but chaotic behaviour often tends to disappear at higher truncation levels. It is therefore interesting that Brooke & Moss (1994) found chaotic solutions (i.e., with positive Lyapunov exponent) in a two-dimensional (axisymmetric) $\alpha\Omega$-dynamo in torus geometry, where α-quenching was the only nonlinearity.

4.7.5 Feedback from Small-scale Motions

As the mean magnetic field increases, the turbulence becomes anisotropic (Krause & Rüdiger 1975, Schumann 1976) and this affects the various turbulent transport coefficients. A typical example is α-quenching

$$\alpha(\langle \mathbf{B} \rangle) = \alpha_0(1 - \alpha_B \langle \mathbf{B} \rangle^2) \qquad (4.7.5)$$

(Rüdiger 1974). The coefficient α_B is estimated as $1/B_{eq}^2$, meaning that quenching becomes efficient once the mean magnetic field becomes comparable to the equipartition value, $B_{eq}^2 = \mu_0 \rho u_t^2$. A solution for a spherical dynamo model with uniform α_0 using (7.5) has been presented by Rüdiger (1973). The more realistic case of an α-effect that changes sign about the equator ($\alpha_0 \propto \cos\theta$) has been considered by Jepps (1975), and has frequently been used in a number of subsequent investigations (e.g., Yoshimura 1975, Ivanova

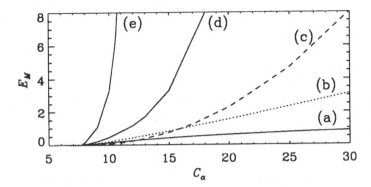

Figure 4.10 The magnetic energy E_M versus C_α for different feedback mechanisms and $\alpha_B = 1$: (a) using (7.5), (b) using (7.6), (c) using (7.7), (d, e) using (7.6) and (7.8) with (d) $\eta_B = 0.5$ and (e) $\eta_B = 1$.

& Ruzmaikin 1977, Krause & Meinel 1988, Brandenburg *et al.* 1989*a*, Schmitt & Schüssler 1989, Rädler & Wiedemann 1989, Rädler *et al.* 1990, Jennings 1991). In these papers an *ad hoc* extension of the above expression to arbitrarily strong magnetic fields is used by taking

$$\alpha(\langle\mathbf{B}\rangle) = \alpha_0/(1 + \alpha_B\langle\mathbf{B}\rangle^2). \tag{4.7.6}$$

In a bifurcation diagram (figure 4.10) for $\mathcal{A}0$ solutions we compare different feedbacks for α^2-dynamos in a full sphere using standard vacuum boundary conditions. We also present the results for

$$\alpha(\langle\mathbf{B}\rangle) = \alpha_0/(1 + \alpha_B^{1/2}|\langle\mathbf{B}\rangle|) \tag{4.7.7}$$

(cf. Brandenburg *et al.* 1993*b*), and for combined α- and η-quenching with

$$\eta_t = \eta_t^{(0)}/(1 + \eta_B\langle\mathbf{B}\rangle^2), \tag{4.7.8}$$

using different coefficients η_B. The η-quenching mechanism is probably an important nonlinear effect that has been known about for a long time (Roberts & Soward 1975), but it has been neglected in many dynamo models.

A feedback via the current helicity, $\langle\mathbf{J}'\cdot\mathbf{B}'\rangle$, in the form

$$\alpha = \alpha_K + \alpha_M = -\tau_{\mathrm{cor}}\langle\omega'\cdot\mathbf{u}'\rangle + \tau_{\mathrm{cor}}\langle\mathbf{J}'\cdot\mathbf{B}'\rangle/\langle\rho\rangle \tag{4.7.9}$$

has been proposed by Vainshtein (1972). Following Kleeorin & Ruzmaikin (1981), the magnetic part α_M satisfies an explicitly time dependent equation of the form

$$\frac{\partial\alpha_M}{\partial t} = Q(\langle\mathbf{J}\rangle\cdot\langle\mathbf{B}\rangle + q\alpha\langle\mathbf{B}\rangle^2) + \nu_\alpha\nabla^2\alpha_M, \tag{4.7.10}$$

where Q, q, and ν_α are constant parameters. The physically interesting case is where $\nu_\alpha < \eta_t$, corresponding to an adjustment time of the magnetic α-effect that is longer than the magnetic diffusion time (cf. Zel'dovich, Ruzmaikin & Sokoloff 1983). Numerical simulations indicate, however, that $\alpha_M \ll \alpha_K$ (Brandenburg *et al.* 1990*c*).

Magnetic buoyancy also acts as a feedback (Leighton 1969). It leads to a radial transport of the average magnetic field outwards (Moss, Tuominen & Brandenburg 1990). However, Kitchatinov & Pipin (1993) found that in the sun magnetic buoyancy is negligible compared to α-quenching.

4.7.6 Feedback from Large-scale Motions

The feedback from the large-scale motions does not involve poorly known quenching coefficients, and the large-scale velocity $\langle u \rangle$ is obtained by solving the momentum equation (here in a nonrotating frame),

$$\frac{D \langle u \rangle}{Dt} = -\left\langle \frac{1}{\rho} \nabla p \right\rangle + g + \varpi \Omega^2 + \left\langle \frac{1}{\rho} J \times B \right\rangle - \frac{1}{\rho} \nabla \cdot (\rho Q), \qquad (4.7.11)$$

simultaneously with the induction equation (7.1). Here, $D/Dt = \partial/\partial t + \langle u \rangle \cdot \nabla$ is the advective derivative, ρ density, p pressure, g gravity, $\varpi = (\sin\theta, \cos\theta, 0)\varpi$ in spherical coordinates, Q is the Reynolds stress tensor, and $\Omega = \langle u_\phi \rangle / \varpi$. Simple α^2-dynamos with a feedback from $\langle J \rangle \times \langle B \rangle$ have been considered by Malkus & Proctor (1975) and Proctor (1977), and this is sometimes referred to as the 'Malkus–Proctor mechanism'. Recently, Barker (1993) and Barker & Moss (1993) found that the axisymmetric solutions are unstable, and that the field evolves towards $S1$.

It has been proposed (Schüssler 1981, Yoshimura 1981) that the feedback from the large-scale magnetic field might be responsible for driving the solar torsional oscillations (Howard & LaBonte 1981). However, it has now been argued that it is in fact the feedback from the small-scale magnetic field, $\langle J' \times B' \rangle$, that modifies the Λ-effect and drives in this way the torsional oscillation of the sun (Kitchatinov 1988, Rüdiger & Kitchatinov 1990, Rüdiger 1992). The torsional oscillation, which has an 11 year period, is accompanied by a cyclic variation of the meridional flow, and has also been found from sunspot proper motions (Tuominen & Virtanen 1984). The feedback from $\langle J' \times B' \rangle$ has also been studied by Kleeorin *et al.* (1990), who investigated the possibility of a large-scale instability.

In practice, both large-scale and small-scale feedbacks have to be considered and it is not *a priori* clear which of the two dominate. These questions have been discussed in Brandenburg *et al.* (1990*b*) and Rüdiger & Kitchatinov (1990). In general, small-scale feedbacks (α- and Λ-quenching) dominate over the large-scale feedback if the normalised correlation length $\xi = \ell/R$ is large enough (around 0.1), corresponding to the lower parts of the CZ.

4.8 THE SOLAR DYNAMO AND DIFFERENTIAL ROTATION

Rüdiger & Kitchatinov (1993) pointed out that all stars, including the sun, must be considered as rapid rotators in the sense that the inverse Rossby number, $\Omega^* = 2\Omega\tau_{cor}$, is large compared to unity. In this case the tensor structure of α_{ij} and η_{ijk} becomes important, and the magnitude of the tensor components is generally reduced (Kitchatinov, Pipin & Rüdiger 1994). The same is true for the Λ-effect and the turbulent viscosity, N_{ijkl} (Kitchatinov & Rüdiger 1993). These results have triggered new investigations of the solar dynamo and differential rotation.

4.8.1 Dynamo Action in the Overshoot Layer

Using the full expressions for the anisotropic α- and η-tensors, Rüdiger & Brandenburg (1993) constructed a dynamo model for the solar overshoot layer. This was formally achieved by assuming that only the stratification in the turbulence intensity contributes to α. The differential rotation was taken from helioseismological inversions. Three fundamental problems have been addressed in this model, with the following results:

(1) The cycle period is close to the solar value of 22 years. This is due to three different contributions. Firstly, because of rapid rotation, η_t is reduced by a factor $\Omega^* \approx 5$ relative to the reference value $\eta_0 \equiv u_t^2\tau_{cor}/3$. Secondly, in the overshoot layer, $u_t = 20\,\mathrm{m/s}$ and $\tau_{cor} = 10\,\mathrm{days}$, so $\eta_0 \approx 10^{12}\mathrm{cm}^2/\mathrm{s}$ is smaller than in the bulk of the CZ. Thirdly, the effect of magnetic buoyancy is to increase the cycle period by another factor of two.

(2) The butterfly diagram shows distinct equatorward and poleward branches at low and high latitudes, respectively. The relative strength of the two branches is determined by the latitudinal profile of α. Rüdiger & Brandenburg argue that the traditional $\alpha \sim \cos\theta$ dependence results from including only terms linear in $\Omega \cdot \nabla \ln(\rho u_t) \sim \cos\theta$ (Krause 1967). In particular in the overshoot layer the gradient in u_t is large and therefore higher order terms, such as $\cos^3\theta$, are possible. Writing $\alpha \sim \cos\theta(1 - \alpha_U\cos^2\theta)$, with α_U as a free parameter, a butterfly diagram compatible with the solar case is obtained when $\alpha_U = 1$. Similar α profiles have been proposed by Yoshimura (1975), Schmitt (1987), Tuominen, Rüdiger & Brandenburg (1988), Belvedere, Proctor & Lanzafame (1991), and Prautzsch (1993).

(3) The phase relation between the poloidal and toroidal fields has been a notorious problem, because a positive $\partial\Omega/\partial r$ generates B_ϕ from B_r with the same sign, i.e., $B_rB_\phi > 0$. This is in contrast to the observed phase relation (Stix 1976, Yoshimura 1976). However, if the toroidal field is evaluated at the bottom of the CZ (where sunspots are anchored), and the poloidal field at the top, then B_r and B_ϕ are almost in antiphase; see figure 4.11. At higher latitudes, we do have $B_rB_\phi < 0$, because there $\partial\Omega/\partial r < 0$.

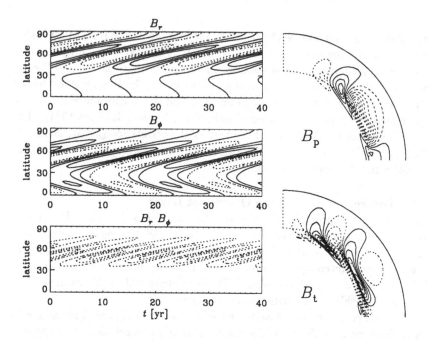

Figure 4.11 The field geometry for the model of Rüdiger & Brandenburg (1993). *Left*: butterfly diagrams for B_r, B_ϕ and $B_r B_\phi$. *Right*: poloidal field lines and contours of toroidal field strength in a meridional plane.

4.8.2 Dynamo Action in the Convection Zone

There are a number of problems associated with dynamo action in the bulk of the CZ: the period is too short and it is not easy to produce butterfly diagrams compatible with the sun. There is, however, one important effect that can significantly change the magnetic field geometry: an anisotropic transport of magnetic field such that poloidal fields are transported poleward and toroidal fields equatorward (Kitchatinov 1991, 1993). In figure 4.12 we show such a model where we have also restored the effects of the density stratification (using a solar mixing length model) so leading to a positive α in the bulk of the CZ and a negative α in the overshoot layer (for the northern hemisphere). Otherwise the model is similar to that of Rüdiger & Brandenburg. However, the value of η_t has been lowered by a factor of ten to make the dynamo supercritical and to obtain a cycle period closer to 22 years.

A promising way to distinguish between an overshoot layer dynamo and a 'distributed' dynamo is to compare dynamo activity in stars with different rotation rates. Preliminary investigations seem to favour the distributed dynamo, although a number of unresolved questions still remain.

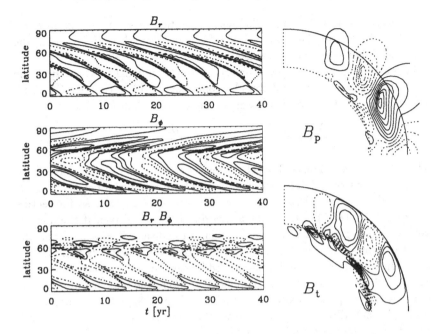

Figure 4.12 The field geometry for a model with dynamo action in the entire CZ.

4.8.3 Differential Rotation and Meridional Flow

In the previous section we assumed the differential rotation to be given. This is dynamically inconsistent, because nonuniform rotation always generates a meridional flow which, in turn, feeds back on Ω. The evolution of Ω is governed by

$$\frac{\partial}{\partial t}(\rho \varpi^2 \Omega) = -\nabla \cdot (\rho \varpi^2 \Omega \langle \mathbf{u}_p \rangle - \varpi \langle B_\phi \mathbf{B}_p \rangle + \mathbf{1}_\phi \cdot \mathbf{Q}), \qquad (4.8.1)$$

where $\mathbf{Q}$ was introduced in section 4.6.2, and differential rotation is generated by nondiffusive contributions to $\mathbf{Q}$ (Λ-effect). An important control parameter is the Taylor number, $\mathrm{Ta} = 4\Omega_0^2 R^4 / \nu_t^2$. With stress-free boundary conditions, $Q_{r\phi} = 0$, the angular momentum is conserved, i.e., $\Omega_0 \equiv \int \rho \varpi^2 \Omega \, dV / \int \rho \varpi^2 \, dV = \text{constant}$, and the value of Ta is thus determined by the initial condition.

If Ta is small, the meridional flow is weak and the approximate solution is $\nabla \cdot (\rho \mathbf{Q}) \approx 0$. For $\mathrm{Ta} \lesssim 10^5$ it is possible to match the observed differential rotation by a suitable choice of Λ-effect parameters (Tuominen & Rüdiger 1989). This is no longer true if Ta is large, because then a strong meridional circulation is driven. Köhler (1970) showed that, in the incompressible case, the meridional flow driven by differential rotation achieves a maximum for a certain Taylor number ($Ta \approx 2 \times 10^6$), and above this critical value the meridional flow speed decreases again.

In order to discuss the basic drivers of the meridional flow, we split the velocity into its poloidal and toroidal components, $\langle u \rangle = \langle u_p \rangle + \varpi \Omega 1_\phi$, and take the curl of the poloidal part of the momentum equation (7.11)

$$ 1_\phi \cdot \nabla \times \left(\frac{D \langle u_p \rangle}{Dt} \right) = 1_\phi \cdot \langle \nabla T \times \nabla S \rangle + \varpi \frac{\partial \Omega^2}{\partial z} + 1_\phi \cdot \nabla \times [\cdots], \quad (4.8.2) $$

where the terms in square brackets represent Lorentz and viscous forces, S is the specific entropy, T temperature, and we have assumed a perfect gas, i.e., $p = \rho \mathcal{R} T$ and $S = c_v \ln p - c_p \ln \rho$, where $\mathcal{R} = c_p - c_v$; see also the lecture of Fearn (chapter 7). Note that a meridional flow is driven if the angular velocity has a gradient in the z-direction. In addition, if $S \neq$ constant and if the contours of S and T are not parallel, then this too drives a meridional flow. In stellar interiors this is the Eddington–Sweet circulation, but the same mechanism also works in a stellar CZ, where the temperature gradient is necessarily superadiabatic (e.g., Moss & Vilhu 1983). In the presence of magnetic fields the Lorentz force too affects the meridional flow either directly or indirectly via the toroidal velocity.

For sufficiently large values of Ta the meridional flow decreases with ν_t and a state is achieved where the terms on the right-hand side of (8.2) are in approximate balance. In the barotropic case $\langle \nabla T \times \nabla S \rangle = 0$, and consequently $\partial \Omega / \partial z \rightarrow 0$ (cf. Durney 1976). This is the Taylor–Proudman theorem, originally formulated for the incompressible case. In the sun $Ta = 3 \times 10^7$ (with $\nu_t = 5 \times 10^{12} \text{ cm}^2/\text{s}$), which would be sufficiently large for the Taylor–Proudman theorem to apply, and yet the solar Ω contours do not lie on cylinders, as inferred from helioseismology (Brown & Morrow 1987). However, a number of other Ω-profiles are also compatible with the data (Wilson 1992). Gough *et al.* (1993) found that even nearly cylindrical Ω-contours in the outer tangent cylinder, $\varpi > 0.7R$, are compatible with the first five terms of the Legendre expansion of the frequency splittings. In the following we discuss theoretical possibilities for obtaining noncylindrical Ω contours.

In the sun the Lorentz force cannot be responsible for $\partial \Omega / \partial z \neq 0$, because this would imply significant variations of Ω with the solar cycle, stronger than those observed (3–8%). Since viscous forces are assumed to be weak, the only remaining term is $\langle \nabla T \times \nabla S \rangle$ (Durney & Spruit 1979, Stix 1989). This baroclinic term has contributions from the mean stratification as well as from the fluctuating parts of S and T. The second term, $\langle \nabla T' \times \nabla S' \rangle$, is only important in the upper part of the CZ where entropy fluctuations are large. In order that the average stratification contributes significantly to the baroclinic term, S must strongly vary with latitude. This would cause latitu-

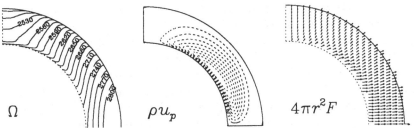

Figure 4.13 Contours of Ω, streamlines of ρu_p, and vectors of the flux $4\pi r^2 F$ for a model with anisotropic eddy conductivity.

dinal convective fluxes that greatly exceed those in the radial direction (e.g., Brandenburg *et al.* 1992). Durney (1991) argues that this would still be compatible with the observations, because any nonuniform flux pattern imprinted deep in the CZ will be screened at the surface (Spruit 1977). According to Durney (1987) the latitudinal gradient of S is essentially determined by the rotation law which, in turn, is determined by the Reynolds stresses. For a detailed analysis of his approach see also Durney (1989, 1993).

Following Rüdiger (1989) and Kitchatinov *et al.* (1994), a latitudinal gradient of S is caused by an anisotropic eddy conductivity tensor which, in the case of rapid rotation, takes the form

$$\chi_{ij} \sim \chi_t (\delta_{ij} + \hat{\Omega}_i \hat{\Omega}_j). \qquad (4.8.3)$$

Using the model of Brandenburg *et al.* (1992), with $\mathrm{Ta} = 3 \times 10^7$, $V^{(0)} = -0.5$ and $V^{(1)} = H^{(1)} = 0.6$ for the Λ-effect parameters, we now obtain Ω contours that deviate from cylindrical contours; see figure 4.13. Similar results have recently been obtained by Kitchatinov *et al.* (1994). To summarise, an anisotropic eddy conductivity is crucial in order to obtain realistic Ω contours that are not constant on cylinders.

4.9 FUTURE SIMULATIONS
Over the last few years there have been substantial improvements of mean-field models, and sometimes we seem to be close to a more coherent understanding of solar and stellar activity. However, there still remain fundamental problems regarding the applicability of mean-field theory, and we would therefore like to see solar and stellar activity being simulated on the computer without relying on mean-field theory.

In the near future it should be possible to perform dynamo simulations in a much larger domain which will allow us to study the development of the inverse cascade in more detail. It is probably important to include the latitudinal change in Ω (using a β-plane or an imposed differential rotation) to make the two horizontal directions in the box sufficiently distinct. The hope is that the large-scale field, that results from the inverse cascade, orients itself preferentially into the east–west direction. Finally, it is crucial that the large-scale magnetic field is able to produce sufficiently strong, systematically oriented, toroidal flux tubes from which bipolar regions can emerge in accordance with Hale's polarity law. For this to happen it is important that strong flux tubes can build up before they eventually die away. This requires the presence of an overshoot layer where the turbulent diffusion is weak, and a very fine mesh so that the numerical dissipation is small enough.

Acknowledgements

I thank the Isaac Newton Institute and the members of the Dynamo Theory group for a very stimulating atmosphere. I am especially grateful to Peter Fox, Andrew Gilbert, David Moss, Åke Nordlund, Mike Proctor, Günther Rüdiger, Bob Stein, Ilkka Tuominen and Nigel Weiss for comments and important discussions. This work was supported in part by the SERC.

REFERENCES

Babcock, H.W. 1961 The topology of the sun's magnetic field and the 22-year cycle. *Astrophys. J.* **133**, 572–587.

Barker, D.M. 1993 Numerical simulations of 3-D $\alpha\Omega$-dynamos. In *Solar and Planetary dynamos* (ed. M.R.E. Proctor, P.C. Matthews & A.M. Rucklidge), pp. 27–34. Cambridge University Press.

Barker, D.M. & Moss, D. 1993 Alpha-quenched alpha–lambda dynamos and the excitation of nonaxisymmetric magnetic fields. In *The Cosmic Dynamo* (ed. F. Krause, K.-H. Rädler & G. Rüdiger), pp. 147–151, IAU Symposium No. 157. Kluwer.

Batchelor, G.K. 1950 On the spontaneous magnetic field in a conducting liquid in turbulent motion. *Proc R. Soc. Lond.* A **201**, 405–416.

Belvedere, G., Proctor, M.R.E. & Lanzafame, G. 1991 The latitude belts of solar activity as a consequence of a boundary-layer dynamo. *Nature* **350**, 481–483.

Brandenburg, A., Krause, F., Meinel, R., Moss, D. & Tuominen, I. 1989a The stability of nonlinear dynamos and the limited role of kinematic growth rates. *Astron. Astrophys.* **213**, 411–422.

Brandenburg, A., Krause, F. & Tuominen, I. 1989b Parity selection in nonlinear dynamos. In *Turbulence and Nonlinear Dynamics in MHD Flows* (ed. M. Meneguzzi, A. Pouquet & P.-L. Sulem), pp. 35–40, Elsevier.

Brandenburg, A., Tuominen, I. & Moss, D. 1989c On the nonlinear stability of dynamo models. *Geophys. Astrophys. Fluid Dynam.* **49**, 129–141.

Brandenburg, A., Meinel, R., Moss, D. & Tuominen, I. 1990a Variation of even and odd parity in solar dynamo models. In *Solar Photosphere: Structure, Convection and Magnetic Fields* (ed. J.O. Stenflo), pp. 379–382. Kluwer.

Brandenburg, A., Moss, D., Rüdiger, G. & Tuominen, I. 1990b The nonlinear solar dynamo and differential rotation: a Taylor number puzzle? *Solar Phys.* **128**, 243–252.

Brandenburg, A., Nordlund, Å., Pulkkinen, P., Stein, R.F. & Tuominen, I. 1990c 3-D simulation of turbulent cyclonic magnetoconvection. *Astron. Astrophys.* **232**, 277–291.

Brandenburg, A., Jennings, R.L., Nordlund, Å., Stein, R.F. & Tuominen, I. 1991 The role of overshoot in solar activity: a direct simulation of the dynamo. In *The Sun and Cool Stars: Activity, Magnetism, Dynamos* (ed. I. Tuominen, D. Moss & G. Rüdiger), pp. 86–88. Lecture Notes in Physics, vol. 380. Springer–Verlag.

Brandenburg, A., Moss, D. & Tuominen, I. 1992 Stratification and thermodynamics in mean-field dynamos. *Astron. Astrophys.* **265**, 328–344.

Brandenburg, A., Jennings, R.L., Nordlund, Å., Rieutord, M., Stein, R. F. & Tuominen, I. 1993a Magnetic structures in a dynamo simulation. *J. Fluid Mech.* (submitted).

Brandenburg, A., Krause, F., Nordlund, Å., Ruzmaikin, A.A., Stein, R.F. & Tuominen, I. 1993b On the magnetic fluctuations produced by a large scale magnetic field. *Astrophys. J.* (submitted).

Brooke, J. & Moss, D. 1994 Nonlinear dynamos in torus geometry: transition to chaos. *Mon. Not. R. Astron. Soc.* **266**, 733–739.

Brown, T.M. & Morrow, C.A. 1987 Depth and latitude dependence of solar rotation. *Astrophys. J.* **314**, L21–L26.

Cattaneo, F. & Hughes, D.W. 1988 The nonlinear breakup of a magnetic layer: instability to interchange modes. *J. Fluid Mech.* **196**, 323–344.

Cattaneo, F., Hughes, D.W. & Weiss, N.O. 1991 What is a stellar dynamo? *Mon. Not. R. Astron. Soc.* **253**, 479–484.

Chan, K.L. 1992 Compressible convection: deeper layers. In *Cool Stars, Stellar Systems, and the Sun* (ed. M.S. Giampapa & J.A. Bookbinder), pp. 165–167. ASP Conf. Ser., vol. 26.

Chan, K.L. & Sofia, S. 1986 Turbulent compressible convection in a deep atmosphere. III. Tests on the validity and limitation of the numerical approach. *Astrophys. J.* **307**, 222–241.

Charbonneau, P. & MacGregor, K.B. 1992 Angular momentum transport in magnetised stellar radiative zones. I. Numerical solutions to the core spin-up model problem. *Astrophys. J.* **387**, 639–661.

Choudhuri, A.R. 1992 Stochastic fluctuations of the solar dynamo. *Astron. Astrophys.* **253**, 277–285.

D'Silva, S. 1993 Can equipartition fields produce the tilts of bipolar magnetic regions? *Astrophys. J.* **407**, 385–397.

Durney, B.R. 1976 On the constancy along cylinders of the angular velocity in the solar convection zone. *Astrophys. J.* **204**, 589–596.

Durney, B.R. 1987 The generalisation of mixing length theory to rotating convection zones and applications to the sun. In *The Internal Solar Angular Velocity* (ed. B.R. Durney & S. Sofia), pp. 235–261. Reidel.

Durney, B.R. 1989 On the behaviour of the angular velocity in the lower part of the solar convection zone. *Astrophys. J.* **338**, 509–527.

Durney, B.R. 1991 Observational constraints on theories of the solar differential rotation. *Astrophys. J.* **378**, 378–397.

Durney, B.R. 1993 On the solar differential rotation: meridional motions associated with a slowly varying angular velocity. *Astrophys. J.* **407**, 367–379.

Durney, B.R. & Spruit, H.C. 1979 On the dynamics of stellar convection zones: the effect of rotation on the turbulent viscosity and conductivity. *Astrophys. J.* **234**, 1067–1078.

Durney, B.R., De Young, D.S. & Passot, T.P. 1990 On the generation of the solar magnetic field in a region of weak buoyancy. *Astrophys. J.* **362**, 709–721.

Ferrière, K. 1993 The full alpha-tensor due to supernova explosions and superbubbles in the galactic disk. *Astrophys. J.* **404**, 162–184.

Fox, P.A., Theobald, M.L. & Sofia, S. 1991 Compressible magnetic convection: formulation and two-dimensional models. *Astrophys. J.* **383**, 860–881.

Frisch, U., Pouquet, A., Léorat, J. & Mazure, A. 1975 Possibility of an inverse cascade of magnetic helicity in hydrodynamic turbulence. *J. Fluid Mech.* **68**, 769–778.

Frisch, U., She, Z.S. & Sulem, P.-L. 1987 Large scale flow driven by the anisotropic kinetic alpha effect. *Physica D* **28**, 382–392.

Galanti, B., Sulem, P.-L. & Gilbert, A.D. 1991 Inverse cascades and time-dependent dynamos in MHD flows, *Physica D* **47**, 416–426.

Gilbert, A.D. Otani, N.F. & Childress, S. 1993 Simple dynamical fast dynamos. In *Solar and Planetary dynamos* (ed. M.R.E. Proctor, P.C. Matthews & A.M. Rucklidge), pp. 129–136. Cambridge University Press.

Gilbert, A.D. & Sulem, P.-L. 1990 On inverse cascades in alpha effect dynamos. *Geophys. Astrophys. Fluid Dynam.* **51**, 243–261.

Gilman, P.A. 1983 Dynamically consistent nonlinear dynamos driven by convection in a rotating spherical shell. II. Dynamos with cycles and strong feedbacks. *Astrophys. J. Suppl.* **53**, 243–268.

Gilman, P.A. 1992 What can we learn about solar cycle mechanisms from observed velocity fields. In *The Solar Cycle* (ed. K.L. Harvey), pp. 241–255. ASP Conf. Ser., vol. 27.

Gilman, P.A. & Miller, J. 1981 Dynamically consistent nonlinear dynamos driven by convection in a rotating spherical shell. *Astrophys. J. Suppl.* **46**, 211–238.

Glatzmaier, G.A. 1984 Numerical simulations of stellar convective dynamos. I. The model and method. *J. Comput. Phys.* **55**, 461–484.

Glatzmaier, G.A. 1985 Numerical simulations of stellar convective dynamos. II. Field propagation in the convection zone. *Astrophys. J.* **291**, 300–307.

Golub, L., Davis, J.M. & Krieger, A.S. 1979 Anticorrelation of X-ray bright points with sunspot number, 1970–1978. *Astrophys. J.* **229**, L145–L150.

Gough, D.O., Kosovichev, A.G., Sekii, T., Libbrecht, K.G. & Woodard, M.F. 1993 The form of the angular velocity in the solar convection zone. In *Seismology of the Sun and Stars* (ed. T. Brown), pp. 213–216, Proc. GONG Conf., ASP Conf. Ser., vol. 42.

Howard, R. & LaBonte, B.J. 1981 Surface magnetic fields during the solar activity cycle. *Solar Phys.* **74**, 131–145.

Hoyng, P. 1988 Turbulent transport of magnetic fields. III. Stochastic excitation of global magnetic field modes. *Astrophys. J.* **332**, 857–871.

Hurlburt, N.E. & Toomre, J. 1988 Magnetic fields interacting with nonlinear compressible convection. *Astrophys. J.* **327**, 920–932.

Hurlburt, N.E., Toomre, J. & Massaguer, J.M. 1984 Two-dimensional compressible convection extending over multiple scale heights. *Astrophys. J.* **282**, 557–573.

Hurlburt, N.E., Proctor, M.R.E., Weiss, N.O. & Brownjohn, D.P. 1989 Time-dependent compressible magnetoconvection Part 1. Travelling waves and oscillations. *J. Fluid Mech.* **207**, 587–628.

Ivanova, T.S. & Ruzmaikin, A.A. 1977 A nonlinear magnetohydrodynamic model of the solar dynamo. *Sov. Astron.* **21**, 479–485.

Jennings, R.L. 1991 Symmetry beaking in a nonlinear $\alpha\omega$-dynamo. *Geophys. Astrophys. Fluid Dynam.* **57**, 147–189.

Jennings, R.L., Brandenburg, A., Nordlund, Å. & Stein, R.F. 1992 Evolution of a magnetic flux tube in two dimensional penetrative convection. *Mon. Not. R. Astron. Soc.* **259**, 465–473.

Jepps, S.A. 1975 Numerical models of hydromagnetic dynamos. *J. Fluid Mech.* **67**, 625–646.

Jones, C.A. & Galloway, D.J. 1993 Alpha-quenching in cylindrical magnetoconvection. In *Solar and Planetary dynamos* (ed. M.R.E. Proctor, P.C. Matthews & A.M. Rucklidge), pp. 161–170. Cambridge University Press.

Jones, C.A., Weiss, N.O. & Cattaneo, F. 1985 Nonlinear dynamos: a generalisation of the Lorentz equations. *Physica D* **14**, 161–174.

Kaisig, M., Rüdiger, G. & Yorke, H.W. 1993 The alpha-effect by supernova explosions. *Astron. Astrophys.* **274**, 757–764.

Kazantsev, A.P. 1968 Enhancement of a magnetic field by a conducting fluid. *Sov. Phys. JETP* **26**, 1031–1034.

Kida, S., Yanase, S. & Mizushima, J. 1991 Statistical properties of MHD turbulence and turbulent dynamo. *Phys. Fluids A* **3**, 457–465.

Kim, Y.C. & Fox, P.A. 1993 (in preparation).

Kitchatinov, L.L. 1988 Nonlinear dynamo effects for an inhomogeneously turbulent rotating fluid. *Astron. Nachr.* **309**, 197–211.

Kitchatinov, L.L. 1991 Turbulent transport of magnetic fields in a highly conducting rotating fluid and the solar cycle. *Astron. Astrophys.* **243**, 483–491.

Kitchatinov, L.L. 1993 Turbulent transport of magnetic fields and the solar dynamo. In *The Cosmic Dynamo* (ed. F. Krause, K.-H. Rädler & G. Rüdiger), pp. 13–17. Kluwer.

Kitchatinov, L.L. & Pipin, V.V. 1993 Mean-field buoyancy. *Astron. Astrophys.* **274**, 647–652.

Kitchatinov, L.L. & Rüdiger, G. 1993 Λ-effect and differential rotation in stellar convection zones. *Astron. Astrophys.* **276**, 96–102.

Kitchatinov, L.L., Pipin, V.V. & Rüdiger, G. 1994 Turbulent viscosity, magnetic diffusivity, and heat conductivity under the influence of rotation and magnetic field. *Astron. Nachr.* (submitted).

Kleeorin, N.I. & Ruzmaikin, A.A. 1981 Properties of a nonlinear solar dynamo model. *Geophys. Astrophys. Fluid Dynam.* **17**, 281–296.

Kleeorin, N.I., Rogachevskii, I.V. & Ruzmaikin, A.A. 1990 Magnetic force reversal and instability in a plasma with advanced magnetohydrodynamic turbulence. *Sov. Phys. JETP* **70**, 878–883.

Köhler, H. 1970 Differential rotation caused by anisotropic turbulent viscosity. *Solar Phys.* **13**, 3–18.

Kraichnan, R.H. 1979 Consistency of the α-effect turbulent dynamo. *Phys. Rev. Lett.* **42**, 1677–1680.

Krause, F. 1967 Eine Lösung des Dynamoproblems auf der Grundlage einer linearen Theorie der magnetohydrodynamischen Turbulenz. Habilitationsschrift, University of Jena.

Krause, F. 1976 Mean-field magnetohydrodynamics of the solar convection zone. In *Basic Mechanisms of Solar Activity* (ed. V. Bumba & J. Kleczek), pp. 305–321. IAU Symp. 71. D. Reidel.

Krause, F. & Meinel, R. 1988 Stability of simple nonlinear α²-dynamos. *Geophys. Astrophys. Fluid Dynam.* **43**, 95–117.

Krause, F. & Rädler, K.-H. 1980 *Mean-Field Magnetohydrodynamics and Dynamo Theory*. Akademie-Verlag, Berlin.

Krause, F. & Rüdiger, G. 1975 On the turbulent decay of strong magnetic fields and the development of sunspot areas. *Solar Phys.* **42**, 107–119.

Leighton, R.B. 1969 A magneto-kinematic model of the solar cycle. *Astrophys. J.* **156**, 1–26.

Malkus, W.V.R. & Proctor, M.R.E. 1975 The macrodynamics of α-effect dynamos in rotating fluids. *J. Fluid Mech.* **67**, 417–443.

Meinel, R. & Brandenburg, A. 1990 Behaviour of highly supercritical α-effect dynamos. *Astron. Astrophys.* **238**, 369–376.

Meneguzzi, M. & Pouquet, A. 1989 Turbulent dynamos driven by convection. *J. Fluid Mech.* **205**, 297–312.

Meneguzzi, M., Frisch, U. & Pouquet, A. 1981 Helical and nonhelical turbulent dynamos. *Phys. Rev. Lett.* **47**, 1060–1064.

Moffatt, H.K. 1970 Turbulent dynamo action at low magnetic Reynolds number. *J. Fluid Mech.* **41**, 435–452.

Moffatt, H.K. 1972 An approach to a dynamic theory of dynamo action in a rotating conducting fluid. *J. Fluid Mech.* **53**, 385–399.

Moffatt, H.K. 1978 *Magnetic Field Generation in Electrically conducting fluids.* Cambridge University Press.

Molchanov, S.A., Ruzmaikin, A.A. & Sokoloff, D.D. 1985 Kinematic dynamo in random flow. *Sov. Phys. Usp.* **28**, 307–327.

Molchanov, S.A., Ruzmaikin, A.A. & Sokoloff, D.D. 1988 Short-correlated random flow as a fast dynamo. *Sov. Phys. Dokl.* **32**, 569–570.

Moss, D. & Vilhu, O. 1983 Models of stellar differential rotation on the lower main sequence. *Astron. Astrophys.* **119**, 47–53.

Moss, D., Tuominen, I. & Brandenburg, A. 1990 Nonlinear dynamos with magnetic buoyancy in spherical geometry. *Astron. Astrophys.* **228**, 284–294.

Moss, D., Tuominen, I. & Brandenburg, A. 1991 Nonlinear nonaxisymmetric dynamo models for cool stars. *Astron. Astrophys.* **245**, 129–135.

Moss, D., Brandenburg, A., Tavakol, R. K. & Tuominen, I. 1992 Stochastic effects in mean-field dynamos. *Astron. Astrophys.* **265**, 843–849.

Nicklaus, B. & Stix, M. 1988 Corrections to first order smoothing in mean-field electrodynamics. *Geophys. Astrophys. Fluid Dynam.* **43**, 149–166.

Nordlund, Å. & Stein, R.F. 1990 3-D simulations of solar and stellar convection and magnetoconvection. *Comp. Phys. Comm.* **59**, 119–125.

Nordlund, Å., Brandenburg, A., Jennings, R.L., Rieutord, M., Ruokolainen, J., Stein, R. F. & Tuominen, I. 1992 Dynamo action in stratified convection with overshoot. *Astrophys. J.* **392**, 647–652.

Novikov, V.G., Ruzmaikin, A.A. & Sokoloff, D.D. 1983 Kinematic dynamo in a reflection-invariant random field. *Sov. Phys. JETP* **58**, 527–532.

Panesar, J.S. & Nelson, A.H. 1992 Numerical models of 3-D galactic dynamos. *Astron. Astrophys.* **264**, 77–85.

Parker, E.N. 1955 The formation of sunspots from the solar toroidal field. *Astrophys. J.* **121**, 491–507.

Parker, E.N. 1984 Magnetic buoyancy and the escape of magnetic fields from stars. *Astrophys. J.* **281**, 839–845.

Passot, T. & Pouquet, A. 1987 Numerical simulation of compressible homogeneous flows in the turbulent regime. *J. Fluid Mech.* **181**, 441–466.

Peyret, R. & Taylor, T.D. 1986 *Computational Methods for Fluid Flow.* Springer–Verlag, New York.

Pouquet, A., Frisch, U. & Léorat, J. 1976 Strong MHD helical turbulence and the nonlinear dynamo effect. *J. Fluid Mech.* **77**, 321–354.

Prautzsch, T. 1993 The dynamo mechanism in the deep convection zone of the sun. In *Solar and Planetary dynamos* (ed. M.R.E. Proctor, P.C. Matthews & A.M. Rucklidge), pp. 249–256. Cambridge University Press.

Priest, E.R. 1982 *Solar Magnetohydrodynamics.* D. Reidel.

Proctor, M.R.E. 1977 Numerical solutions of nonlinear α-effect dynamo equations. *J. Fluid Mech.* **80**, 769–784.

Pulkkinen, P., Tuominen, I., Brandenburg, A., Nordlund, Å. & Stein, R.F. 1991 Simulation of rotational effects on turbulence in the solar convective zone. In *The Sun and Cool Stars: Activity, Magnetism, Dynamos* (ed. I. Tuominen, D. Moss & G. Rüdiger), pp. 98–100. Lecture Notes in Physics, vol. 380. Springer–Verlag.

Pulkkinen, P., Tuominen, I., Brandenburg, A., Nordlund, Å. & Stein, R.F. 1993a Rotational effects on convection simulated at different latitudes. *Astron. Astrophys.* **267** 265–275.

Pulkkinen, P., Tuominen, I., Brandenburg, A., Nordlund, Å. & Stein, R.F. 1993b Reynolds stresses derived from simulations. In *The Cosmic Dynamo* (ed. F. Krause, K.-H. Rädler & G. Rüdiger), pp. 123–127, IAU Symposium No. 157. Kluwer.

Rädler, K.-H. & Wiedemann, E. 1989 Numerical experiments with a simple nonlinear mean-field dynamo model. *Geophys. Astrophys. Fluid Dynam.* **49**, 71–80.

Rädler, K.-H., Wiedemann, E., Brandenburg, A., Meinel, R. & Tuominen, I. 1990 Nonlinear mean-field dynamo models: stability and evolution of three-dimensional magnetic field configurations. *Astron. Astrophys.* **239**, 413–423.

Richtmyer, R.D. & Morton, K.W. 1967 *Difference Methods for Initial-value Problems.* John Wiley & Sons.

Roberts, P.H. & Soward, A.M. 1975 A unified approach to mean field electrodynamics. *Astron. Nachr.* **296**, 49–64.

Rüdiger, G. 1973 Behandlung eines einfachen magnetohydrodynamischen Dynamos mittels Linearisierung. *Astron. Nachr.* **294**, 183–186.

Rüdiger, G. 1974 The influence of a uniform magnetic field of arbitrary strength on turbulence. *Astron. Nachr.* **295**, 275–284.

Rüdiger, G. 1980a Rapidly rotating α^2-dynamo models. *Astron. Nachr.* **301**, 181–187.

Rüdiger, G. 1980b Reynolds stresses and differential rotation I. On recent calculations of zonal fluxes in slowly rotating stars. *Geophys. Astrophys. Fluid Dynam.* **16**, 239–261.

Rüdiger, G. 1982 A heuristic approach to a non-local theory of turbulent channel flow. *Zeitschr. Angewandt. Math. Mech.* **6**, 95–101.

Rüdiger, G. 1989 *Differential Rotation and Stellar Convection: Sun and Solar-type Stars.* Gordon & Breach.

Rüdiger, G. 1992 Solar torsional oscillations as due to magnetic suppression and deformation of turbulence. In *Cool Stars, Stellar Systems, and the Sun* (ed. M.S. Giampapa & J.A. Bookbinder), pp. 185–187. ASP Conf. Ser., vol. 26.

Rüdiger, G. & Brandenburg, A. 1993 A solar dynamo in the overshoot layer: cycle period and butterfly diagram. *Astrophys. J.* (submitted).

Rüdiger, G. & Elstner, D. 1994 Non-axisymmetry vs. axisymmetry in dynamo-excited stellar magnetic fields. *Astron. Astrophys.* **281**, 46–50.

Rüdiger, G. & Kitchatinov, L.L. 1990 The turbulent stresses in the theory of the solar torsional oscillation. *Astron. Astrophys.* **236**, 503–508.

Rüdiger, G. & Kitchatinov, L.L. 1993 Alpha-effect and alpha-quenching. *Astron. Astrophys.* **269**, 581–588.

Ruzmaikin, A.A. 1981 The solar cycle as a strange attractor. *Comments Astrophys.* **9**, 85–96.

Saar, S.H., Piskunov, N.E. & Tuominen, I. 1992 Magnetic surface images of the BY Dra star HD 82558. In *Cool Stars, Stellar Systems, and the Sun* (ed. M.S. Giampapa & J.A. Bookbinder), pp. 255–258. ASP Conf. Ser., vol. 26.

Schmidt-Voigt, M. 1989 Time-dependent MHD simulations for cometary plasmas. *Astron. Astrophys.* **210**, 433–454.

Schmitt, D. 1987 An $\alpha\omega$-dynamo with an α-effect due to magnetostrophic waves. *Astron. Astrophys.* **174**, 281–287.

Schmitt, D. 1993 The solar dynamo. In *The Cosmic Dynamo* (ed. F. Krause, K.-H. Rädler & G. Rüdiger), pp. 1–12. Kluwer.

Schmitt, D. & Schüssler, M. 1989 Non-linear dynamos. I. One-dimensional model of a thin layer dynamo. *Astron. Astrophys.* **223**, 343–351.

Schrijver, C.J. 1987 Solar active regions: radiative intensities and large scale parameters of the magnetic field. *Astron. Astrophys.* **180**, 241–252.

Schumann, U. 1976 Numerical simulation of the transition from three- to two-dimensional turbulence under a uniform magnetic field. *J. Fluid Mech.* **74**, 31–58.

Schüssler, M. 1981 The solar torsional oscillation and dynamo models of the solar cycle. *Astron. Astrophys.* **94**, L17–L18.

Schüssler, M. 1983 Stellar dynamo theory. In *Solar and Stellar Magnetic Fields: Origins and Coronal Effects* (ed. J.O. Stenflo), pp. 213–234. D. Reidel.

Schüssler, M. 1987 Magnetic fields and the rotation of the solar convection zone. In *The Internal Solar Angular Velocity* (ed. B.R. Durney & S. Sofia), pp. 303–320. D. Reidel.

Sokoloff, D.D., Nesme-Ribes, E. 1993 The Maunder minimum, a mixed parity dynamo mode? *Astron. Astrophys.* (in press).

Spiegel, E.A. & Weiss, N.O. 1980 Magnetic activity and variation in the solar luminosity. *Nature* **287**, 616–617.

Spiegel, E.A. & Zahn, J.-P. 1992 The solar tachocline. *Astron. Astrophys.* **265**, 106–114.

Spruit, H.C. 1977 Appearance at the solar surface of disturbances in the heat flow associated with differential rotation. *Astron. Astrophys.* **55**, 151–153.

Steenbeck, M. & Krause, F. 1969 Zur Dynamotheorie stellarer und planetarer Magnetfelder I. Berechnung sonnenähnlicher Wechselfeldgeneratoren. *Astron. Nachr.* **291**, 49–84.

Steenbeck, M., Krause, F. & Rädler, K.-H. 1966 A calculation of the mean electromotive force in an electrically conducting fluid in turbulent motion, under the influence of Coriolis forces. *Z. Naturforsch* **21a**, 369–376. (English transl.: *The turbulent dynamo...*, by P.H. Roberts & M. Stix, Tech. Note 60, NCAR, Boulder, Colorado (1971).)

Stix, M. 1976 Differential rotation and the solar dynamo. *Astron. Astrophys.* **47**, 243–254.

Stix, M. 1981 Theory of the solar cycle. *Solar Phys.* **74**, 79–101.

Stix, M. 1989 The sun's differential rotation. *Rev. Mod. Astron.* **2**, 248–266.

Tao, L., Cattaneo, F. & Vainshtein, S.I. 1993 Evidence for the suppression of the alpha-effect by weak magnetic fields. In *Solar and Planetary dynamos* (ed. M.R.E. Proctor, P.C. Matthews & A.M. Rucklidge), pp. 303–310. Cambridge University Press.

Tavakol, R.K. 1978. Is the sun almost-intransitive? *Nature* **276**, 802–803.

Tuominen, I. & Rüdiger, G. 1989 Solar differential rotation as a multiparameter turbulence problem. *Astron. Astrophys.* **217**, 217–228.

Tuominen, I. & Virtanen, H. 1984 Oscillatory motions in the sun corresponding to the solar cycle. *Astron. Nachr.* **305**, 225–228.

Tuominen, I., Rüdiger, G. & Brandenburg, A. 1988 Observational constraints for solar type dynamos. In *Activity in Cool Star Envelopes* (ed. O. Havnes, B.R. Pettersen, J.H.M.M. Schmitt & J.E. Solheim), pp. 13–20. Kluwer.

Vainshtein, S.I. 1972 Nonlinear problem of the turbulent dynamo. *Sov. Phys. JETP* **34**, 327–331.

Vainshtein, S.I. & Cattaneo, F. 1992 Nonlinear restrictions on dynamo action. *Astrophys. J.* **393**, 165–171.

Vainshtein, S.I. & Zel'dovich, Ya.B. 1972 Origin of magnetic fields in astrophysics. *Sov. Phys. Usp.* **15**, 159–172.

Valdettaro, L. & Meneguzzi, M. 1991 Turbulent dynamos driven by convection inside spherical shells. In *The Sun and Cool Stars: Activity, Magnetism, Dynamos* (ed. I. Tuominen, D. Moss & G. Rüdiger), pp. 80–85, IAU Coll. 130, Lecture Notes in Physics. Springer–Verlag.

Virtanen, H. 1989 Solar observational hydrodynamics from the sunspot group statistics. Licentiate thesis, Univ. of Helsinki.

Ward, F. 1965 The general circulation of the solar atmosphere and the maintenance of the equatorial acceleration. *Astrophys. J.* **141**, 534–547.

Weisshaar, E. 1982 A numerical study of α^2-dynamos with anisotropic α-effect. *Geophys. Astrophys. Fluid Dynam.* **21**, 285–301.

Wilson, P.R. 1992 Helioseismology data and the solar dynamo. *Astrophys. J.* **399**, 294–299.

Yoshimura, H. 1975 A model of the solar cycle driven by the dynamo action of the global convection in the solar convection zone. *Astrophys. J. Suppl.* **29**, 467–494.

Yoshimura, H. 1976 Phase relation between the poloidal and toroidal solar-cycle general magnetic fields and location of the origin of the surface magnetic fields. *Solar Phys.* **50**, 3–23.

Yoshimura, H. 1981 Solar-cycle Lorentz force waves and the torsional oscillations of the sun. *Astrophys. J.* **247**, 1102–1112.

Zel'dovich, Ya.B., Ruzmaikin, A.A. & Sokoloff, D.D. 1983 *Magnetic fields in Astrophysics.* Gordon & Breach.

Zhang, K. & Busse, F.H. 1989 Magnetohydrodynamic dynamos in rotating spherical shells. *Geophys. Astrophys. Fluid Dynam.* **49**, 97–116.

CHAPTER 5

Energy Sources for Planetary Dynamos

W. V. R. MALKUS

Department of Mathematics
Massachusetts Institute of Technology
Cambridge, MA 02139, USA

5.1 INTRODUCTION

In the belief that only unkind gods would arrange two energy sources for planetary dynamos as equally important, this re-exploration of plausible sources seeks to eliminate rotational energy in favor of convection. Recent experiments and theory of the 'elliptical' instabilities in a rotating fluid due to precessional and tidal strains provide quantitative results for velocity fields and energy production. The adequacy of these flows to produce a dynamo on both terrestrial and giant planets is assessed in the context of 'strong field' scaling. With little ambiguity it is concluded that Mercury, Venus, and Mars can not have a dynamo of tidal or precessional origin. The case for today's Earth is marginal. Here precessional strains (accidentally comparable to tidal strains) also are potential sources of inertial instabilities. The ancient Earth with its closer Moon, as well as all the giant planets, have tides well in excess of those needed to critically maintain dynamos. Hence the project proposed here proves to be successful only in part — an Earth in the distant future will not be able to sustain the geodynamo with its rotational energy. On the other hand, convection remains a possible dynamo energy source, with such a large number of undetermined processes and parameters that it is unfairly easy to establish conditions for its inadequacy. A large literature explores its adequacy. A brief review of this literature, in both a 'strong-field' and 'weak-field' context, advances several cautionary restraints to be employed on that day when the limits of validity of a quantitative dynamo-convection theory are to be determined.

In the following section, the minimum conditions on strains and flow are recalled which permit the growth of a magnetic field in an electrically conducting homogeneous fluid. Section 5.2 starts with a re-assessment of the flows observed in rotating spheroids due to tidal and precessional forcing. The relatively new awareness that these body forces can lead to sub-harmonic in-

161

M.R.E. Proctor & A.D. Gilbert (eds.)
Lectures on Solar and Planetary Dynamos, 161–179
©1994 Cambridge University Press.

ertial bifurcation is outlined with the simplest theoretical and experimental examples. In section 5.3 these findings are applied to the planets using the conditions of section 5.1 and the conventional 'strong field' scaling, to establish limits of applicability. Section 5.4 explores recent contributions to the large literature on planetary convection as a dynamo source. Here, hope is advanced that some unique qualitative consequence of theory may be reflected in planetary observations. The energy consumption by small scales of motion is assessed as a possible guide in delimiting dynamo-convection behavior. In final paragraphs of section 5.4 recent thoughts on compositional convection are explored. Energetic considerations raise doubts about any type of convection as a geodynamo source, but the many determining material parameters for the Earth are so uncertain that a hopeful literature can persist indefinitely. In a final discussion several paths towards testable dynamo theory are suggested.

5.1.1 Integral Requirements to Sustain a Planetary Dynamo

From the magnetic diffusion equations

$$\frac{\partial \mathbf{B}}{\partial t} = \nabla \times (\mathbf{v} \times \mathbf{B}) + \lambda \nabla^2 \mathbf{B}, \quad \nabla \cdot \mathbf{B} = 0, \tag{5.1.1}$$

George Backus (1958) derived the inequality

$$\frac{1}{2}\frac{d}{dt} <\ln |\mathbf{B}|^2> \leq \left(m(t) - \frac{\pi^2 \lambda}{r_0^2} \right), \tag{5.1.2}$$

where $\mathbf{B}$ is the magnetic field, λ is the magnetic diffusivity, $<\cdot>$ indicates an average over all space, r_0 is the radius of the smallest sphere enclosing the fluid velocity field $\mathbf{v}$, and $m(t)$ is the largest principal value of the rate-of-strain tensor $S_{ij} \equiv \frac{1}{2}(\partial v_i/\partial x_j + \partial v_j/\partial x_i)$.

Hence the condition

$$\frac{m(t)r_0^2}{\lambda} \geq \pi^2 \tag{5.1.3}$$

must be met by all homogeneous dynamos.

Childress (1969), by a different route, established the condition on a magnetic Reynolds number

$$Rm \equiv \frac{U r_0}{\lambda} \geq \pi, \tag{5.1.4}$$

where U is the largest velocity within r_0, while Busse (1978) determined that

$$\frac{(\mathbf{v} \cdot \mathbf{r})_{\max}}{r_0} \geq \left(\frac{\lambda}{r_0} \right) \left(2 \frac{E_P}{E_M} \right)^{1/2}, \tag{5.1.5}$$

where E_P/E_M is the ratio of magnetic energy in the poloidal component of a planetary field to the total magnetic energy.

All these conditions will provide guidance and restraint in the following assessments.

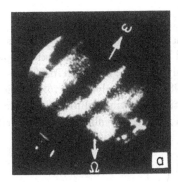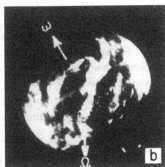

Figure 5.1 Precessional flow in a laboratory spheroid: (*a*) laminar nonlinear flow and (*b*) turbulent 'spin-over'.

5.2 TIDAL AND PRECESSIONAL RESONANT BIFURCATIONS

5.2.1 Earlier Studies
Planetary scientists are familiar with the resonant response of stably strati-fied atmospheres to tidal forcing (e.g., Lindzen 1991). In such circumstance those quasi-two-dimensional Hough solutions to the Laplace tidal equations which have the same period and phase velocity as the forcing function grow until limited by dissipative processes. However, study of the sub-harmonic bifurcation of the three-dimensional Poincaré modes in near-neutral regions of an atmosphere or core fluid has received negligible attention. In this sec-tion, the 'triad' interactions which are responsible for such bifurcation will be discussed, and the properties of the principal modes, the 'spin-over' modes, will be derived and exhibited in laboratory experiment.

  In early studies of tidal and precessional flow induced in rotating oblate spheroids (Poincaré 1910, Bondi & Lyttleton 1953, Roberts & Stewartson 1963, 1965, Busse 1968, Malkus 1968, Suess 1969 and recently Vanyo 1991), the singular behavior of the viscous boundary layer was the central concern. Figure 5.1(*a*) exhibits the shear layers induced at critical latitudes in a labo-ratory flow by precession. The dramatic instability seen in figure 5.1(*b*) was interpreted as a break down of the shear layers. The analytical resolution of such shear layer break-down proved intractable. Only in the last few years has

it become clear that figure 5.1(b) exhibited a quite independent bifurcation
due to the global strain caused by precession. The origin, orientation and
magnitude of this strain will be derived below. Before so doing, the simplest
example of such a process, 'tidal distortion in a cylinder', will be explored.

5.2.2 Elliptical Bifurcation

Emerging from earlier studies of the instability of parallel shear flows, the
general nature of the three-dimensional bifurcations of an elliptical flow in
an infinite domain was clarified by Bayly (1986). The principal 'spin-over'
mode was isolated by Waleffe (1988, 1990) in his analytical reformulation of
these fully non-linear, viscous, exponentially growing solutions. A compact
restatement for the spin-over bifurcation follows. An elliptical flow is written

$$\mathbf{U} = \begin{pmatrix} 0 & -\gamma - \epsilon & 0 \\ \gamma - \epsilon & 0 & 0 \\ 0 & 0 & 0 \end{pmatrix} \begin{pmatrix} x \\ y \\ z \end{pmatrix} = \gamma \times \mathbf{r} + \mathbf{D} \cdot \mathbf{r}, \qquad (5.2.1)$$

which is a complete two-dimensional solution, with vorticity 2γ, of the Navier–
Stokes equations, where γ is the magnitude of the constant vector rotation
γ, ϵ is the strain rate, and $\mathbf{D}$ is the strain matrix. The complete perturbation
vorticity equation for departures, $\mathbf{u}$, from the elliptical state (2.1) is

$$\frac{\partial \omega}{\partial t} + \mathbf{U} \cdot \nabla \omega = \gamma \times \omega + \mathbf{D} \cdot \omega + 2\gamma \frac{\partial \mathbf{u}}{\partial z} - \nabla \times (\omega \times \mathbf{u}) + \nu \nabla^2 \omega. \qquad (5.2.2)$$

For a $\mathbf{u}$ of the spin-over form

$$\mathbf{u} = \frac{1}{2} \omega(t) \times \mathbf{r}, \qquad (5.2.3)$$

representing solid rotation about an arbitrary axis, (2.2) reduces to

$$\frac{\partial \omega}{\partial t} = \mathbf{D} \cdot \omega, \qquad (5.2.4)$$

which has an exponentially growing eigenmode with ω along the stretching
axis ($-45°$ in the x-z plane).

5.2.3 Experiments and Theory in a Bounded Domain

Since this growing mode is a complete non-linear solution, and does not alter
the basic elliptical flow, the energy for its growth must be flowing in myste-
riously from infinity. Could such solutions, or slight modifications of them,
exist in a bounded domain? The suggestion found in earlier experiments
(Suess 1969) was that an inertial instability occurred. McEwan (1970) observed similar resonant forcing and (1971) interpreted the phenomenon as a
triad instability, while Gledzer et al. (1975) specifically addressed the linear,

inviscid problem of its origin in the elliptical flow. Steady-state experiments in a tidally distorted elastic cylinder (Malkus 1989) confirmed the initial instability of these solutions if

$$\beta \equiv \frac{\epsilon}{\gamma} \geq E^{1/2} \equiv \left(\frac{\nu}{\gamma r_0^2}\right)^{1/2}, \qquad (5.2.5)$$

where ν is the kinematic viscosity of the fluid in the cylinder of radius r_0. The experiments also established that the growing modes drew energy for their growth from the basic rotation of the flow, but that their exponential growth was terminated by an explosive collapse when about half of the rotational energy had been transferred to them. A period of spin-up then occurred during which the rotation was restored, to be followed again by exponential growth and collapse. No stable finite amplitude solutions for the most unstable modes have been observed for values of β (defined in (2.5)) from a few percent above its critical value to several times critical. Even under such just post-critical conditions growth continued unabated until half the energy of rotation was transferred. Figure 5.2(a) exhibits the growing spin-over mode in a rotating fluid cylinder ten centimeters in diameter, with a 'tidal' distortion of two percent. In figure 5.2(b) one sees the 'collapse' of this mode, in (c) one sees the weak Taylor columns accompanying re-spin up, while (d) exhibits the next intermittent growth, but with opposite symmetry. Figure 5.3 is a brief record of rotation of the central twenty percent of the fluid with a β similar to that visualized in figure 5.2.

5.2.4 Amplitude Limiting Processes

It is implausible that the small tidal amplitude in a planet could lead to such disaster as recorded in figure 5.3. Detuning of principal modes, geometry, stratification, turbulent convection, the bifurcation which produces magnetic fields, any of these may be responsible for limiting the amplitude of this otherwise unbridled growth.

Detuning of the spin-over mode can occur when the body responsible for the tidal distortion rotates about the contained fluid. Since this is always the case for planetary tides, the effect deserves first attention. In the simple cylindrical model, (2.2) would then include a Coriolis term $\nabla \times (2\omega_m \times u)$, where ω_m is the rotation of the tidal distortion relative to a rest frame. Detuning has been explored in the case of unbounded flow by Craik (1989) (who has made numerous contributions to this topic and to its hydromagnetic extensions, (Craik 1988, Craik & Allen 1992)). In his thesis Kerswell (1992) has considered detuning in bounded flows, e.g., oblate spheroids, and has concluded that modes whose phase velocities match the angular motion of the tidal distortion are the first to grow. These growth rates are close to those of the spin-over mode when the angular motion is small compared to

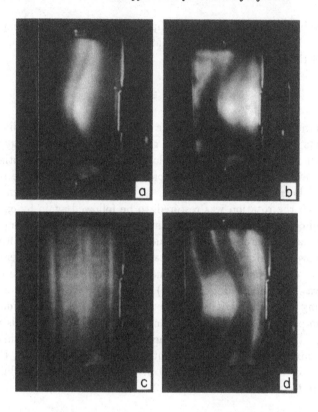

Figure 5.2 Elliptical instability: (*a*) growth, (*b*) collapse, (*c*) re-spinup and (*d*) growth.

the planetary rotation rate. (At the other extreme, as one anticipates, growth
rates vanish for a 'frozen' tide.)

Study of the tidal excitation of modes in oblate spheroidal geometry is more
demanding than in cylindrical geometry. In both cases the inviscid eigenmodes are exact solutions of the Poincaré equation, thoroughly discussed by
Greenspan (1968). These three-dimensional vortex waves have rotation as
their 'restoring force', whereas gravity is, of course, the restoring force for the
surface and internal waves due to density stratification. The vortex waves
are infinitely more numerous than the gravity waves, but are restricted to the
frequency range of plus or minus twice the angular rotation rate of the fluid.
Tidal forcing is at approximately twice the rotation rate; hence it does not
directly resonate with any one of the Poincaré modes, but can excite a pair of
modes to form a resonant triad. For cylindrical geometry these modes and the
associated dispersion relations are easily determined. For oblate spheroidal

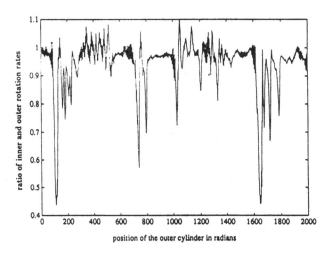

Figure 5.3 Post bifurcation rotation of the central 20% of the fluid.

geometry the modes are products of associated Legendre polynomials, and their dispersion relations emerge from more lengthy computation. However, in both geometries the spin-over modes and their phase velocities are analytically established. Details for oblate spheroidal geometry and precessional forcing will be given in a following paragraph.

The role of stable stratification in cylindrical geometry has been studied by Kerswell (1993a) who finds that radial stratification can increase the growth rates for elliptical instability by enriching the modal frequency spectrum. A parallel study for oblate spheroid geometry has not been completed. However, in this case a complete set of spin-over modes had been found (Malkus 1967) which are 'shellular' and uninfluenced by radial stratification. Hence, it is anticipated that planetary stable stratification will not reduce the growth rate of the principal elliptic bifurcations, but may reduce the critical condition for such bifurcation.

Convection occurs when the thermal gradient exceeds the adiabatic lapse rate. No theoretical study has been made yet of the joint problem of elliptical and convective instability to determine modification of eigenstructures and eigenvalues. Initial experiments in cylinders suggest that both the transport of heat and momentum are enhanced by the interaction, but no experiments have been done in spheroids.

Lastly the role of an (imposed) magnetic field in elliptic bifurcation has been established by Kerswell (1993a) as invariably stabilizing. Despite the large class of hydromagnetic waves added to the inertial spectrum by the Lorentz force, growth rates are reduced, at least a bit. Indeed, this may

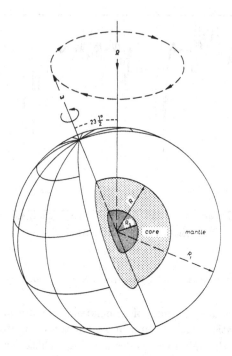

Figure 5.4 The precessing Earth.

relate to the role of magnetic fields in the planets, and will be investigated in the next section.

5.2.5 Precessional Strains in Oblate Spheroids

With the Earth's fluid core in mind, Poincaré (1910) advanced an elegantly simple resolution of the core fluid response to the Earth's precession. Figure 5.4 pictures the precessing earth and labels the several vectors. Poincaré's complete solution for the steady motion in the rotating frame containing Ω and ω is given by

$$\mathbf{v} = \omega \times \mathbf{r} + \nabla A, \tag{5.2.6}$$

where

$$\omega = \omega_z \left(\mathbf{1}_z - \frac{(2+\eta)(\Omega \times \mathbf{1}_z) \times \mathbf{1}_z}{\eta \omega_z + 2(\mathbf{1}_z \cdot \Omega)(1+\eta)} \right), \tag{5.2.7}$$

$$A = \frac{\eta \omega_z (\Omega \times \mathbf{1}_z) \cdot \mathbf{r}(\mathbf{1}_z \cdot \mathbf{r})}{\eta \omega_z + 2(\mathbf{1}_z \cdot \Omega)(1+\eta)}, \tag{5.2.8}$$

ω approaches $\omega_z \mathbf{1}_z$ as Ω approaches zero, and where

$$r^2 + \eta(\mathbf{1}_z \cdot \mathbf{r})^2 = 1 \tag{5.2.9}$$

is the shape of the 'container', with η as a measure of the ellipticity of the oblate spheroid. The velocity field (2.6) is a non-linear, viscous solution to the full Navier–Stokes equation except for a thin viscous boundary layer, addressed in section 5.2.1. Bifurcations due to the body strains given by (2.6) have not been addressed in previous literature. However, the similarity between (2.6) and (2.1) suggests similar phenomena. If the plane containing Ω and ω is chosen as the x-z plane, equations (2.6–8) can be rewritten as

$$\mathbf{v} = (\omega_z \mathbf{1}_z + \omega_x \mathbf{1}_x) \times \mathbf{r} + \nabla(ayz), \tag{5.2.10}$$

where to satisfy $\mathbf{v} \cdot \hat{\mathbf{n}} = 0$ on the boundary, equation (2.9),

$$a = \frac{-\eta \Omega_x \omega_z}{\eta \omega_z + 2(1 + \eta)\Omega_z}, \tag{5.2.11}$$

and

$$\omega_x = -\frac{(2+\eta)}{\eta} a. \tag{5.2.12}$$

Equation (2.11) reveals that when $\omega_z \gg \Omega_z$ (for the Earth $\Omega_z/\omega_z \simeq 10^{-7}$), the strain, a, appears to be independent of η. Actually the prescribed Ω depends on η, the inclination angle $(23\frac{1}{2}°)$, the mass and distance of the body responsible for the (mutual) precession. Equation (2.6) may now be written in the form of (2.1) as

$$\mathbf{v} = \begin{pmatrix} 0 & -\omega_z & 0 \\ \omega_z & 0 & a - \omega_x \\ 0 & a + \omega_x & 0 \end{pmatrix} \begin{pmatrix} x \\ y \\ z \end{pmatrix} = \omega \times \mathbf{r} + \mathbf{D} \cdot \mathbf{r}, \tag{5.2.13}$$

where

$$\mathbf{D} \equiv \begin{pmatrix} 0 & 0 & 0 \\ 0 & 0 & a \\ 0 & a & 0 \end{pmatrix} \tag{5.2.14}$$

represents a strain in a plane orthogonal to the plane containing the tidal strain. As in the case of a tidal strain which rotates relative to the inertial frame, as discussed in section 5.2.4 the simple spin-over mode of equation (2.3) may not grow initially. A pair of the many other modes, with phase velocity matching the Ω drift will be first to bifurcate. The structure of this linear eigenvalue problem is discussed in Kerswell (1993b, c). Figure 5.1(b) provides assurances that the principal spin-over mode dominates the finite amplitude behavior for $a > E^{1/2}$.

5.3 APPLICATION TO THE PLANETS

5.3.1 Minimum Conditions Required for a Dynamo

Can tides meet the integral requirements to sustain a dynamo which were set forth in section 5.1?

Table 5.1, from Novotny (1983), (for precessional parameters see Dolginov 1977 and Vanyo 1992) lists planetary tides h, the principal origin of those tides, the (estimated) radius of the conducting core, its mean density, the period of rotation, and the average observed magnetic dipole field strength extrapolated to the core radius. (The fields H_P for Uranus and Neptune are from Ness *et al.* (1986, 1989).)

Using the Earth as a first example, the condition set by Backus (1.3), requires that the strain due to tides, $h|\omega|/r_0 \geq 4 \times 10^{-12}$ s^{-1} (from table 5.1), and/or due to precession, $\Omega_x = 10^{-11} \sin 23.5° \approx 4 \times 10^{-12}$ s^{-1}, must exceed $\lambda \pi^2/r_0^2$. For the Earth, the magnetic diffusivity is estimated at 10^4 cm^2/s, plus or minus a factor of three. Hence Backus' condition is met in the entire body of the flow by both tide and precession separately, but only marginally for the largest proposed λ.

The Childress condition (1.4) requires determining a maximum velocity. Certainly, the tidal velocity, $|\omega|h \simeq 1.5 \times 10^{-3}$ cm/s, which is radial, is less than the maximum velocity which will be induced by such a tide (and very much less than in the experiments reported in section 5.2.1). Using this velocity, the Childress condition is exceeded by a factor of ten.

Lastly, the Busse condition (1.5) is met with this same radial tidal velocity, $|\omega|h$, by a factor of twenty, even in the 'weak field' limit $E_P \simeq E_m$.

From table 5.1, none of the other terrestrial planets meet any of these three conditions, even for markedly different magnetic diffusivity. Yet all the great planets exceed the conditions with large margins. The following paragraphs explore possible consequences of magnetic reaction to the growth of a spin-over mode.

5.3.2 'Strong Field' Scaling

Dynamo theorists have come to believe that almost any three-dimensional flow of sufficiently large magnetic Reynolds number will produce a dynamo. The large-scale bifurcation of spin-over, figure 5.2(a), would seem a reasonable candidate. A magnetic field could stop the growth by reducing the rotational constraints responsible for the phase and structure of the Poincaré modes. Hence, for scaling purposes one can picture a magnetostrophic radial balance of the Coriolis force due to a toroidal flow, V_T, and the radial component of the Lorentz force, $(1/4\pi\rho)(\nabla \times \mathbf{H}) \times \mathbf{H} \cdot \mathbf{1}_r$, written as

$$2|\omega|V_T \simeq \frac{1}{4\pi\rho}H_T^2/r, \qquad (5.3.1)$$

Planet	r_0 (km)	ρ (gm/cm^3)	T (days)
Mercury	1840	7.60	58.6
Venus	3150	10.60	243.09
Earth	3485	10.615	1.00
Mars	1500	7.50	1.03
Jupiter	66400	1.33	0.41
Saturn	46800	.70	0.43
Uranus	18000	1.31	0.45
Neptune	15100	1.66	0.65

Planet	H_P (Gauss)	h (m)	P.B.
Mercury	8×10^{-3}	.59	Sun (.59)
Venus	1.6×10^{-4}	.11	Sun (.11)
Earth	1.9	.15	Moon (.10)
Mars	7.4×10^{-3}	.01	Sun (.01)
Jupiter	5.3	19.45	Io (15.48)
Saturn	.47	2.83	Titan (1.44)
Uranus	.65	1.15	Ariel (.66)
Neptune	.53	6.97	Triton (6.97)

Table 5.1 Physical parameters and tides of planetary cores: radius of the core r_0, mean core density ρ, rotational period T, average observed magnetic dipole field strength extrapolated to the core radius H_P, height of the static tides h, the main perturbing body (P.B.) and the amplitude of its tidal waves (in brackets). (After Novotny 1983.)

where H_T is a representative toroidal component of the magnetic field, and r ($< r_0$) is some characteristic length scale also to be determined. In this 'strong field' balance it is presumed that other radial pressure gradients have, at most, comparable effect. The scaling estimate (3.1) arises from different reasoning and models in current reviews (e.g., Roberts 1988, Soward 1991, Roberts & Soward 1992) of dynamo theory.

From the magnetic diffusion equation (1.1) a balance for dissipation due to H_T is

$$\lambda H_T / r^2 = \frac{V_T}{r} H_P, \qquad (5.3.2)$$

where H_P is a representative poloidal component of the magnetic field, and that part of the field which can be observed outside the conducting fluid.

When the magnetic Reynolds number

$$Rm \equiv \frac{V_T r_0}{\lambda} \gg Rm_c, \qquad (5.3.3)$$

where Rm_c is the critical value for magnetic growth (and certainly larger than the Childress minimum (1.4)), the diffusive length scale r can be estimated from

$$Rm_c = \frac{V_T r}{\lambda} = \left(\frac{V_T r_0}{\lambda}\right)\left(\frac{r}{r_0}\right). \tag{5.3.4}$$

Hence, from equations (3.1,2,4) one finds

$$H_T = (2|\omega|4\pi\rho\lambda)^{1/2} Rm_c^{1/2} \tag{5.3.5}$$

and

$$H_P = (2|\omega|4\pi\rho\lambda)^{1/2} Rm_c^{-1/2}, \tag{5.3.6}$$

similar to Roberts (1988). Equations (3.5) and (3.6) are independent of r_0 and V_T (as long as V_T meets the minimum conditions of (1.4)). The product fields

$$H_T H_P = 2|\omega|4\pi\rho\lambda \tag{5.3.7}$$

were the unique consequence of the 'macrodynamic' dynamo scaling reported in Malkus & Proctor (1975). However, the estimated energy consumption of such dynamos from equation (3.2) is

$$\Phi \equiv \left(\frac{V_T H_P H_T}{r}\right)\left(\frac{4}{3}\pi r_0^3\right) = \left(\frac{V_T^2}{Rm_c}\right)(2|\omega|4\pi\rho)\left(\frac{4}{3}\pi r_0^3\right), \tag{5.3.8}$$

which is independent of λ, but requires an estimate of V_T and Rm_c. For $Rm \gg Rm_c$ one expects that $\mathbf{V} \times \mathbf{H} \simeq 0$; that is the flow will be almost parallel to the magnetic field. In this case from equations (3.5, 6),

$$\frac{H_T}{H_P} = Rm_c \simeq \frac{V_T}{V_P}. \tag{5.3.9}$$

Since V_P is at least comparable to the poloidal tidal velocity $|\omega|h$, then from (3.9),

$$V_T \simeq Rm_c|\omega|h. \tag{5.3.10}$$

If then Rm_c is at least greater than π^2

$$\Phi > (|\omega|h\pi)^2 (2|\omega|4\pi\rho)(4\pi r_0^3/3) \simeq 3 \times 10^{19} \text{ erg/s} \tag{5.3.11}$$

for Earth values from table 5.1.

For the Earth and the giant planets which pass the minimum criteria for a tidal energy source, equation (3.6) for their poloidal surface fields indicates that all will have an H_P of order 1 Gauss. The dependence on $\lambda^{1/2}$ may account for the variations reported in table 5.1. However, none of the scaling arguments above rule out a convective source and its accompanying zonal flow V_T, if such a source can provide sufficient energy.

5.4 CONVECTIVE ENERGY SOURCES FOR PLANETARY DYNAMOS

5.4.1 Previous Reviews

For several decades the origin of the dynamo process has been sought as a consequence of thermal convection in the sun and planets. The special phenomenon of compositional convection is believed to be most important in the case of the Earth, and is discussed in a final subsection.

This section will be brief for two reasons: there are excellent articles and reviews available, e.g., Busse (1978), Roberts & Gubbins (1987) and Zhang & Busse (1990); secondly, this author has not contributed to magneto-convection literature for many years (Malkus 1959).

5.4.2 Laboratory Paradigms

As in the phenomena of elliptical bifurcation, an observable convective process in a rotating laboratory experiment suggests the possible behavior of convection in the planetary setting. The lovely experiment of Busse & Carrigan (1976) is shown in figure 5.5, and confirms the theoretical expectations of its eigenstructure. Early study of the conditions under which the finite-amplitude convection produced a magnetic field suggested a 'weak field' balance in which the Coriolis force remained much larger than the Lorentz force. This balance significantly reduces the energy requirement to maintain the dynamo and led (Busse 1978) to an estimate for the toroidal magnetic field of

$$H_T \leq |\omega| r_0 \left(\frac{\rho \kappa}{\lambda}\right)^{1/2} \times 10^{-4}, \qquad (5.4.1)$$

where κ is the thermal diffusivity. This is a small field indeed, and the H_P would be smaller still. It is not impossible that a 'weak field' regime exists in three of the terrestrial planets, yet recent models and computation (Roberts 1988) suggest the basic instability of such regimes. Numerical studies by Zhang & Busse (1989) offer the most profound insight into the evolutionary details of a magneto-convective process. Depending on the value of the Ekman number E, the Prandtl number ν/κ, and thermal contrasts, they find both steady (DC) and periodic (AC) dynamos close to the critical values for their bifurcations. Perhaps clever modeling can lead to numerical results appropriate to a planetary scale, and one day permit quantitative and qualitative assessments supporting magneto-convection.

5.4.3 The Solar Dynamo

The solar convection is so energetic that it does not permit the kind of dynamo pictured in section 5.4.2 (Weiss 1990). However, the Sun has a strong oscillatory dynamo with a twenty-two year period. Recent studies (see, e.g.,

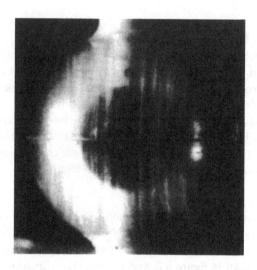

Figure 5.5 Convection columns in a rotating spherical fluid shell heated from the outside and cooled from within (from Busse & Carrigan 1976).

Prautzsch 1993) explore the possibility that the principal solar magnetic fields may be generated in, and initially confined to, the stably stratified shear layer at the base of the solar convection zone. Perhaps the strains in that shear zone are the actual source of the magnetic field, which would then become a secondary rather than primary consequence of the intense convection above?

5.4.4 Energetic Requirements for α^2 and $\alpha\omega$ Dynamos

The turbulent transport of heat and momentum usually requires many small scales of motion. If these small scales have helicity, as they would in a rotating system, they can initiate a large scale magnetic field (α^2 dynamos). If there is large scale shear in the flow, this furthers the potential instability ($\alpha\omega$ dynamos). As H.K. Moffatt has pointed out, in private communication, the energetic requirements for a large scale field, H_T, are increased by the square of the ratio of the small to large scale motions in such dynamos. Hence in many planetary physical contexts these α^2 and $\alpha\omega$ dynamos are energetically quite 'impractical' and should be taken at best as pedagogical idealizations. A geophysical example is discussed in a following paragraph.

5.4.5 Compositional Convection in the Earth

A geodynamo driven by thermal convection alone has been doubted for at least three decades.

The reasons are based on estimates of total heat flux from the core fluid to the mantle, and the estimates of the minimum flux needed to maintain an adiabatic thermal gradient in the core fluid. The difference of these two heat fluxes is available to produce fluid motion, but at a low efficiency (in the Earth approximately five percent). Values for total flux from core to mantle dance around 10^{20} erg/s (Roberts & Gubbins 1987; Gubbins, Masters & Jacobs 1979; Loper & Roberts 1983), while estimates of the heat flux needed to maintain the adiabat include this value but are centered at 6×10^{19} ergs/s. Five percent of the difference is less than a tenth of the energy needed to maintain the 'strong field' dynamo of equation (3.11). Again, both heat flux estimates have plus or minus factors of approximately three, hence a thermally convective geodynamo remains faintly possible. A resolution was proposed by Braginsky (1963), following the studies of Verhoogen (1961) on energy released by the growth of the solid inner core. (A simplified and amechanistic re-assessment of the Verhoogen proposal can be read in Buffett *et al.* 1992.) Braginsky suggested that the buoyant gravitational energy released by the solidification of iron from an iron rich alloy would be much more efficient than thermal buoyancy in producing motion. Estimates in Loper & Roberts (1983) are that the overall efficiency is approximately twenty five percent, five times as effective as thermal convection alone. Doubts about maintaining an adiabatic gradient in the core fluid may be removed by the many studies now completed on this topic.

As in the case of convection and elliptical instability, a laboratory paradigm exists of the release of buoyancy by deposition of a denser liquid component. This elegant and suggestive demonstration (Loper 1978), shows that the deposition is by the growth of dendrites, while the buoyancy is released in isolated plumes of fluid from 'chimneys' in a dendritic mush (Loper & Roberts 1983). A study of the vertical evolution of such plumes has recently been published (Loper & Moffatt 1993). However, little has been written yet of the manner in which many small plumes (of estimated spacing 1–10 km at the inner core boundary) will 'efficiently' produce a large scale magnetic field. To follow Moffatt's estimate that energy requirements in $\alpha\omega$ dynamos increase as the square of the ratio of large to small scale would mitigate against such a process. Given the (parametric) flexibility of this inquiry, one is sure that another path can be found.

5.5 IN TEMPORARY CONCLUSION

In the introduction to this paper the writer may have been misguided regarding the concern of the gods. The geodynamo, more so than any other dynamo, seems fragilely balanced with respect to each proposed energy source. Perhaps buoyancy sources and rotational energy both must be called upon to maintain it. Perhaps, despite its long past, the geodynamo has less future, and will end up like the other fields of the terrestrial planets. Or perhaps, 'weak field' and 'strong field' states snap back and forth at a rate related to the energy available. Certainly the role of the magnetic field is to transport momentum, not heat. Hence experimental and computational exploration of that role in the accessible convective regimes, and accessible 'tidal' bifurcations, can generate that understanding of mechanism so missing in the kinematic studies of the past.

Numerical investigation of non-linear dynamo behavior appears to be, indeed, the current and future path. It is hoped that it will be concentrated on testing hypotheses — with the primary goal of discarding the many misdirected proposals that abound in the study of planetary dynamos.

Acknowledgements

The author wishes to thank F.H. Busse and R.R. Kerswell for their thoughtful reading of an earlier draft of this paper. This study has been supported by grant ATM89-01473 from the National Science Foundation.

REFERENCES

Backus, G.E. 1958 A class of self-sustaining dissipative spherical dynamos. *Ann. Phys. N.Y.* **4**, 372–447.

Bayly, B.J. 1986 Three-dimensional instability of elliptical flow, *Phys. Rev. Lett.* **57**, 2160–2163.

Bondi, H. & Lyttleton, R.A. 1953 On the dynamical theory of the rotation of the earth. II. The effect of precession on the motion of the liquid core. *Proc. Camb. Phil. Soc.* **49**, 498–515.

Braginsky, S.I. 1963 Structure of the F layer and reasons for convection in the Earth's core. *Dokl. Akad. Nauk. SSSR* **149**, 1311–1314. (English transl.: *Sov. Phys. Dokl.* **149**, 8–10, (1963).)

Buffett, B.A., Huppert, H.E., Lister, J.R. & Woods, A.W. 1992 Analytical model for solidification of the Earth's core. *Nature* **356**, 329–331.

Busse, F. 1968 Steady fluid flow in a precessing spheroidal shell. *J. Fluid Mech.* **33**, 739–751.

Busse, F.H. 1978 Magnetohydrodynamics of the Earth's dynamo. *Annu. Rev. Fluid Mech.* **10**, 435–462.

Busse, F.H. & Carrigan, C.R. 1976 Laboratory simulation of thermal convection in rotating planets and stars. *Science* **191**, 81–83.

Childress, S. 1969 Théorie magnetohydrodynamique de l'effet dynamo. Report from Departement Mécanique de la Faculté des Sciences, Univ. Paris.

Craik, A.D.D. 1988 A class of exact solutions in viscous incompressible magnetohydrodynamics. Proc R. Soc. Lond. A 417, 235–244.

Craik, A.D.D. 1989 The stability of unbounded two- and three- dimensional flows subject to body forces: some exact solutions. J. Fluid Mech. 198, 275–292.

Craik, A.D.D. & Allen, H.R. 1992 The stability of three-dimensional time-periodic flows with spatially uniform strain rates. J. Fluid Mech. 234, 613–627.

Dolginov, Sh.Sh. 1977 Planetary magnetism: A survey. Geomag. Aeron. 17, 569. (English transl.: 391–406.)

Gledzer, Ye. B., Dolzhanskiy, F.V., Obukhov, A.M. & Ponomarev, V.M. 1975 An experimental and theoretical study of the stability of motion of a liquid in an elliptical cylinder. Izv., Atmos. & Oceanic Phys. 11, 981–992. (English transl.: 617–622.)

Greenspan, H.P. 1968 The Theory of Rotating Fluids. Cambridge University Press.

Gubbins, D., Masters, T.G. & Jacobs, J.A. 1979 Thermal evolution of the Earth's core. Geophys. J. R. Astr. Soc. 59, 57–99.

Kerswell, R.R. 1992 Elliptical instabilities of stratified, hydromagnetic waves and the Earth's outer core. PhD Dissertation, Department of Mathematics, MIT.

Kerswell, R.R. 1993a Elliptical instabilities of stratified hydromagnetic waves. Geophys. Astrophys. Fluid Dynam. 71, 105–143.

Kerswell, R.R. 1993b Instabilities of tidally and precessionally induced flows. In Solar and Planetary dynamos (ed. M.R.E. Proctor, P.C. Matthews & A.M. Rucklidge), pp. 181–179. Cambridge University Press.

Kerswell, R.R. 1993c The instability of precessing flow. Geophys. Astrophys. Fluid Dynam. 72, 107–144.

Lindzen, R.S. 1991 Some remarks on the dynamics of the Jovian atmospheres. Geophys. Astrophys. Fluid Dynam. 58, 123–131.

Loper D.E. 1978 The gravitationally powered dynamo. Geophys. J. R. Astr. Soc. 54, 389–404.

Loper, D.E. & Moffatt, H.K. 1993 Small-scale hydromagnetic flow in the Earth's core: rise of a vertical buoyant plume. Geophys. Astrophys. Fluid Dynam. 68, 177–202.

Loper, D.E. & Roberts, P.H. 1983 Compositional convection and the gravitationally powered dynamo. In Stellar and Planetary Magnetism, vol. 2 (ed. A.M. Soward), pp. 297–327. Gordon and Breach.

McEwan, A.D. 1970 Inertial oscillations in a rotating fluid cylinder. J. Fluid Mech. 40, 603–640.

McEwan, A.D. 1971 Degeneration of resonantly-excited standing internal gravity waves. *J. Fluid Mech.* **50**, 431–448.

Malkus, W.V.R. 1959 Magnetoconvection in a viscous fluid of infinite electrical conductivity. *Astrophys. J.* **130**, 259–275.

Malkus, W.V.R. 1967 Hydromagnetic planetary waves. *J. Fluid Mech.* **28**, 793–802.

Malkus, W.V.R. 1968 Precession of the Earth as the cause of geomagnetism. *Science* **160**, 259–264.

Malkus, W.V.R. 1989 An experimental study of global instabilities due to the tidal (elliptical) distortion of a rotating elastic cylinder. *Geophys. Astrophys. Fluid Dynam.* **48**, 123–134.

Malkus, W.V.R. & Proctor M.R.E. 1975 The macrodynamics of α-effect dynamos in rotating fluids. *J. Fluid Mech.* **67**, 417–444.

Ness, N.F., Acuna, M.H., Behannon, K.W., Burlaga, L.F., Connerney, J.E.P., Lepping, R.P. & Neubauer, F.M. 1986 Magnetic fields at Uranus. *Science* **233**, 85–89.

Ness, N.F., Acuna, M.H., Burlaga, L.F., Connerney, J.E.P., Lepping, R.P. & Neubauer, F.M. 1989 Magnetic fields at Neptune. *Science* **246**, 1473–1478.

Novotny, O. 1983 Empirical relations between planetary magnetism and tides. In *Stellar and Planetary Magnetism*, vol. 2 (ed. A.M. Soward) pp. 289–293. Gordon and Breach.

Poincaré, H. 1910 Sur la précession des corps deformable. *Bull. Astron.* **27**, 231.

Prautzsch, T. 1993 The dynamo mechanism in the deep convective zone of the Sun. In *Solar and Planetary dynamos* (ed. M.R.E. Proctor, P.C. Matthews & A.M. Rucklidge), pp. 249–256. Cambridge University Press.

Roberts, P.H. 1988 Future of geodynamo theory. *Geophys. Astrophys. Fluid Dynam.* **44**, 3–31.

Roberts, P.H. & Gubbins, D. 1987 Origin of the main field: kinematics. In *Geomagnetism*, vol. 2 (ed. J.A. Jacobs), pp. 185–249. Academic Press.

Roberts, P.H. & Soward, A.M. 1992 Dynamo theory. *Annu. Rev. Fluid Mech.* **24**, 459–512.

Roberts, P.H. & Stewartson, K. 1963 On the stability of a Maclaurin spheroid of small viscosity. *Astrophys. J.* **137**, 777–790.

Roberts, P.H. & Stewartson, K. 1965 Motion of a liquid in a spheroidal cavity of a precessing rigid body. II. *Proc. Camb. Phil. Soc.* **61**, 279–288.

Soward, A.M. 1991 The Earth's dynamo. *Geophys. Astrophys. Fluid Dynam.* **62**, 191–209.

Suess, S. 1969 Effects of gravitational tides on a rotating fluid. PhD dissertation, Department of Planetary Science, UCLA.

Vanyo, J.P. 1991 A geodynamo powered by Luni-Solar precession. *Geophys. Astrophys. Fluid Dynam.* **59**, 209–234.

Verhoogen, J. 1961 Heat balance of the Earth's core. *Geophys. J. R. Astr. Soc.* **4**, 276–281.

Waleffe, F. 1988 3D instability of bounded elliptical flow. *Woods Hole Oceanog. Inst. Rep.* WHOI–89–26, 302–315.

Waleffe, F. 1990 On the three-dimensional instability of strained vortices. *Phys. Fluids A* **2**, 76–80.

Weiss, N.O. 1990 Solar & stellar convection zones. *Comp. Phys. Rep.* **12**, 233–245.

Zhang, K. & Busse, F.H. 1989 Convection driven magnetohydrodynamic dynamos in rotating spherical shells. *Geophys. Astrophys. Fluid Dynam.* **49**, 97–116.

Zhang, K. & Busse, F.H. 1990 Generation of magnetic field by convection in a rotating spherical fluid shell of infinite Prandtl number. *Phys. Earth Planet. Inter.* **59**, 208–222.

CHAPTER 6

Fast Dynamos

ANDREW M. SOWARD

Deptartment of Mathematics and Statistics
University of Newcastle upon Tyne
Newcastle upon Tyne NE1 7RU, UK.

6.1 INTRODUCTION

The notion of a dynamo was clearly defined by Roberts in chapter 1. Essentially, in the absence of motion, magnetic field not maintained by external current sources decays on the diffusion time scale $\tau_\eta = \mathcal{L}^2/\eta$, where $\mathcal{L}$ is the characteristic length of the system. On the other hand, motion may advect magnetic field on the advection time $\tau_\mathcal{V} = \mathcal{L}/\mathcal{V}$, where $\mathcal{V}$ is a typical velocity of the system. Their ratio $R = \tau_\eta/\tau_\mathcal{V}$ defines the magnetic Reynolds number (see (1.2.37)). Basically there is a competition between advection and diffusion. Advection may intensify the magnetic field while diffusion acts to destroy it. Accordingly, a dynamo will only operate when the diffusion time τ_η is at least as long as the advection time $\tau_\mathcal{V}$. In many astrophysical systems, however, the magnetic Reynolds number is large

$$R \gg 1 \qquad (6.1.1)$$

and the time scales separate. Since only advection amplifies the magnetic field, we do not expect magnetic field to grow on a timescale shorter than $\tau_\mathcal{V}$ (see also (3.9) below), but can it grow that fast? This is the central question with which fast dynamo theory is concerned, and it is ideas related to large R which concern us in this chapter.

The question of whether or not fast dynamos exist focuses our attention on the key issues of large-R dynamo theory. Indeed, their existence is implicit in many mean field theories upon which, for example, solar dynamo models depend (e.g., Stix 1987). The mean field development for turbulent velocity fields was explained in section 1.5. For the particular case of pseudo-isotropic turbulence, the parameters α and β measuring the strengths of the α-effect and turbulent diffusivity, respectively, were introduced (1.5.44, 45). The mean field dynamo equation (1.5.70) which results, depends on the mean fluid velocity $\overline{\mathbf{V}}$, and the values of α and the total diffusivity η^{T} ($= \eta + \beta$

181

M.R.E. Proctor & A.D. Gilbert (eds.)
Lectures on Solar and Planetary Dynamos, 181–217
©1994 Cambridge University Press.

(1.5.50)). However, when the corresponding microscale magnetic Reynolds number R_m (1.5.36) for the turbulence is large but the correlation time is short, heuristic arguments suggest that α and β take the values (1.5.54, 55) independent of the magnetic diffusivity. When in addition it is assumed that η is negligible in comparison with β, the remaining mean field dynamo equation (1.5.70) no longer depends on the magnetic diffusivity η. From that point of view, any dynamo predicted by it is fast because only convective time scales remain in the equation! The explicit evaluation of the coefficients α and β in the large R_m limit encounters severe mathematical difficulties (e.g., Moffatt 1974, Parker 1979) because, unlike in the small R_m limit, the perturbation fields $\mathbf{B}'$ produced (1.5.39) are large compared to the mean field $\overline{\mathbf{B}}$. Clearly a resolution of these issues provides a strong motivation for understanding fast dynamos. Indeed, as yet, there is no mathematical proof that fast dynamos exist. Though the numerical and analytical evidence now available (see section 6.6.3 below) strongly supports their existence for a wide range of sufficiently complicated flows.

In this chapter, the turbulent dynamo will not be addressed. Instead attention will be restricted to simple flows for which the dynamo question can be clearly stated. In particular, we will consider either steady flows or time-periodic flows with period P (1.3.28). In the former case, the solution can be represented as a sum of eigenfunctions $\mathbf{B}_\alpha(\mathbf{x})$ each with their own growth rates λ_α (1.3.18). In the latter case, those eigenfunctions are replaced by periodic functions $\mathbf{B}_\alpha(\mathbf{x}, t)$ with the same period P of the flow and the growth rates λ_α (possibly complex) become Floquet exponents. The steady flow case may be regarded as the limit $P \downarrow 0$ of the periodic case. In both cases, however, we can identify the fastest growth rate and define its limiting value as $R \to \infty$ by

$$\lambda_\infty \tau_\mathcal{V} = \lim_{R \to \infty} \left[\sup_\alpha (\lambda_\alpha \tau_\mathcal{V}) \right]. \tag{6.1.2}$$

The dynamo is fast if it grows exponentially on the convective timescale, namely

$$\lambda_\infty \tau_\mathcal{V} > 0. \tag{6.1.3}$$

Otherwise it is slow.

This chapter is organised as follows. In section 6.2 we develop some of the mathematical apparatus useful for large-R flows. We consider linear flows in section 6.3, isolating those magnetic features, particularly flux ropes and sheets, which are robust in the limit $R \to \infty$. Some slow dynamos are mentioned in section 6.4. Idealised rope dynamos are introduced in section 6.5, which isolate important features of fast dynamos towards which section 6.6 develops. Many of the ideas described here have evolved from our earlier reviews, namely Soward & Childress (1986), Soward (1988), Fearn, Roberts & Soward (1988) and Roberts & Soward (1992). Nevertheless, Childress (1992) provides an excellent review, which emphasises some different aspects of fast dynamo theory, particularly on the map related topics described by Bayly in chapter 10 below.

6.2 THE CAUCHY SOLUTION AND REFERENCE FIELDS

6.2.1 Perfectly Conducting Fluids

In order to set the scene we begin with some remarks about well-known features of flow kinematics (Ottino 1989, Tabor 1992). Consider two neighbouring fluid particles, one at $\mathbf{x}$ and the other at $\mathbf{x} + \delta\mathbf{x}$. As they move, their separation $\delta\mathbf{x}$ is governed by

$$D(\delta\mathbf{x})/Dt = \mathbf{V}(\mathbf{x} + \delta\mathbf{x}, t) - \mathbf{V}(\mathbf{x}, t) = \mathbf{S}\delta\mathbf{x}, \tag{6.2.1}$$

where D/Dt is the material derivative and

$$S_{ij} = \partial V_i / \partial x_j \tag{6.2.2}$$

is the velocity gradient tensor. On the other hand, when the fluid is a perfect electrical conductor

$$\eta = 0, \qquad (R = \infty) \tag{6.2.3}$$

and the fluid is incompressible ($\nabla \cdot \mathbf{V} = 0$), the magnetic induction equation (1.2.34) reduces to

$$D\mathbf{B}/Dt = \mathbf{S}\mathbf{B}. \tag{6.2.4}$$

Since the equation coincides with (2.1), we have immediately the particular integral

$$\mathbf{B}(\mathbf{x}, t) = k\delta\mathbf{x}, \qquad (Dk/Dt = 0), \tag{6.2.5}$$

where k is constant for fluid particles.

The above result suggests that in the case of perfectly conducting fluids a Lagrangian formulation is more appropriate. Accordingly we identify the path $\mathbf{x} = \mathbf{x}(\mathbf{a}, t)$ of a fluid particle initially ($t = 0$) at $\mathbf{x} = \mathbf{a}$. Then for our two neighbouring fluid particles, their current separation $\delta\mathbf{x}$ is related to their initial separation $\delta\mathbf{a}$ by

$$\delta\mathbf{x} = \mathbf{J}\delta\mathbf{a} \tag{6.2.6}$$

through the Jacobian matrix

$$J_{ij}(\mathbf{a}, t) = \partial x_i / \partial a_j, \qquad (\mathbf{x} = \mathbf{x}(\mathbf{a}, t)). \tag{6.2.7}$$

We note that direct differentiation with respect to t gives

$$D\mathbf{J}/Dt = \mathbf{S}\mathbf{J}, \qquad (D\mathbf{x}/Dt = \mathbf{V}), \tag{6.2.8}$$

where in the new **a**-coordinate system the material derivation D/Dt is simply $\partial/\partial t$ at fixed **a**. From this point of view the constant k in (2.5) is a function of **a** alone, determined by the initial value

$$\mathbf{b(a)} = \mathbf{B(a,0)} = k\delta\mathbf{a} \qquad (6.2.9)$$

of the magnetic field. Since (2.5) and (2.9) hold for all $\delta\mathbf{a}$, use of (2.6) leads to the Cauchy solution (1.2.58), namely

$$\mathbf{B(x,}t) = \mathbf{Jb(a)}. \qquad (6.2.10)$$

Formally this follows directly upon differentiation with respect to t and use of (2.4, 8). Note also that the incompressibility condition implies that

$$\det \mathbf{J} = 1, \qquad (\nabla \cdot \mathbf{V} = 0). \qquad (6.2.11)$$

The result (2.5) has a simple physical interpretation. It shows that magnetic field lines are material lines frozen to the fluid as it moves. On the one hand, the direction of the field is determined by the direction of the line element $\delta\mathbf{x}$ initially parallel to it, as defined by (2.9). On the other hand, its strength is proportional to the length δl $(= |\delta\mathbf{x}|)$ of line elements,

$$|\mathbf{B(x,}t)| = |k|\delta l. \qquad (6.2.12)$$

Indeed we have

$$(\delta l)^2 = \delta\mathbf{x} \cdot \delta\mathbf{x} = \delta\mathbf{a} \cdot \mathbf{\Lambda}\delta\mathbf{a}, \qquad (6.2.13)$$

where $\mathbf{\Lambda}$ is the symmetric matrix

$$\mathbf{\Lambda} = \mathbf{J}^{\mathrm{T}}\mathbf{J}, \qquad (\Lambda_{ij} = (\partial\mathbf{x}/\partial a_i) \cdot (\partial\mathbf{x}/\partial a_j)). \qquad (6.2.14)$$

Note that the dot denotes the scalar multiplication of vectors, the absence of a dot denotes matrix multiplication and the superscript ^{T} denotes the transpose matrix. Suppose that we identify the direction of the initial line element $\delta\mathbf{a}$ by the vector **e** and write

$$\delta\mathbf{a} = |\delta\mathbf{a}|\mathbf{e}. \qquad (6.2.15)$$

Then the rate at which the line element expands is measured by the Liapunov exponent

$$\lambda(\mathbf{a}, \mathbf{e}) = \limsup_{t\to\infty}\left\{\frac{1}{2t}\ln(\mathbf{e}\cdot\mathbf{\Lambda e})\right\}. \qquad (6.2.16)$$

This in turn determines the growth rate of the magnetic field strength (2.12). Following Childress (1992) we will call the flow a stretching flow if for some **a**, **e** the Liapunov exponent is positive. Accordingly a stretching flow leads to exponential growth of magnetic field. Note, however, that this growth rate is a local property of individual fluid particles. Since exponential stretching corresponds to chaotic trajectories, they may be dense in a finite region. In any case it should be contrasted with the diffusive problem for which there are global growth rates λ_α (1.3.18). This distinction between local and global leads to difficulties of interpretation of frozen field results, for ultimately we are concerned with the limit $\eta \downarrow 0$ and not the singular case $\eta = 0$.

An interesting correspondence can be made with the case of a scalar field $\Phi(\mathbf{x}, t)$ frozen to the flow and given by its initial value

$$\Phi(\mathbf{x}, t) = \phi(\mathbf{a}). \tag{6.2.17}$$

Its gradient is

$$\nabla \Phi = [\mathbf{J}^{-1}]^T \nabla_{\mathbf{a}} \phi, \tag{6.2.18}$$

where we have introduced the notation

$$(\nabla_{\mathbf{a}} \phi)_i = \partial \phi / \partial a_i. \tag{6.2.19}$$

The magnitude of $\nabla \Phi$ is determined from

$$|\nabla \Phi|^2 = [\nabla_{\mathbf{a}} \phi] \cdot \mathbf{\Lambda}^{-1} [\nabla_{\mathbf{a}} \phi], \tag{6.2.20}$$

where

$$\mathbf{\Lambda}^{-1} = \mathbf{J}^{-1} [\mathbf{J}^{-1}]^T, \qquad ((\mathbf{\Lambda}^{-1})_{ij} = \nabla_{\mathbf{x}} a_i \cdot \nabla_{\mathbf{x}} a_j). \tag{6.2.21}$$

Whereas $\mathbf{\Lambda}$ measures extension of line elements, the inverse matrix $\mathbf{\Lambda}^{-1}$ is important because, in contrast, it measures compression.

An alternative representation for the magnetic field $\mathbf{B}$ is

$$\mathbf{B} = \nabla \times \mathbf{A}. \tag{6.2.22}$$

Here the vector potential is not unique. Nevertheless, if we wish, we may define its evolution uniquely by

$$D\mathbf{A}/Dt = -\mathbf{S}^T \mathbf{A}. \tag{6.2.23}$$

Verification of (2.23) follows from taking its curl, use of (2.4) and the incompressibility condition $\nabla \cdot \mathbf{V} = 0$. Together with (2.4) it is readily shown that

$$h = \mathbf{B} \cdot \mathbf{A}, \qquad (Dh/Dt = 0) \tag{6.2.24}$$

is constant for fluid particles, $h = h(\mathbf{a})$. This result is compatible with the solution

$$\mathbf{A} = [\mathbf{J}^{-1}]^T \mathcal{A} \tag{6.2.25}$$

of (2.23) in terms of the initial value

$$\mathcal{A}(\mathbf{a}) = \mathbf{A}(\mathbf{a}, 0) \tag{6.2.26}$$

of the magnetic vector potential $\mathbf{A}(\mathbf{x}, t)$. From this point of view, whereas $\mathbf{B}$ behaves like a material line, $\mathbf{A}$ behaves like a material area. Note that $\mathbf{A}$ is only determined up to an additive gradient, which, as (2.18) shows, also transforms like $\mathbf{A}$. However, in general, if the gauge of $\mathbf{A}$ is specified (say $\nabla \cdot \mathbf{A} = 0$) then an additional term $-\nabla_{\mathbf{x}} \phi$ must be included on the right-hand side of (2.23) in order that the gauge condition can be met.

Note that use of the vector potential $\mathbf{A}$ is closely linked to the adjoint problem, which Bayly & Childress (1989) have usefully employed for certain dynamo problems involving maps. The idea is explained by Bayly in section 10.5 below.

6.2.2 Finite Magnetic Diffusivity

When the fluid has finite electrical conductivity and the magnetic Reynolds number is large, we anticipate that, under certain conditions, the field is close to its frozen field value. This suggests that the representations (2.10) and (2.25) may continue to be helpful. Now, however, $\mathbf{b}$ and $\mathcal{A}$ are to be interpreted as reference fields which are no longer constant but evolve with time;

$$\mathbf{B}(\mathbf{x}, t) = \mathbf{J}\mathbf{b}(\mathbf{a}, t), \qquad \mathbf{A}(\mathbf{x}, t) = [\mathbf{J}^{-1}]^T \mathcal{A}(\mathbf{a}, t). \qquad (6.2.27, 28)$$

These forms satisfy the magnetic induction equation provided that

$$D\mathcal{A}/Dt = \eta(\mathcal{E}_{\mathbf{a}} - \nabla_{\mathbf{a}}\phi), \qquad (\mathbf{b} = \nabla_{\mathbf{a}} \times \mathcal{A}) \qquad (6.2.29)$$

for some ϕ, where

$$\mathcal{E}_{\mathbf{a}} = -\Lambda[\nabla_{\mathbf{a}} \times \mathbf{b}_{\mathbf{a}}], \qquad (\mathbf{b}_{\mathbf{a}} = \Lambda\mathbf{b}) \qquad (6.2.30)$$

or

$$\mathcal{E}_{\mathbf{a}} = -\lambda\mathbf{b} - [\Lambda^{-1}\nabla_{\mathbf{a}}] \times \mathbf{b} \qquad (6.2.31)$$

where

$$\lambda_{ij} = (\partial\mathbf{x}/\partial a_i) \cdot [\nabla_{\mathbf{x}} \times (\partial\mathbf{x}/\partial a_j)]. \qquad (6.2.32)$$

It follows that $h \ (= \mathbf{B} \cdot \mathbf{A} = \mathbf{b} \cdot \mathcal{A})$ satisfies

$$Dh/Dt = -2\eta H - \eta\nabla_{\mathbf{a}} \cdot (\mathbf{b}\phi + \mathcal{A} \times \mathcal{E}_{\mathbf{a}}), \qquad (6.2.33)$$

where

$$H = -\mathbf{b} \cdot \mathcal{E}_{\mathbf{a}} = \mathbf{b}_{\mathbf{a}} \cdot (\nabla_{\mathbf{a}} \times \mathbf{b}_{\mathbf{a}}) = \mathbf{B} \cdot \nabla_{\mathbf{x}} \times \mathbf{B} \qquad (6.2.34)$$

is the magnetic helicity. For an unbounded fluid with $\mathbf{B}$ and $\mathbf{A}$ decaying sufficiently fast at infinity, we have the identity

$$d\mathcal{H}/dt = -2\eta \int H \, dV, \qquad (\mathcal{H} = \int h \, dV), \qquad (6.2.35)$$

where integration is taken throughout all space.

The formulation and results stated above raise a number of important issues. To begin, Woltjer (1958) realised that for perfectly conducting fluids $\mathcal{H}$ is an invariant, which is linked to topological properties of the fluid (e.g., Moffatt 1978, 1992, Moffatt & Tsinober 1992). For finitely conducting fluids (2.35) says that $\mathcal{H}$ only changes in the presence of magnetic helicity H. Note, however, that for $\mathcal{H}$ to grow it is not sufficient to have magnetic helicity. Axisymmetric magnetic fields generally possess magnetic helicity but they are not dynamos. In this context, beware! Some symmetric dynamo configurations are such that the integrals on both sides of (2.35) vanish, in which case the region of consideration must be restricted and the arguments modified.

Now a superficial inspection of (2.29) suggests that the reference field **b** is likely to evolve on the magnetic diffusion time scale τ_η. Nevertheless **B** itself could evolve rapidly on the advective time τ_ν and so be a fast dynamo. A fast dynamo mode, however, must have both **B** and **A** growing exponentially on the convective time scale τ_ν and consequently $\mathcal{H}$ itself. At first sight such a statement appears to be incompatible with (2.35) in which η vanishes in the perfect conductivity limit. Moffatt & Proctor (1985) argued that, in order for a fast dynamo mode to satisfy the helicity requirements of (2.35), it is necessary that the energy of the magnetic field be concentrated on the short order $R^{-1/2}\mathcal{L}$ diffusive length scale. This appears to be a fundamental requirement for fast dynamo models.

As for the Lagrangian formulation (2.29–32) itself, the hope is, of course, that for a limited time, at any rate before the length scales have reduced to order $R^{-1/2}\mathcal{L}$, the right-hand side of (2.29) is small when $R \gg 1$. Accordingly the reference fields **b** and $\mathcal{A}$ evolve slowly. The reason for the length-scale reduction is simple. As time proceeds, the coordinate transformation between the **a**- and **x**-frames becomes highly contorted and the compression matrix Λ^{-1} in (2.31) is likely to increase secularly causing $\mathcal{E}_\mathbf{a}$ to become extremely large. Despite the obvious drawbacks of the formalism, there are useful results that can be obtained from it as we will see in the following sections. Before closing this section we note that, even for a uniform magnetic field, $\mathcal{E}_\mathbf{a}$ may be non-zero. Indeed its component parallel to **b** is responsible for the α-effect in some slow dynamos (see section 6.4). So, for example, in the case of the unit field $\mathbf{b} = (1, 0, 0)$, the α-coefficient is $-\eta H$, where we have noted $\lambda_{11} = H$.

6.3 LINEAR FLOWS

6.3.1 The General Solutions

An important application of the evolving reference field formulation of the previous section is to linear flows of the type

$$V = Sx, \qquad (S = S(t)), \tag{6.3.1}$$

in which the velocity gradient is a function of t alone. In this case the coordinate transformation is particularly simple,

$$x = Ja, \qquad (J = J(t)), \tag{6.3.2}$$

where J is a function of t alone governed by (2.8). It follows from the definitions (2.14, 32) that

$$\Lambda = \Lambda(t), \qquad \lambda = 0. \tag{6.3.3}$$

For this flow a simple general solution for the reference magnetic field $b(a, t)$ (2.27) can be found in the form of the Fourier transform solution (3.9) below of an initial value problem.

Zel'dovich *et al.* (1984) considered the dynamo properties of random linear flows on arbitrary initial magnetic fields. Consider the Fourier decomposition of some initial magnetic field

$$B_0(x) = \int \hat{B}_0(k_0) \exp(ik_0 \cdot x) \, d^3k_0, \qquad (k_0 \cdot \hat{B}_0(k_0) = 0). \tag{6.3.4}$$

Then for any time dependent linear flow, (2.29) admits a solution of the form

$$b(a, t) = \int \hat{b}(k_0, t) \exp(ik_0 \cdot a) \, d^3k_0, \qquad (k_0 \cdot \hat{b}(k_0) = 0), \tag{6.3.5}$$

in which the phase

$$\phi(k_0, a) = k_0 \cdot a = k \cdot x \tag{6.3.6}$$

is frozen to the flow (cf. (2.17))

$$k(k_0, t) = [J^{-1}]^T k_0. \tag{6.3.7}$$

Substitution into (2.31) yields

$$D\hat{b}/Dt = -\eta(k_0 \cdot \Lambda^{-1} k_0)\hat{b} = -\eta|k|^2\hat{b} \tag{6.3.8}$$

and so the solution (3.5), derived this way by Fearn *et al.* (1988), to our initial value problem (3.4) is given by

$$\hat{b}(k_0, t) = \hat{B}_0(k_0) \exp\left\{-\eta \int_0^t |k(k_0, t)|^2 \, dt\right\}. \tag{6.3.9}$$

This simple result shows that the decay of $\hat{b}(k_0, t)$ is controlled by the evolving wave number $|k|$ in physical space. There is one immediate consequence. Whatever the growth rate in real space of a particular mode characterised by the initial value $\hat{B}_0(k_0) \exp(ik_0 \cdot x)$ when the fluid is perfectly conducting, the presence of finite diffusivity η reduces that growth rate. Indeed Vishik (1988, 1989) makes use of this idea to show that for motion which is nowhere a stretching flow, a dynamo is necessarily slow. Essentially he argues that for non-stretching flow the dynamo can only be fast if the reference field $b(a, t)$ grows on the fast time scale. That in turn requires b to vary on the short $R^{-1/2}$ diffusive length scale; then a WKBJ-type approximation can be made. Within the framework of that approximation only the local behaviour of the fluid in the neighbourhood of a fluid particle need be considered. The local flow is linear and a solution of the type (3.9) is obtained in the leading order approximation. We remark that the other extreme of a steady motion which is everywhere a stretching flow, and consequently likely to be a very potent fast dynamo, has been investigated by Bayly (1986).

Linear flow problems have also been considered in the mean field context when an alpha effect has been included. In particular, Gvaramadze *et al.* (1988*a*, *b*) and Gvaramadze & Cheketiani (1989) have obtained solutions when the term $\nabla \times (\alpha B)$ is added to the right-hand side of the magnetic induction equation (1.2.39). For constant α, which they consider, our solution (3.4, 5) continues to hold. The equation (3.8) governing $\hat{b}$, however, is modified by the addition of the term $i\alpha k \times \hat{b}$ to its right-hand side. Consequently, the transform solution is now less easily obtained and not given simply by (3.9). It should be emphasised that these authors also include the effect of compressibility of the flow for which further minor modifications of the governing equations are necessary.

The mathematical apparatus set up here is also applicable to linear maps, particularly the cat map dynamo described by Bayly in section 10.4.1 below. In those problems the map adopted is generally instantaneous and a period of stasis follows during which time the magnetic field diffuses. Accordingly $|k(k_0, t)|$ in the integrand of (3.9) is simply the constant value taken by the modulus of the wave vector during the diffusive period (see, e.g., Gilbert 1993). For a series of maps the integral is replaced by a sum of values obtained after each sequential application of the map (see, e.g., Klapper 1994). It should be emphasised that the cat map is not a map in Euclidean space (Arnol'd *et al.* 1981) but is instead a linear map on a two-torus. Though not physically realisable, the abstraction leads to relatively simple treatments of important dynamo mechanisms. These issues will not be discused further here but the matter is taken up again in section 10.4.1, where the cat map is defined.

In the following sections explicit examples are given of some simple steady linear flows similar to those described by Soward (1988). That development is from a slightly different point of view and though less extensive is complementary, being conceptually simpler.

6.3.2 Pure Straining Motion

In the case of the two-dimensional stretching stagnation point flow

$$\mathbf{V} = T_\nu^{-1}(-x, y, 0) \qquad (6.3.10)$$

the Jacobian matrix is diagonal, $J_{ij} = 0$ $(i \neq j)$, and diagonal elements take the values

$$J_{11} = J^{-1}, \qquad J_{22} = J, \qquad J_{33} = 1, \qquad (J = \exp(t/T_\nu)). \qquad (6.3.11)$$

This flow provides the simplest example of the exponential growth of frozen magnetic field at a material point in a flow field.

Consider the unidirectional magnetic field

$$\mathbf{B} = (0, B(x, t), 0) \qquad (6.3.12)$$

aligned with the stretching y-direction for which the Liapunov exponent (2.16) is $\lambda(\mathbf{a}, \mathbf{1}_y) = T_\nu^{-1}$ for all $\mathbf{a}$. According to our results individual components $\hat{B}$ of its Fourier decomposition,

$$B(x, t) = \int \hat{B}(k, t) \exp(ikx) \, dk_0, \qquad (k = k_0 J), \qquad (6.3.13)$$

evolve according to

$$\hat{B}(k, t) = \hat{B}_0(k_0) J \exp[-\tfrac{1}{2}R^{-1}(J^2 - 1)], \qquad (6.3.14)$$

where R is the magnetic Reynolds number

$$R = (\eta k_0^2 T_\nu)^{-1} \qquad (6.3.15)$$

based on the initial wavelength $2\pi/k_0$. For large R, the result (3.14) shows that $\hat{B}$ grows exponentially until the local wavelength $2\pi/k$ is of the same order as the diffusive length scale

$$l_\eta = (2\eta T_\nu)^{1/2} = k_0^{-1}(2/R)^{1/2}. \qquad (6.3.16)$$

That occurs at time

$$t_{\max} = \tfrac{1}{2}T_\nu \ln R \qquad (6.3.17)$$

when the magnetic field is amplified considerably to the value

$$\hat{B}_{\max} = \hat{B}_0 R^{1/2}. \qquad (6.3.18)$$

Subsequently, $t \geq t_{max}$, as the local length scale continues to decrease super-exponential decay follows.

From (3.13, 14) the evolution of an arbitrary initial unidirectional magnetic field with total flux

$$\Phi = \int B_0(x)\, dx = 2\pi \widehat{B}_0(0) \tag{6.3.19}$$

is given by

$$B(x,t) = \int \widehat{B}_0(J^{-1}k) \exp[-\tfrac{1}{4}(kl_\eta)^2(1 - J^{-2}) + ikx]\, dk. \tag{6.3.20}$$

Solutions of this general type were derived by Clark (1964, 1965). As t increases, so does J, and the dominant contribution to B originates from increasingly longer initial length scales, $k_0^{-1} = O(J)$ (see (3.13, 14). Eventually as $t \to \infty$ (and $J \to \infty$) only the initial mean field remains. The total magnetic flux is preserved and is confined within the flux sheet

$$B_s(x) = (\Phi/\sqrt{\pi}l_\eta) \exp[-(x/l_\eta)^2]. \tag{6.3.21}$$

Of course, this steady state solution persists for all time if $B_0(x) = B_s(x)$.

6.3.3 Flux Ropes and Sheets
The most general form of a linear flow is

$$\mathbf{V} = \mathbf{D}\mathbf{x} + \boldsymbol{\Omega} \times \mathbf{x}. \tag{6.3.22}$$

Consider axes chosen parallel to the principal axes of the constant symmetric rate of strain tensor $\mathbf{D}$, with eigenvalues $\lambda_1(< 0)$, $\lambda_2(> 0)$ and λ_3. In general, this flow will not support permanent magnetic features in the form of flux ropes and sheets. One special case, however, deserves attention, namely that of constant vorticity $2\boldsymbol{\Omega}$ ($= 2\Omega \mathbf{1}_y$) parallel to the y-axis corresponding to the positive eigenvalue λ_2.

The important features of the kinematics take place in the plane perpendicular to $\boldsymbol{\Omega}$ and so we will denote vector components in that xz-plane by the subscript $\perp$. There the velocity is

$$\mathbf{V}_\perp = \mathbf{S}_\perp \mathbf{x}_\perp, \qquad \mathbf{S}_\perp = \begin{pmatrix} \lambda_1 & \Omega \\ -\Omega & \lambda_3 \end{pmatrix}, \tag{6.3.23}$$

where the velocity gradient matrix $\mathbf{S}_\perp$ satisfies

$$\mathbf{S}_\perp + \lambda_2 \mathbf{I}_\perp + \zeta\, \mathbf{S}_\perp^{-1} = 0, \qquad (\zeta = \lambda_1\lambda_3 + \Omega^2). \tag{6.3.24}$$

Consequently its eigenvalues s_1, s_3 are the roots of

$$s^2 + \lambda_2 s + \zeta = 0, \qquad (s_1 + s_3 = \lambda_1 + \lambda_3 = -\lambda_2). \tag{6.3.25}$$

The corresponding Jacobian matrix which solves (2.8) is

$$\mathbf{J}_\perp(t) = \dot{J}(t)\mathbf{I}_\perp - \zeta J(t)\mathbf{S}_\perp^{-1}, \qquad (6.3.26)$$

where the dot denotes the time derivative and

$$J(t) = [\exp(s_1 t) - \exp(s_3 t)]/(s_1 - s_3). \qquad (6.3.27)$$

Incompressibility implies that

$$\det \mathbf{J}_\perp = J_{22}^{-1} = \exp(-\lambda_2 t). \qquad (6.3.28)$$

Provided that

$$\det \mathbf{S}_\perp \equiv \zeta > 0, \qquad (6.3.29)$$

both $\Re(s_1)$ and $\Re(s_3)$ are negative and all fluid particles approach the rotation y-axis.

The flow described can support magnetic field

$$\mathbf{B} = (0, B(\mathbf{x}_\perp, t), 0). \qquad (6.3.30)$$

Use of the results (3.4–9) shows that the general solution can be expressed in the form

$$B(\mathbf{x}_\perp, t) = \int \hat{B}_0(\mathbf{J}_\perp^T \mathbf{k}_\perp) \exp[-\eta \mathbf{k}_\perp \cdot \mathbf{M}_\perp \mathbf{k}_\perp + i \mathbf{k}_\perp \cdot \mathbf{x}_\perp] \, d^2 \mathbf{k}_\perp, \qquad (6.3.31)$$

where again, as we defined in (3.4), $\hat{B}_0(\mathbf{k}_\perp)$ is the Fourier transform of the initial field ($\mathbf{J}_\perp(0) = \mathbf{I}_\perp$, $\mathbf{M}_\perp(0) = 0$) and we have noted that $J_{22} \, d^2 \mathbf{k}_{\perp 0} = (\det \mathbf{J}_\perp)^{-1} d^2 \mathbf{k}_{\perp 0} = d^2 \mathbf{k}_\perp$. The integral representation of $\mathbf{M}_\perp$ emerging from (3.8) and (3.9) is simplified using the transition property $\mathbf{J}_\perp(t) = \mathbf{J}_\perp(\tau)\mathbf{J}_\perp(t-\tau)$. Specifically $\mathbf{J}_\perp^{-1}(-t) = \mathbf{J}_\perp(t)$ enables us to write

$$\mathbf{M}_\perp(t) = \int_0^t \mathbf{\Lambda}_\perp^{-1}(-\tau) \, d\tau, \qquad (\mathbf{\Lambda}_\perp^{-1}(-t) = \mathbf{J}_\perp(t)\mathbf{J}_\perp^T(t)). \qquad (6.3.32)$$

For converging flows, $\zeta > 0$, $J(t)$ tends to zero as t tends to infinity. Consequently the asymptotic behaviour of the magnetic field (3.31) is determined by the limiting values

$$\mathbf{J}_\perp(\infty) = 0, \qquad \mathbf{M}_\perp(\infty) = \mathbf{M}_\perp, \qquad (6.3.33)$$

where

$$\mathbf{M}_\perp = \tfrac{1}{2}\lambda_2^{-1}[\mathbf{I}_\perp + \zeta(\mathbf{S}_\perp^T \mathbf{S}_\perp)^{-1}] \qquad (6.3.34)$$

is the symmetric matrix which solves $\mathbf{M}_\perp^T \mathbf{S}_\perp^T + \mathbf{S}_\perp \mathbf{M}_\perp = -\mathbf{I}_\perp$ and can be expressed in the form

$$\mathbf{M}_\perp = \frac{1}{2\lambda_2 \zeta} \begin{pmatrix} \lambda_3 & \Omega \\ \Omega & -\lambda_1 \end{pmatrix} \begin{pmatrix} -\lambda_2 & 2\Omega \\ 2\Omega & \lambda_2 \end{pmatrix}. \qquad (6.3.35)$$

To construct the explicit form of the solution defined by (3.31, 33, 35), we rotate axes through an angle θ about the y-axis in a positive sense and set

$$
\begin{aligned}
\mathbf{X}_\perp &= \mathbf{R}_\perp \mathbf{x}_\perp, \\
\mathbf{K}_\perp &= \mathbf{R}_\perp \mathbf{k}_\perp,
\end{aligned}
\qquad
\mathbf{R}_\perp = \begin{pmatrix} \cos\theta & -\sin\theta \\ \sin\theta & \cos\theta \end{pmatrix}.
\tag{6.3.36}
$$

The new X- and Z-axes are aligned with the principal axes of the symmetric matrix $\mathbf{S}_\perp^T \mathbf{S}_\perp$, when

$$
\tan 2\theta = 2\Omega/\lambda_2.
\tag{6.3.37}
$$

Relative to these new axes $\mathbf{S}_\perp^T \mathbf{S}_\perp$, $\boldsymbol{\Lambda}_\perp$ and $\mathbf{M}_\perp$ are all diagonal and $\mathbf{M}_\perp$ in particular is given by

$$
\mathbf{R}_\perp \mathbf{M}_\perp \mathbf{R}_\perp^T = \frac{1}{4\eta} \begin{pmatrix} l_1^2 & 0 \\ 0 & l_3^2 \end{pmatrix},
\tag{6.3.38}
$$

where

$$
l_1 = (2\eta)^{1/2}[-\lambda_3 \sin^2\theta - \lambda_1 \cos^2\theta]^{-1/2},
\tag{6.3.39}
$$

$$
l_3 = (2\eta)^{1/2}[-\lambda_1 \sin^2\theta - \lambda_3 \cos^2\theta]^{-1/2}.
\tag{6.3.40}
$$

The evaluation of (3.31) is now simple and yields the flux rope solution

$$
B = \frac{\Phi}{S} \exp\left\{ -\left(\frac{X}{l_1}\right)^2 - \left(\frac{Z}{l_3}\right)^2 \right\}.
\tag{6.3.41}
$$

Here

$$
\Phi = \int B \, d^2\mathbf{x}_\perp = (2\pi)^2 \hat{B}_0(0)
\tag{6.3.42}
$$

is the total magnetic flux which remains constant for all time, and

$$
S = \pi l_1 l_3 = 2\pi\eta\zeta^{-1/2} \cos 2\theta
\tag{6.3.43}
$$

is the area of the ellipse with semi major axes given by the diffusion lengths l_1 and l_3. Note that (3.41) is a steady solution of the magnetic induction equation for all values of ζ. When $\zeta < 0$ the area S defined by (3.43) does not exist implying that l_3^2 (say) is negative. It means that the solution (3.41) is unbounded at infinity and is not the solution of our initial value problem with

$$
B(\mathbf{x}_\perp, t) \to 0 \qquad \text{as} \qquad |\mathbf{x}_\perp| \to \infty.
\tag{6.3.44}
$$

Though unbounded at infinity, the special case $\zeta = 0$ with $l_3 = \infty$ and $l_1 = (2\eta/\lambda_2)^{1/2}$ has physical significance. It corresponds to a y-directed flux sheet in the plane $X = 0$ and is characterised by

$$
\Omega^2 = -\lambda_1 \lambda_3, \qquad (\lambda_3 > 0).
\tag{6.3.45}
$$

where $\tan\theta = \pm\sqrt{-\lambda_3/\lambda_1}$ for $\pm\Omega > 0$. Our earlier sheet solution (3.21) corresponds to the special case $\Omega = \lambda_3 = 0$.

There are several features of the solution worth noting. Firstly, when there is no rotation $\Omega = 0$, we have $\mathbf{M}_\perp = -\frac{1}{2}\mathbf{D}^{-1}$ and the principal axes of the rope coincide with those of $\mathbf{D}$ ($\theta = 0$). This standard solution, which occurs when both λ_1 and λ_3 are negative, is given, for example, by Moffatt (1978). For the axisymmetric case, $\lambda_1 = \lambda_3$, Galloway & Zheligovsky (1993) report additional unsteady non-axisymmetric flux rope solutions. Secondly with rotation ($\Omega \neq 0$) and λ_1 negative, the rope can also exist when λ_3 is positive provided that $\lambda_3 < -\Omega^2/\lambda_1$. This condition, $\zeta > 0$, ensures that all fluid particles approach the rotation axis. This latter case is of particular interest because numerical experiments simulating hydrodynamic turbulence (Vincent & Meneguzzi 1991) indicate that vortex filaments aligned roughly in the y-direction form in this environment with only weak stretching in that direction ($|\lambda_1| \doteq |\lambda_3| \gg \lambda_2 > 0$). Some recent work on MHD turbulence (see Brandenburg, chapter 4) also emphasises the importance of this configuration. Finally we note that the limiting case $\zeta = 0$ with

$$\lambda_3 \doteq -\lambda_1 \doteq \Omega \gg \lambda_2 \qquad (6.3.46)$$

corresponds to a flux sheet aligned to the direction $(1,0,1)$ of the almost linear shear

$$\mathbf{V} \doteq \Omega(z - x)(1, 0, 1). \qquad (6.3.47)$$

By itself this unidirectional flow cannot support a flux sheet. Nevertheless the weak inflow due to the expanding y-direction ($\lambda_2 > 0$) is just capable of confining it. Of course, in an evolving flow, for which λ_1, λ_3 and Ω change slowly with time, flux ropes can be supported as long as $\zeta > 0$. As $-\lambda_1\lambda_3$ increases through the value Ω^2, the rope first expands to form a sheet and then ceases to be confined being torn apart by the advection.

Since our earlier result (3.14) shows that fluctuating fields are rapidly destroyed even in highly conducting fluids, we cannot over-emphasise the importance of flux ropes and sheets. In isolation they are robust features which preserve their strength (see (3.42)). Indeed, even when magnetic diffusion is ignored as it is in some fast dynamo models, magnetic flux across a finite area is likely to provide the most reliable measure of fast dynamo activity, in as much as the result is insensitive to the addition of small magnetic diffusion.

6.3.4 Linear Shear

In the previous two sections we have discussed examples of stretching flows. The linear shear introduced in (3.47) by itself is a non-stretching flow and typifies a wide class of simple flows. A convenient representation for the velocity is

$$\mathbf{V} = T_\nu^{-1}(y, 0, 0), \qquad (6.3.48)$$

for which the diagonal elements of the Jabobian matrix are unity, while all the off-diagonal elements are zero except for J_{12} ($= t/T_\nu$). For this flow the Liapunov exponent $\lambda(\mathbf{a}, \mathbf{e})$ is zero for all $\mathbf{a}$, $\mathbf{e}$.

As in section 6.3.2, we consider the effect of the flow (3.48) on a Fourier component of magnetic field initially aligned normal to the flow in the y-direction, namely

$$\mathbf{B}_0(x) = [0, \, \hat{B}_0 \sin(k_0 x), \, 0]. \tag{6.3.49}$$

Moffatt & Kamkar (1983) showed that subsequently the field is given by

$$\mathbf{B}(\mathbf{x}, t) = \hat{B}(t) \sin(\mathbf{k} \cdot \mathbf{x})[t/T_\nu, \, 1, \, 0], \tag{6.3.50}$$

where

$$\mathbf{k} = k_0(1, -t/T_\nu, 0) \tag{6.3.51}$$

and

$$\hat{B}(t) = \hat{B}_0 \exp[-R^{-1}\{(t/T_\nu) + \tfrac{1}{3}(t/T_\nu)^3\}], \tag{6.3.52}$$

in which the magnetic Reynolds number R is still defined by (3.15). For large R, the magnetic field only grows linearly during the frozen field stage. That growth ends at time

$$t_{\max} = O(T_\nu R^{1/3}), \tag{6.3.53}$$

when the magnetic field has been amplified considerably to the value

$$B_{\max} = O(\hat{B}_0 R^{1/3}). \tag{6.3.54}$$

Subsequently the field decays rapidly.

The importance of the above result is that it explains the mechanism of flux expulsion in regions of steady flow following closed streamlines, like the case of magnetic flux expelled by a closed eddy investigated numerically by Weiss (1966). Certainly the mechanism is central to many slow dynamo models. Indeed numerical simulations of slow dynamos, according to Dr D.J. Galloway (personal communication), are also slow to settle down to their asymptotic states. He argues that this is probably due in part to the long time of order $T_\nu R^{1/3}$ for flux to be expelled. Our earlier result (3.17) suggests that fields adjust on the much shorter time of order $T_\nu \ln R$ in stretching flows.

6.4 SLOW DYNAMOS

6.4.1 Nearly Aligned Fields
Though Vishik (1988) has shown that non-stretching flows are never fast dynamos, it is of interest to know how such dynamos operate because it provides insight into large-R dynamo mechanisms. The steady linear shear discussed in the previous section is the prototype model of such a flow. The

most important feature illustrated by the result (3.50) is the tendency of magnetic field to align itself with the flow. The mechanism, by which aligned field is created and possibly amplified from the transverse magnetic field, is the well known ω-effect of many astrophysical dynamos. There are two possibilities. Firstly, if the transverse field is spatially fluctuating the aligned field created is eventually destroyed on the flux expulsion time scale $T_V R^{1/3}$. Secondly, if the transverse field has a mean part, then a strong aligned field is created, which like flux ropes, may be robust on relatively long time scales. It is this second case which interests us here. In particular, our main concern is the mechanism by which the transverse magnetic field is maintained. In all cases, described below, an α-effect can be isolated leading to dynamos of $\alpha\omega$-type.

Suppose that the dominant contributions to the magnetic field and flow are aligned

$$\mathbf{V} = M\mathbf{h}, \qquad \mathbf{B} = K\mathbf{h}, \qquad (6.4.1,2)$$

where $\mathbf{h}$ is solenoidal ($\nabla \cdot \mathbf{h} = 0$) and M and K are scalars constant on stream/field lines (($\mathbf{h} \cdot \nabla)M = (\mathbf{h} \cdot \nabla)K = 0$). The objective of higher order theory (Soward 1990) is the determination of an evolution equation for the field amplitude K. The key idea is that when the magnetic field (4.2) diffuses, the associated electric current, $\mathbf{J} = \mu^{-1}\nabla \times \mathbf{B}$, produces an emf with a component parallel to $\mathbf{B}$ of magnitude $-\eta\lambda\mathbf{B}$. Here

$$\lambda = \frac{\mathbf{h} \cdot \nabla \times \mathbf{h}}{|\mathbf{h}|^2} = \frac{\mathbf{V} \cdot \nabla \times \mathbf{V}}{|\mathbf{V}|^2} = \frac{\mathbf{B} \cdot \nabla \times \mathbf{B}}{|\mathbf{B}|^2} \qquad (6.4.3)$$

provides a measure of the twist of the magnetic field lines and is linked to the magnetic helicity, $H = \lambda|\mathbf{B}|^2$ (see (2.34)) and also to the helicity of the flow (1.4.4). This component of the emf produces an α-effect of magnitude $-\eta\lambda$, which is responsible for generation of smaller transverse magnetic field. Since this α-effect is proportional to η, any dynamo which results is necessarily slow.

6.4.2 Helical Dynamos

The simplest example of nearly aligned slow dynamos is provided by the screw dynamo investigated by Ruzmaikin et al. (1988) and Gilbert (1988). The model has astrophysical applications to the jets of active galaxies and in that context Reshetnyak et al. (1991) and Shukurov & Sokoloff (1993) have considered the evolution of a localised magnetic dynamo wave packet. Relative to cylindrical polar coordinates (s, ϕ, z) they considered the unbounded axisymmetric flow

$$\mathbf{V} = s\omega\mathbf{1}_\phi + (U - c)\mathbf{1}_z. \qquad (6.4.4)$$

Unlike the Ponomarenko (1973) dynamo described in section 1.4.1, ω and U are continuous functions of s, rather than piecewise constant. Since dynamo models are generally propagating waves, we have added the constant axial velocity c in the definition (4.4), chosen so that a particular mode is brought to rest in the moving frame. By this device the dominant magnetic field aligns itself with the flow.

Following Lortz (1968), we introduce helical (s, ξ)-coordinates, where

$$\xi = m\phi + kz, \qquad (6.4.5)$$

and define h in (4.2) by

$$h = q(-ks1_\phi + m1_z), \qquad (q = |h|^2). \qquad (6.4.6)$$

It has the Beltrami property

$$\nabla \times h = \lambda h, \qquad (\lambda = -2mkq). \qquad (6.4.7)$$

Lortz (1968) noted that dynamo solutions were possible, which are functions of s and ξ alone. For them the velocity and magnetic field take the form

$$V = Mh + \nabla\psi \times h, \qquad B = Kh + \nabla\chi \times h. \qquad (6.4.8,9)$$

For our particular choice of velocity (4.4) M and ψ are functions of s alone.

Now unless the pitch $(U - c)/\omega$ of the streamlines (4.4) is constant independent of s, the vector h is only aligned to V on isolated critical surfaces $s = s_0$ characterised by $V \times h = 0$, namely

$$m\omega_0 + k(U_0 - c) = 0, \qquad (6.4.10)$$

where the subscript zero denotes the critical values. There the secondary flow $(\nabla\psi \times h)_0$ vanishes. Off these surfaces, the secondary flow $\nabla\psi \times h$ transverse to h shuffles the aligned field. By that we mean that, if K has fluctuating dependence on ξ, the transverse shear reduces the radial s length scale leading to enhanced ohmic dissipation. Essentially, this is the flux expulsion mechanism described in section 6.3.4. To minimise this effect the dynamo mode selects a surface on which the radial derivative of the pitch vanishes

$$m\omega_0' + kU_0' = 0, \qquad (\lambda_0 = 2U_0'\omega_0'/(U_0'' + s_0^2\omega_0'')), \qquad (6.4.11)$$

where the prime denotes the radial derivative. For fixed m and k (4.11) determines the location of the critical surface, while in turn (4.10) fixes the wave speed c. It is the departure of the pitch from its $-m/k$ value at order $(s - s_0)^2$, together with ohmic dissipation, which localises the dynamo. The resulting $\alpha\omega$-dynamo equations are

$$D\chi/Dt = -\eta\lambda_0 K + \eta\mathcal{L}(\chi) \qquad (6.4.12)$$

$$DK/Dt = (s_0^2 q_0)^{-1/2}(U_0'^2 + s_0^2 \omega_0'^2)^{1/2} \partial \chi / \partial \xi + \eta \mathcal{L}(K), \qquad (6.4.13)$$

where

$$\mathcal{L} \equiv \partial^2/\partial s^2 + (s_0^2 q_0)^{-1} \partial^2/\partial \xi^2, \qquad (6.4.14)$$

$$D/Dt = \partial/\partial t + (m\omega_0'' + kU_0'')\tfrac{1}{2}(s - s_0)^2 \partial/\partial \xi \qquad (6.4.15)$$

(see Soward 1990). The small transverse field $\nabla \chi \times h$ is due to the action of the α-effect, $-\eta \lambda_0$, as explained in section 6.4.1. Maintenance of the strong aligned field is due to the shear $(0, s_0\omega_0', U_0')$ acting on that weak transverse field as quantified by (4.13).

Our choice of magnitude for m and k is arranged so that the ξ length scale is of order unity. When $s_0\omega_0' = O(U_0')$ so that $s_0\lambda_0$ is of order unity (see (4.11)), the magnetic field for the fastest growing mode spirals about the h-lines on the small transverse length scale $s_0 q_0^{1/2}$, where

$$q_0^{1/2} = (k^2 s_0^2 + m^2)^{-1/2} = O(R^{-1/3}). \qquad (6.4.16)$$

Here the magnetic Reynolds number is

$$R = s_0^2(U_0'^2 + s_0^2 \omega_0'^2)^{1/2}/\eta. \qquad (6.4.17)$$

The ratio of the transverse to aligned field is of order $R^{-1/3}$. The mode will grow relatively quickly on the time scale $R^{-2/3}(s_0^2/\eta)$ provided the meridional shear characterised by $m\omega_0'' + kU_0''$ in (4.15) is not too large. This destructive flux expulsion mechanism decreases the growth rate and when big enough causes the mode to decay. A more careful analysis shows that the crucial parameter is the dynamo number $D = s_0\lambda_0 R$ (see (1.4.53)) rather than R. Reassuringly D vanishes in the two-dimensional limits $\omega_0' = 0$ and $U_0' = 0$ (see (4.11)).

The Ponomarenko (1973) dynamo described in section 1.4.1 is a degenerate case which requires a separate analysis. Gilbert (1988) showed that the mode is similar to that described above, except that it is concentrated on the shorter $s_0 R^{-1/2}$ diffusive length scale containing the shear discontinuity and grows on the fast $R^{-1}(s_0^2/\eta)$ convective time scale. The singular feature, which leads to fast growth, is the fact that, however thin the dynamo layer becomes, it always contains all the shear. Smoothing out the shear into a region of finite width necessarily leads to a slow dynamo.

6.4.3 Hybrid Eulerian–Lagrangian Approach

A refinement of the reference field approach of section 6.2.2 proves useful in the case of almost aligned fields. As we have already seen in our formulation (4.1, 2) for the dominant field, we do not expect much variation along the field line. In that spirit we wish only to identify material lines and not fluid particles. From this point of view, we are interested in a transformation

$$\mathbf{x} = \mathbf{x}(\mathbf{a}, t) \qquad (6.4.18)$$

which under the frozen field representation takes the reference field line to the actual field line position. Of course, $\mathbf{x}$ could be the actual new position of a fluid particle but, in general, it is more convenient to identify a point on the field line close to the original position, as we do for the Braginsky (1964) model (see section 1.4.3). Since $(\partial\mathbf{x}/\partial t)_a$ is no longer equal to $\mathbf{V}$, we define their difference to be

$$\mathbf{V} - (\partial\mathbf{x}/\partial t)_a = \mathbf{J}v \qquad (6.4.19)$$

by analogy with the frozen field transformation (2.10) for $\mathbf{B}$. With this choice, v defines the advective velocity in the a-frame in the sense that

$$D/Dt \equiv (\partial/\partial t)_{\mathbf{x}} + \mathbf{V}\cdot\nabla_{\mathbf{x}} = (\partial/\partial t)_a + v\cdot\nabla_a. \qquad (6.4.20)$$

Under this generalisation some of the earlier results of section 6.2 must be modified. In particular (2.8) becomes

$$D\mathbf{J}/Dt = \mathbf{S}\mathbf{J} - \mathbf{J}\mathcal{S}, \qquad (\mathcal{S}_{ij} = \partial v_i/\partial a_j). \qquad (6.4.21)$$

Accordingly for frozen fields ($\eta = 0$), which satisfy (2.23), it follows immediately that the reference potential (2.25) obeys

$$D\mathcal{A}/Dt = -\mathcal{S}^{\mathrm{T}}\mathcal{A}. \qquad (6.4.22)$$

Armed with this refinement of (2.23), it is easy to make the appropriate modifications to the diffusive equation (2.29) and so obtain

$$(\partial\mathbf{b}/\partial t)_a = \nabla_a \times (v \times \mathbf{b}) + \eta\,\nabla_a \times \mathcal{E}_a. \qquad (6.4.23)$$

It reduces to the conventional form (1.2.34) of the magnetic induction equation, when $\mathbf{x} = \mathbf{a}$.

Soward (1972) pointed out that the above formulation is ideally suited to the Braginsky (1964) dynamo described in section 1.4.3. For simplicity we restrict attention to steady flows, for which $(\partial\mathbf{x}/\partial t)_a = 0$, but note that the theory has more general applicability. Within the framework of the Braginsky scaling (1.4.46) a velocity $\overline{\mathbf{V}}$ can be identified (see below (1.4.52)), which has closed streamlines, C. Those streamlines are close to azimuthal circles, $(s, z) =$ const., and are a distance at most of order $R^{-1/2}\mathcal{L}$ from them. Accordingly, the circles in the a-frame map into the real x-frame by the transformation

$$\mathbf{x} = \mathbf{a} + \mathrm{O}(R^{-1/2}\mathcal{L}). \qquad (6.4.24)$$

By construction the velocity v in the a-frame is almost azimuthal with a small meridional part

$$v = v_\phi \mathbf{1}_\phi + R^{-1}v_M \qquad (6.4.25)$$

so that

$$\widetilde{\mathbf{V}} = (\mathbf{J}\mathbf{1}_\phi)v_\phi, \qquad \overline{\mathbf{V}}_M = \mathbf{J}\mathbf{v}_M, \qquad (6.4.26)$$

where $R^{-1}\overline{\mathbf{V}}_M$ is the effective meridional velocity introduced below (1.4.47). Indeed, it is in this a-frame that the Braginsky equations (1.4.48, 49) should be interpreted. His magnetic field is the azimuthal average of b and not B. It means that, in real space, his averages are taken about the distorted contours C. When we consider the value of $\mathcal{E}_a$ in cylindrical polar coordinates some adjustments need to be made to (2.31). Nevertheless, all changes introduced by the transformation in (4.24) are negligible except for the α-effect

$$\alpha = -\eta(2\pi)^{-1} \int_{-\pi}^{\pi} \lambda_{\phi\phi} \, d\phi = \mathrm{O}(R^{-1}\eta/\mathcal{L}), \qquad (6.4.27)$$

whose explicit value is given by (1.4.52). Here the value of $\lambda_{\phi\phi}$ is closely linked to λ defined by (4.3) with $\mathbf{V}$ replaced by $\widetilde{\mathbf{V}}$. Braginsky's (1964) choice of scaling leads to a slow dynamo, which evolves on the diffusion time.

6.5 ROPE DYNAMOS

In section 6.3 we identified magnetic flux in ropes as robust features as the magnetic diffusivity η tends to zero. It is, therefore, not surprising that early models illustrating how a fast dynamo might operate were developed manipulating magnetic flux tubes assumed frozen to the fluid ($\eta = 0$). Refining an earlier idea of Alfvén (1950), Vainshtein & Zel'dovich (1972) considered a closed tube of magnetic flux $\Phi_0 = \int_S \mathbf{B}\cdot\mathbf{n}\,dS$, where S is the tube cross section. By stretching the tube to twice its length, the field strength is doubled and the cross section halved. The tube is then twisted to form a figure of eight and folded to merge with itself. By this device, the total cross section, which contains two cross sections of the original rope, is restored to its original area, but now contains double the original flux. After n repetitions of this process each taking time T_V the total flux becomes

$$\Phi_n = 2^n \Phi_0 \qquad (6.5.1)$$

and leads to the flux growth rate

$$\Gamma = (nT_V)^{-1} \ln(\Phi_n/\Phi_0) = T_V^{-1} \ln 2. \qquad (6.5.2)$$

This is also the growth rate of material lines, which may be identified with the topological entropy Γ_E. Of course, small structures develop, which eventually diffuse away, but the basic field doubling is unaffected. The underlying stretch-twist-fold mechanism (STF-map), upon which the rope dynamo depends, is central to the notion of a fast dynamo.

Variations of this theme were proposed by Finn & Ott (1988) and we will give details of one of them. It depends in part on a non-dynamo mechanism, which we will call the SPF-map. If differs from the STF-map in as much as there is no twist. Instead the two halves of the loop are pinched off (P) and folded. Except near the pinching, where there is some field line complexity, a double loop structure is again formed but flux is directed in opposite directions in the two loops. The unsigned flux $\int_S |\mathbf{B} \cdot \mathbf{n}| \, dS$ again doubles its strength so that the topological entropy Γ_E is again given by (5.2). On the other hand, the total flux now vanishes and continues to vanish under subsequent applications of the map. The process is analogous to the baker's non-dynamo discussed by Bayly in section 10.4.2. Finn & Ott's (1988) idea is essentially to apply the following sequence. First, the STF-map is applied to the loop. Second, of the two loops formed, the STF-map is applied to one, while the SPF-map is applied to the other. At the end of this two stage operation taking total time $T_\mathcal{V}$, the unsigned flux is increased four fold but the total signed flux is only doubled. Under repeated application of the map, the flux growth rate continues to be given by (5.2) but the topological entropy becomes

$$\Gamma_E = T_\mathcal{V}^{-1} \ln 4. \tag{6.5.3}$$

A useful measure of the relation of the signed to unsigned flux is the cancellation exponent introduced by Ott *et al.* (1992). This simple model illustrates the fact that there is no uniquely defined growth rate or for that matter an eigenfunction. This contrasts dramatically with their existence for dissipative problems ($\eta \neq 0$) (see (1.3.16)). Another model of Finn & Ott's (1988) involving maps with uneven stretching illustrates the intermittent and fractal structure of the evolving field. For the baker's map generalisation, see Bayly, section 10.4.3 below.

Let us consider the possible effect of adding ohmic diffusion to our two stage dynamo model. Suppose that the initial length scale is $\mathcal{L}$ and that after the double map the length scale is reduced by a factor $1/2$ corresponding to an area reduction of $1/4$ so that after the n-th map the length scale is

$$\delta(n) = 2^{-n}\mathcal{L}. \tag{6.5.4}$$

Let us assume that after the length scale reaches the diffusive length scale $\mathcal{L}R^{-1/2}$ any finer scale features are averaged on that length scale. This diffusive cut-off is reached after N complete double maps, where

$$N = \tfrac{1}{2}(\ln R)/(\ln 2). \tag{6.5.5}$$

After that time $NT_\mathcal{V}$, the field continues to grow at the flux growth rate Γ defined by (5.2), but the structure remains the same. In other words, an eigenfunction has emerged for the diffusive problem with a well defined growth rate $\gamma = \Gamma$. The transition from frozen field behaviour to the final diffusion controlled state occurs when the magnetic flux is $\Phi_N = R^{1/2}\Phi_0$ whereas the unsigned flux is $R\Phi_0$. Since both now evolve at the same growth rate, the unsigned flux remains larger by a factor $R^{1/2}$. This means that most of the energy resides in the fluctuating field. This model is very simplistic and does not address matters like intermittency and filling factors (see chapter 4), which perhaps pertain more to the other Finn & Ott (1988) model involving uneven stretching. Nevertheless the model does isolate features that fast dynamo models need to address and the emergence of a single growth rate is confirmed by Finn & Ott's (1990) numerical results. Accordingly the flux growth rate, correctly measured, in the frozen field stage may provide a robust measure of the growth rate for the complete diffusive problem in the limit $\eta \downarrow 0$. Indeed, in some average or weak sense, an eigenfunction can be found (cf. the field from the above map after N iterates), but not one which is convergent pointwise (see Soward 1993a, b)!

6.6 SPATIALLY PERIODIC DYNAMOS

6.6.1 ABC Flows
The importance of exponential stretching of field lines by straining motion was identified by Arnol'd et al. (1981), though comparable ideas were developed earlier by Arnol'd (1972) in the context of the hydrodynamic stability of ideal fluid flow. The study of Arnol'd et al. (1981) was soon followed up by Zel'dovich et al. (1984). For perfectly conducting fluids ($\eta = 0$), the Cauchy solution (2.10) ensures that there is exponential growth of the magnetic field in stretching flows with positive Liapunov exponent (2.16). The fast dynamo problem is different and relates to the limit $\eta \downarrow 0$. Here the main concern is linked to inevitable shrinking of the magnetic field length scale, which we mentioned in the previous section. Indeed, Bayly & Childress (1988) have demonstrated that it is possible to construct two-dimensional unsteady stretching flows which amplify the magnetic energy exponentially, even when $\eta = 0$. Nevertheless, when $\eta \neq 0$, Cowling's theorem ensures that the resulting two-dimensional magnetic field necessarily decays as $t \to \infty$. The failure of this dynamo is due to the cancellation of oppositely directed fields on the short diffusive length scale (cf. the SPF-map). For fast dynamo action, on the other hand, we require our motion to exhibit some of the features isolated by the STF-map.

The stretching flows that we discuss below are all spatially periodic on the length $2\pi\mathcal{L}$ parallel to each of our coordinate axes. We use $\mathcal{L}$ and T_V as our unit of length and time so that η becomes the inverse magnetic Reynolds number R^{-1}. Generally our flows are built from ±Beltrami waves of the type

$$\mathbf{V} = (\sin z, \pm \cos z, 0), \qquad (\nabla \times \mathbf{V} = \pm\mathbf{V}). \qquad (6.6.1)$$

The superposition of three mutually orthogonal +Beltrami waves gives the ABC-flow (1.4.7). In the case of two-dimensional motion ($A = 0$) the horizontal xy-components can be expressed in terms of a stream function in the form,

$$\mathbf{V} = (\partial_y \psi, \ -\partial_x \psi, \ K\psi) \tag{6.6.2}$$

with

$$\psi = C \sin y + B \cos x \tag{6.6.3}$$

and $K = 1$. Closely related flows are obtained for $K \neq 1$ and alternative choices of ψ. When the flow is steady, motion lies on the stream surfaces $\psi = $ const., and is generally not a stretching flow. Indeed, exponential stretching only occurs at the isolated stagnation points of the horizontal flow. In section 6.6.2, dynamos based on such flows are discussed. They are not quite fast, but see (6.17) below. If, on the other hand, the flow is time dependent ($B = B(t)$, $C = C(t)$) regions of stretching flow form, where the particle paths are chaotic. These, generally, appear to give fast dynamos and further details are explained in section 6.6.3. Finally at the end of that section, we discuss the steady three-dimensional case in which A, B and C are all non-zero. Again fast dynamo action appears likely in the finite regions of stretching flow which occur.

6.6.2 Slow Dynamos in Steady Two-dimensional Flows

A generalised α-effect.   The main advantage of two-dimensional steady flows is that separable solutions of the magnetic induction equation can be sought of the form

$$\mathbf{B} = \Re\{\tilde{\mathbf{B}}(x,y) \exp[ik(z - ct) + \gamma t]\}, \tag{6.6.4}$$

in which the wave number k, phase speed c and growth rate γ are all real. We may define horizontal and z-averages over the periodicity lengths by

$$\overline{\mathbf{B}} = (2\pi)^{-2} \int_{-\pi}^{\pi} \int_{-\pi}^{\pi} \mathbf{B} \, dx dy, \qquad <\!\mathbf{B}\!> = (k/2\pi) \int_{-\pi/k}^{\pi/k} \mathbf{B} \, dz. \tag{6.6.5}$$

The horizontal average of (6.4) has only horizontal components, $\overline{\mathbf{B}} = \overline{\mathbf{B}}_{\mathrm{H}}$. From it we may define the mean helicity of the horizontally averaged magnetic field by

$$H = <\!\overline{\mathbf{B}}_{\mathrm{H}} \cdot \nabla \times \overline{\mathbf{B}}_{\mathrm{H}}\!>. \tag{6.6.6}$$

The horizontal average of the magnetic induction equation for the particular mode (6.4) yields

$$[(\gamma + R^{-1}k^2) - ick]\overline{\overline{\mathbf{B}}} = ik\mathbf{1}_z \times \overline{\tilde{\mathcal{E}}}_{\mathrm{H}}. \tag{6.6.7}$$

From this we may deduce that the horizontal average of the emf $\mathcal{E}$ ($= \mathbf{V} \times \mathbf{B}$) is given by

$$\overline{\mathcal{E}}_H = \alpha \overline{\mathbf{B}}_H, \qquad (6.6.8)$$

where $\boldsymbol{\alpha}$ is the 2×2 α-tensor with the unique real representation

$$\alpha_{ij} = 2(\gamma + R^{-1}k^2)H^{-1} < |\overline{\mathbf{B}}_H|^2 \delta_{ij} - \overline{B}_i \overline{B}_j > - c\epsilon_{3ij}. \qquad (6.6.9)$$

This unusual form of the α-tensor is dependent on the explicit form of the mode (6.4). In general, it is only defined once the complete solution is already known and different modes define different $\boldsymbol{\alpha}$. For this reason, we call it a generalised α-effect with the functional dependence

$$\boldsymbol{\alpha} = \boldsymbol{\alpha}(k; R). \qquad (6.6.10)$$

This is still oversimplistic because, even with k and R specified, there is a spectrum of modes. If, however, we restrict attention to the fastest growing mode the functional form (6.10) is generally unambiguous.

For long wave length modes with small k, (6.10) may be approximated by its Taylor series expansion

$$\alpha(k) = \alpha(0) + k(\partial\alpha/\partial k)(0) + \dots . \qquad (6.6.11)$$

The first term is the usual α-effect and the second is linked to the concept of eddy diffusivity. For the conventional α-effect to be a useful concept, the Taylor series (6.11) must be rapidly convergent. However, for the large values of k, which give the modes with fastest growth rates, (6.11) is inappropriate. It should be added that, though (6.9) has a complicated tensorial structure, for the symmetric flows, which we consider below, $\boldsymbol{\alpha}$ is diagonal with $c = 0$.

The G.O. Roberts dynamo (1972). He considered the two-dimensional ABC flow (6.2) ($K = 1$) with $B = C$ in (6.3). After a rotation of 45° about the z-axis, a change of origin and a change of length scale the flow can be cast again in the form (6.2) but now with

$$\psi = \sin x \sin y, \qquad K = \sqrt{2}. \qquad (6.6.12)$$

The character of this flow was discussed below (1.4.6). It suffices to say that spiralling vortices are bounded by the planes $x = m\pi$ and $y = n\pi$ (m, n integers). Within the vortices the flow is non-stretching and there, like in the case of the Ponomarenko dynamo (section 6.4.2), only slow dynamo action can occur (see Soward 1990). Any rapid field generation is, therefore, linked to the bounding planes $\psi = 0$, which connect the stagnation points with exponential stretching. The symmetries of the flow are such that $\overline{\mathbf{B}}_H$ is a Beltrami wave of the type (6.1), for which

$$\overline{\overline{\mathbf{B}}}_H = (1, -i), \qquad (c = 0). \qquad (6.6.13)$$

Accordingly the α-tensor (6.9) is isotropic giving

$$\alpha = \alpha \mathbf{I} \qquad (H = -k), \qquad (6.6.14)$$

where the scalar α satisfies

$$\gamma + R^{-1}k^2 = -\alpha k. \qquad (6.6.15)$$

To understand the dynamo mechanism itself, consider the effect of the flow (6.12) on the uniform magnetic field $\overline{\mathbf{B}}_H = \mathbf{1}_x$. On the one hand, the eddies expel the flux and concentrate it on sheets $y = n\pi$ ($n = $ integer) (see section 6.3.4). On the other hand, at the stagnation points, where motion diverges from the planes, tongues of magnetic flux are plucked out on the mutually orthogonal planes $x = m\pi$ ($m = $ integer). Both the sheet and tongue boundary layers have width of order $R^{-1/2}$. If, instead of being uniform, the applied field oscillates with z taking the form $\overline{\mathbf{B}}_H = \mathbf{1}_x \cos kz$, the basic flux sheet may be regarded as stacked flux ropes of alternating direction. Then in the spirit of the rope dynamo (section 6.5), the plucked tongues constitute the stretch mechanism, folded at the ends of the tongues. Now the upwelling and downwelling in neighbouring vortices reorganises the flux in the tongues to help reinforce the y-directed field $\overline{\mathbf{B}}_H = \mathbf{1}_y \sin kz$, which completes the description of our dynamo generated mean field $\overline{\mathbf{B}}_H$ defined by (6.4) and (6.13). Essentially, the spatial periodicity of the system enables the twist operation in the rope dynamo to be replaced by a simple shear in the z-direction with $\partial V_z/\partial \psi = \sqrt{2}$ everywhere but particularly across the cell boundaries $\psi = 0$, where $V_z = 0$. In spirit this is G.O. Robert's (1972) original idea. The notion, as developed in our boundary layer context, has been generalised to other configurations (see section 6.6.3 below), for which Bayly and Childress (1988) have introduced the name stretch–fold–shear (SFS) map.

In the long length scale limit $k \to 0$, the z and t dependence can be ignored and Childress (1979) explained how $\alpha(0)$ can be evaluated by computing $\overline{\mathcal{E}}_H$ simply by analysing the contributions from the boundary layers induced by the action of the flow on a uniform magnetic field. It gives

$$\alpha(0) = -(2/R)^{1/2}\nu, \qquad (\nu = 0.5327\ldots), \qquad (6.6.16)$$

where the constant ν was evaluated numerically by Anufriev & Fishman (1982), Perkins & Zweibel (1987) and analytically by Soward (1987, also 1989). With (6.15) the result suggests that fast dynamo action will occur when the z length scale k^{-1} is comparable to the boundary layer width of order $R^{-1/2}$. Unfortunately the Taylor series expansion (6.11) is not valid for such large k. The reason for the failure can be traced to the long time of order $\ln R$ taken for fluid in the boundary layers, where $\psi = O(R^{-1/2})$, to pass the stagnation points. The long z-displacements of order $R^{-1/2}\ln R$ of fluid particles in the boundary layers mean that the ropes need to exhibit long z-length scales of comparable size, $k = O(R^{1/2}/\ln R)$, in order for the SFS mechanism to operate efficiently. With shorter vertical length scales considerable flux cancellation results from the shear, again of the flux expulsion type. The functional dependence of $\alpha(k; R)$ in this range was determined by Soward (1987). Order of magnitude estimates suggest the fastest growth rate is

$$\gamma = O\left(\ln(\ln R)/\ln R\right) \qquad (6.6.17)$$

occuring when $R^{-1/2}k = O\left((\ln R)^{-1/2}\right)$. Though not of order unity, it is a near miss and illustrates how even a negligible amount of exponential stretching can be very effective. Incidentally, the negative value of $\alpha(0)$ given by (6.16) implies by (6.15) that growing modes are characterised by negative k. By (6.14) they have negative helicity, so that $\overline{\mathbf{B}}_{\mathrm{II}}$ is a $-$Beltrami wave in contrast to the motion which is composed of $+$Beltrami waves.

Cats' eyes flow. Childress & Soward (1989) looked at the ABC flow (1.4.7) with $A = 0$, $B + C = O(1)$ and $C - B = O(\delta)$, where $0 < \delta \ll 1$. It consists of arrays of cats' eyes along lines $y = (n + \frac{1}{2})\pi$ and connected at X-type stagnation points of the horizontal velocity close to $x = (2m + \frac{1}{2}[1 + (-1)^n])\pi$ ($m, n = $ integers). Inside the cats' eyes the flow continues to consist of spiralling vortices. Outside in channels, width of order δ, motion proceeds indefinitely in either the positive or negative direction as dictated by the circulation at the boundaries of the cats' eyes. As in the G.O. Roberts dynamo, it is there in boundary layers of width $R^{-1/2}$ containing the streamlines connecting the X-type stagnation points that the dynamo activity takes place. The crucial parameter is

$$\beta = \delta R^{1/2}, \qquad (6.6.18)$$

which measures the ratio of the channel to boundary layer widths. For $\beta \gg 1$ the boundary layers on either side of the channel are non-overlapping and distinct. Childress & Soward (1989) only looked at the small k limit. The symmetries of the flow are such that

$$\overline{\overline{\mathbf{B}}}_{\mathrm{H}} = (1, ik^{-1}H), \qquad (c = 0) \qquad (6.6.19)$$

(cf. (6.13)) giving a diagonal α-tensor ($\alpha_{12} = \alpha_{21} = 0$) with

$$\alpha_{11}(0) = O(R^{-1/2}\beta^{-2}), \qquad \alpha_{22}(0) = O(R^{-1/2}\beta^3).$$

From (6.9) and (6.19) the mean helicity H satisfies

$$(H/k)^2 = \alpha_{11}(0)/\alpha_{22}(0) = \beta^{-5}, \qquad (6.6.20)$$

showing that the mean field is aligned predominantly in the x-direction as a result of a highly anisotropic α-tensor. The root mean value

$$[\alpha_{11}(0)\alpha_{22}(0)]^{1/2} = O(R^{-1/2}\beta^{1/2}) \qquad (6.6.21)$$

is the important parameter determining the dynamo growth rate. In magnitude it is larger than the value (6.16) for the G.O. Roberts dynamo. Still in the parameter range for which fast dynamo action might occur the approximation again breaks down.

We mention that a comparable problem, in which mean flow is added to the G.O. Roberts motion, has been investigated by Soward & Childress (1990).

6.6.3 Fast Dynamo Models

Pulsed waves. A simple time-periodic flow, which has led to interesting results, is the successive pulsing of two +Beltrami waves

$$\mathbf{V}(x, y, t) = \begin{cases} (0, \; \sin x, \; \cos x) & (0 \le t < \tau), \\ (\sin y, \; 0, \; -\cos y) & (\tau \le t < 2\tau). \end{cases} \qquad (6.6.22)$$

Bayly & Childress (1987, 1988) investigated numerically an almost identical two-dimensional flow for order one values of the period $P = 2\tau$ and convincingly demonstrated fast dynamo action. (See also the related investigation of Gilbert & Bayly (1992) involving random helical flows.) The dynamo properties of the flow with small and large P also raise significant issues. For the case $\tau \ll 1$ the particle paths follow the streamlines of the ABC flow ($A = 0$, $B = C = \frac{1}{2}$) corresponding to the G.O. Roberts cells. There is, however, a thin web containing the cell boundaries, whose thickness tends to zero with τ, in which the particle paths are chaotic. The possibility of fast dynamo action in such webs for closely related flows is discussed by Childress (1993a). As P increases the chaotic regions enlarge until they fill all space, when P is of order unity. For large τ, though the particle paths are highly chaotic, each individual pulse brings considerable order because almost all the magnetic field produced is aligned to the shearing direction of the pulse. Of course, that field may alternate its direction on very short length scales but it does render the problem amenable to asymptotic analysis. Indeed, Soward (1993a, b) shows that a simple mean field analysis, which we outline below, gives the correct growth rate in the large-τ limit $P \to \infty$.

In the frozen field case ($\eta = 0$), the mean field limit is quite simple. Consider the action of the first pulse on a uniform z-independent mean field $\overline{\mathbf{B}}_{\mathrm{H}}$. As a result the transverse field $\overline{B}_x$ remains unchanged, whereas the y-component increases linearly with t and takes the value $\overline{B}_y + \tau\overline{B}_x \cos x$ at the end of the first pulse. By this stretch-fold (SF) process strong fluctuating field is produced. When the z-dependence of the mean field, $\mathbf{B}_H = \Re[\overline{\overline{\mathbf{B}}}_H \exp(ikz)]$, is taken into account the vertical z-shear (S) reorganises the magnetic field so that the mean field $\overline{\overline{\mathbf{B}}}_H$ at the end of the first pulse becomes

$$\overline{\overline{\mathbf{B}}}_H(\tau) = J_0(\zeta)\overline{\overline{\mathbf{B}}}_H(0) - i\tau J_1(\zeta)\overline{\overline{B}}_x(0)\mathbf{1}_y, \qquad \zeta = k\tau, \qquad (6.6.23)$$

where we have used

$$J_0(\zeta) = <\exp(i\zeta\cos x)>, \qquad J_1(\zeta) = -J_0'(\zeta). \qquad (6.6.24)$$

Likewise, after the second pulse we have

$$\overline{\overline{\mathbf{B}}}_H(2\tau) = J_0(\zeta)\overline{\overline{\mathbf{B}}}_H(\tau) + i\tau J_1(\zeta)\overline{\overline{B}}_y(\tau)\mathbf{1}_x. \qquad (6.6.25)$$

In this way there is clearly the possibility of field amplification, when $\zeta \neq 0$. The effect of our double SFS-map can be reproduced by double application of the single map

$$TZ = (AZ)^*, \qquad (6.6.26)$$

where

$$A = \begin{pmatrix} -i\tau J_1(\zeta) & J_0(\zeta) \\ J_0(\zeta) & 0 \end{pmatrix}, \qquad (\tau \gg 1) \qquad (6.6.27)$$

and the star denotes complex conjugate. The fastest growing mode is the self-similar structure

$$\overline{\overline{\mathbf{B}}}_H(2n\tau) = (\tau\Lambda)^{2n}Z, \qquad (n = \text{integer}), \qquad (6.6.28)$$

where Z satisfies the non-standard (in view of the complex conjugation in (6.26)) eigenvalue problem
$$TZ = i(\tau\Lambda)Z. \qquad (6.6.29)$$

As in (6.19), the solution gives the helical magnetic field

$$Z = (1, ik^{-1}H), \qquad (6.6.30)$$

where

$$k^{-1}H = -J_0(\zeta)/(\tau\Lambda), \qquad \Lambda = J_1(\zeta)/[1 - (k^{-1}H)^2]. \qquad (6.6.31)$$

The corresponding growth rate is

$$\gamma = \tau^{-1}\ln(\tau|\Lambda|), \qquad (c = 0) \qquad (6.6.32)$$

which is fast for all $\Lambda \neq 0$ as $\tau \to \infty$. In our large-τ limit $k^{-1}H$ is small of order τ^{-1} and so the mean field of the eigenfunction (6.30) is almost aligned to the x-direction. Accordingly the maximum of our eigenvalue $\Lambda = J_1(\zeta) + O(\tau^{-1})$ is given by

$$\Lambda_{max} = J_1(\zeta_{max}) = 0.5819\ldots, \qquad (6.6.33)$$

when the corresponding value of ζ is

$$\zeta_{max} = 1.8412\ldots, \qquad J_1'(\zeta_{max}) = 0. \qquad (6.6.34)$$

From (6.31) it follows, as in our previous examples, that this amplifying mode has negative mean helicity, H.

Many of the ideas which we explained in section 6.5 for rope dynamos, apply to our SFS-model (see also Bayly, section 10.4.4 below). With each application of the map the length scale is reduced by a large factor of order τ^{-1} (as opposed to $1/2$ in (5.4)). With the inclusion of ohmic dissipation the diffusive length scale is reached after $N+1$ applications of the map, where $N = \frac{1}{2}(\ln R)/(\ln \tau)$ (cf. (5.5)). At this stage diffusion cuts off any further reduction in length scale and an eigenfunction emerges. Nevertheless (6.32) continues to give the growth rate of the dissipative problem in the limit $R \rightarrow \infty$, and moreover Soward (1993a, b) gives the weak solution for the eigenvector. It is weak because there is no pointwise convergence, an aspect which is clear from the Lagrangian viewpoint. Indeed, Soward (1993c) relates the strongest singularities of this generalised eigenvector to the stagnation points of the flow (6.22) at $(x,y) = (m\pi, n\pi)$ (integer m, n). Many of the ideas upon which the construction of the weak solution is based have their counterparts in the recent investigations by Childress (1993b) and Gilbert (1993).

Other pulsed Beltrami wave solutions are discussed by Bayly in section 10.3 below.

Temporally smooth flows. We mention briefly other examples of two-dimensional periodic flows, for which separable **B**-field solutions can be sought proportional to $\exp(ikz)$. Klapper (1992) considered a motion composed of piecewise linear flows of the type $(-x, y, xy)$ fitted together to resemble the G.O. Roberts cellular motion. To it is added the weak time-dependent flow $(0, \epsilon \sin t, 0)$ with $\epsilon \ll 1$. This leads to a chaotic web, as in the small-τ limit for the pulsed waves described above, which now contains the ($\epsilon = 0$) cell boundaries. The fast dynamo action in this web is analysed by maps from one cell boundary to the next taking account of the chaos initiated in the corner regions containing the stagnation points of the unperturbed flow ($\epsilon = 0$). Finite diffusivity is treated using a stochastic Wiener bundle method.

Numerical calculations at finite magnetic Reynolds number were initiated by Otani (1988, 1993). One family of his flows can be cast in the form (6.2, 3) with

$$B + C = 1, \qquad B - C = \epsilon \cos \omega t \qquad (6.6.35)$$

and $\epsilon = 1$. Each flow is characterised by constant values of K and ω. An interesting comparison is made between his numerical results for the growth rate at finite R and a frozen field estimate based on a calculation similar to (6.23–32) within the framework of the rough assumption that the flow is pulsed rather than continuously varying. The small-ϵ limit of (6.35) is the time dependent version of the cats' eyes flow investigated by Childress & Soward (1989). The small ϵ, large ω limit has also been discussed from an analytic point of view by Childress (1993a). Perhaps the most convincing evidence, however, for fast dynamo action at large R is provided by the numerical results of Galloway & Proctor (1992). One of their flows is of the ABC-type (6.2) with

$$\psi = \sin Y(t) + \cos X(t) \qquad (6.6.36)$$

but generalised to the moving frame

$$X(t) = x + \epsilon \cos \omega t, \qquad Y(t) = y + \epsilon \sin \omega t. \qquad (6.6.37)$$

Ponty et al. (1993) have linked the size of the chaotic regions of this flow to the efficiency of the fast dynamo. They have used a Melnikov function to estimate the dependence of its size on the frequency ω. Such methods are also central to the analytic approach of Childress (1993a).

Dynamical effects have been considered in a limited way by Gilbert, Otani & Childress (1993), who let the amplitudes of given flow harmonics respond to the Lorentz force.

Three-dimensional steady flows.   As a preliminary step to render the slow G.O. Roberts (1972) dynamo described in section 6.6.2 fast, Gilbert & Childress (1990) and Gilbert (1992) added the small z-dependent flow, namely the +Beltrami wave $\epsilon(\sin \sqrt{2}z, \cos \sqrt{2}z, 0)$, to the G.O. Roberts motion (6.12). This is an ABC flow with $B = C \gg A = O(\epsilon)$ with $\epsilon \ll 1$. In one respect this is like the Soward & Childress (1990) two-dimensional model with a z-dependent mean horizontal flow. More significantly, like Klapper's (1992) time-dependent flow, a thin steady chaotic web (Zaslavsky, Sagdeev & Chernikov 1988) is introduced containing the G.O. Roberts cell boundary. Gilbert (1992) restricts attention to the frozen field limit. The evolution of an initial field is traced by maps and the mean value of the vertical B_z-field is determined on prescribed cross sections. Numerical results indicate clear exponential growth dependent on the symmetries of the initial field.

From a more general point of view separable kinematic dynamo solutions for ABC-flows can be sought in the form

$$\mathbf{B}(\mathbf{x}, t) = \Re\{\tilde{\mathbf{B}}(\mathbf{x}) \exp[i\mathbf{k} \cdot \mathbf{x} + \lambda t]\}, \qquad (6.6.38)$$

where $\tilde{\mathbf{B}}(\mathbf{x})$ is a complex vector with the same spatial periodicity as the flow, while the wave vector $\mathbf{k}$ is a Floquet exponent. The symmetric ABC flow

($A = B = C$) has attracted much attention. The particle path structure, revealed by the comprehensive investigation of Dombre *et al.* (1986), consists of regions of stretching and non-stretching flow. The non-stretching flow consists of spiralling vortices aligned to each of the mutually orthogonal directions $\mathbf{1}_x$, $\mathbf{1}_y$, $\mathbf{1}_z$. The stretching flow resides in the chaotic region between the twisted and contorted interlocking vortices. Following the pioneering numerical studies of Arnol'd & Korkina (1983), results at larger values of R have been obtained by Galloway & Frisch (1984, 1986) and Lau & Finn (1993). In all these studies, solutions were sought, for simplicity, with the same spatial periodicity as the flow ($\mathbf{k} = 0$ in (6.38)). Galanti, Sulem & Pouquet (1992, 1993) have considered the temporal evolution of fields essentially composed of non-zero values of $\mathbf{k}$ compatible with the periodicity of a square box side $2n\pi$ (n = integer). These authors also considered the feedback of the Lorentz force on the motion (Galanti *et al.* 1992). Results for frozen fields are given by Gilbert (1992). Childress (1979) identified the importance of the streamlines connecting the stagnation points, particularly lines like $x = y = z$. In the immediate vicinity of this one-dimensional manifold, motion consists of an axisymmetric swirling flow. As the streamlines approach the stagnation points they diverge but then converge on to new two-dimensional unstable manifolds. Isolated streamlines on these manifolds connect with other stagnation points, but, in general, the streamlines wander indefinitely throughout the chaotic region to produce a very complicated surface. Nevertheless, by consideration of an idealised axisymmetric model, Childress (1979) (also Ghil & Childress 1987) showed how flux ropes might form on the one-dimensional manifolds and provide a potent source of dynamo action. The idea is simply that the swirl in the $R^{-1/2}$ radius ropes can produce azimuthal magnetic field which is amplified by a factor of order $R^{1/2}$ when the fluid expands onto the unstable manifold. Estimates from the idealised model suggests that an order one isotropic α-effect might result. Certainly there is support for the idea that the magnetic flux is concentrated in those ropes from the numerical results of Galloway & Frisch (1984, 1986) (see also Gilbert 1991, Galloway & O'Brian 1993). Indeed, some recent results (Zheligovsky 1992) for Beltrami flows in bounded spheres, in which there are some comparable heteroclinic connections between the stagnation points, also point to the existence of flux ropes. Some progress towards a large magnetic Reynolds number dynamo theory of flux ropes and sheets was reported by Childress & Soward (1985). Whether or not such boundary layer theories pertain to the fast dynamo remains unclear and a complete theory remains a challenging unsolved problem.

Acknowledgements
Many of the ideas presented have resulted from collaborations with Steve
Childress: an earlier unabridged version of his (1992) lecture notes proved
particularly stimulating. Others with whom I have had beneficial discussions
include Bruce Bayly, Axel Brandenburg, Dave Galloway, Andrew Gilbert,
Vasilii Gvaramadze, Isaac Klapper, Keith Moffatt, Edward Ott, Mike Proc-
tor, Paul Roberts, Alexander Ruzmaikin, Anvar Shukurov, Dmitry Sokoloff,
Misha Vishik and Vladimir Zheligovsky.

REFERENCES

Alfvén, H. 1950 Discussion of the origin of the terrestrial and solar magnetic
fields. *Tellus* **2**, 74–82.

Anufriev, A.P. & Fishman, V.M. 1982 Magnetic field structure in the two-
dimensional motion of a conducting fluid. *Geomag. Aeron.* **22**, 245–248.

Arnol'd, V.I. 1972 Notes on the three-dimensional flow pattern of a perfect
fluid in the presence of a small perturbation of the initial velocity field.
Prikl. Matem. Mekh. **36**(2), 255–262. (English transl.: *Appl. Math. Mech.*
36, 236–242.)

Arnol'd, V.I. & Korkina, E.I. 1983 The growth rate of a magnetic field in a
three dimensional steady incompressible flow. *Moscow Univ. Math. Bull.*
38(3), 50–54.

Arnol'd, V.I., Zel'dovich, Ya.B., Ruzmaikin, A.A., & Sokoloff, D.D. 1981 A
magnetic field in a stationary flow with stretching in Riemannian space.
Zh. Eksp. Teor. Fiz., **81**, 2052–2058. (English transl.: *Sov. Phys. JETP*
54, 1083–1086, (1981).)

Bayly B.J. 1986 Fast magnetic dynamos in chaotic flows. *Phys. Rev. Lett.* **57**,
2800–2803.

Bayly, B.J. & Childress, S. 1987 Fast-dynamo action in unsteady flows and
maps in three dimensions. *Phys. Rev. Lett.* **59**, 1573–1576.

Bayly, B.J. & Childress, S. 1988 Construction of fast dynamos using unsteady
flows and maps in three dimensions. *Geophys. Astrophys. Fluid Dynam.* **44**,
207–240.

Bayly, B.J. & Childress, S. 1989 Unsteady dynamo effects at large magnetic
Reynolds number. *Geophys. Astrophys. Fluid Dynam.* **49**, 23–43.

Braginsky, S.I. 1964 Self-excitation of a magnetic field during the motion of a
highly conducting fluid. *Zh. Eksp. Teor. Fiz. SSSR* **47**, 1084–1098. (English
transl.: *Sov. Phys. JETP* **20**, 726–735 (1965).)

Childress, S. 1979 Alpha-effect in flux ropes and sheets. *Phys. Earth Planet.
Inter.* **20**, 172–180.

Childress, S. 1992 Fast dynamo theory. In *Topological Aspects of the Dynamics
of Fluids and Plasmas* (ed. H.K. Moffatt, G.M. Zaslavsky, P. Comte &

M. Tabor). NATO ASI Series E: Applied Sciences, vol. 218, pp. 111–147. Kluwer.

Childress, S. 1993*a* On the geometry of fast dynamo action in unsteady flows near the onset of chaos. *Geophys. Astrophys. Fluid Dynam.* **73**, 75–90.

Childress, S. 1993*b* Note on perfect fast dynamo action in a large-amplitude SFS map. In *Solar and Planetary dynamos* (ed. M.R.E. Proctor, P.C. Matthews & A.M. Rucklidge), pp. 43–50. Cambridge University Press.

Childress, S. & Soward, A.M. 1985 On the rapid generation of magnetic fields. In *Chaos in Astrophysics* (ed. J.R. Buchler, J.M. Perdang & E.A. Spiegel). NATO ASI Series C: Mathematical and Physical Sciences, vol. 161, pp. 223–244. Reidel.

Childress, S. & Soward, A.M. 1989 Scalar transport and alpha-effect for a family of cat's-eye flows. *J. Fluid Mech.* **205**, 99–133.

Clark, A. 1964 Production and dissipation of magnetic energy by differential fluid motions. *Phys. Fluids* **7**, 1299–1305.

Clark, A. 1965 Some exact solutions in magnetohydrodynamics with astrophysical applications. *Phys. Fluids* **8**, 644–649.

Dombre, T., Frisch, U., Greene, J.M., Hénon, M., Mehr, A. & Soward, A.M. 1986 Chaotic streamlines in ABC flows. *J. Fluid Mech.* **167**, 353–391.

Fearn, D.R., Roberts, P.H. & Soward, A.M. 1988 Convection, stability and the dynamo. In *Energy, Stability and Convection* (ed. G.P. Galdi & B. Straughan). Pitman Research Notes in Mathematics Series, vol. 168, pp. 60–324. Longman.

Finn, J.M. & Ott, E. 1988 Chaotic flows and fast magnetic dynamos. *Phys. Fluids* **31**, 2992–3011.

Finn, J.M. & Ott, E. 1990 The fast kinematic magnetic dynamo and the dissipationless limit. *Phys. Fluids B* **2**, 916–926.

Galanti, P., Sulem, P.L. & Pouquet, A. 1992 Linear and non-linear dynamos associated with ABC flows. *Geophys. Astrophys. Fluid Dynam.* **66**, 183–208.

Galanti, P., Sulem, P.L. & Pouquet, A. 1993 Influence of the period of an ABC flow on its dynamo action. In *Solar and Planetary dynamos* (ed. M.R.E. Proctor, P.C. Matthews & A.M. Rucklidge), pp. 99–103. Cambridge University Press.

Galloway, D.J. & Frisch, U. 1984 A numerical investigation of magnetic field generation in a flow with chaotic streamlines. *Geophys. Astrophys. Fluid Dynam.* **29**, 13–18.

Galloway, D.J. & Frisch, U. 1986 Dynamo action in a family of flows with chaotic streamlines. *Geophys. Astrophys. Fluid Dynam.* **36**, 53–83.

Galloway, D.J. & O'Brian, N.R. 1993 Numerical calculations of dynamos for ABC and related flows. In *Solar and Planetary dynamos* (ed. M.R.E. Proctor, P.C. Matthews & A.M. Rucklidge), pp. 105–113. Cambridge University

Press.

Galloway, D.J. & Proctor, M.R.E. 1992 Numerical calculations of fast dynamos in smooth velocity fields with realistic diffusion. *Nature* **356**, 691–693.

Galloway, D.J. & Zheligovsky, V.A. 1993 On a class of non-axisymmetric flux rope solutions to the electromagnetic induction equation. *Geophys. Astrophys. Fluid Dynam.* (submitted).

Ghil, M. & Childress, S. 1987 *Topics in Geophysical Fluid Dynamics, Atmospheric Dynamics, Dynamo Theory, and Climate Dynamics.* Springer.

Gilbert, A.D. 1988 Fast dynamo action in the Ponomarenko dynamo. *Geophys. Astrophys. Fluid Dynam.* **44**, 241–258.

Gilbert, A.D. 1991 Fast dynamo action in a steady chaotic flow. *Nature* **350**, 483–485.

Gilbert, A.D. 1992 Magnetic field evolution in steady chaotic flows. *Phil. Trans. R. Soc. Lond. A* **339**, 627–656.

Gilbert, A.D. 1993 Towards a realistic fast dynamo: models based on cat maps and pseudo-Anosov maps. *Proc R. Soc. Lond. A* **443**, 585–606.

Gilbert, A.D. & Bayly B.J. 1992 Magnetic field intermittency and fast dynamo action in random helical flows. *J. Fluid Mech.* **241**, 199–214.

Gilbert, A.D. & Childress, S. 1990 Evidence for fast dynamo action in a chaotic web. *Phys. Rev. Lett.* **65**, 2133–2136.

Gilbert, A.D., Otani, N.F. & Childress, S. 1993 Simple dynamical fast dynamos. In *Solar and Planetary dynamos* (ed. M.R.E. Proctor, P.C. Matthews & A.M. Rucklidge), pp. 129–136. Cambridge University Press.

Gvaramadze, V.V. & Cheketiani, O.G. 1989 The stationary distribution of a magnetic field in a spiral turbulence with linear deformations. *Magn. Gidrodin.* **25**(4), 9–14. (English transl.: *Magnetohydrodynamics* **25**, 420–425 (1990).)

Gvaramadze, V.V., Lominadze, J.G., Ruzmaikin, A.A. & Sokoloff, D.D. 1988*a* Magnetic field in a turbulent flow of compressible fluid with a linear profile. *Magn. Gidrodin.* no. 2, 8–10. (English transl.: *Magnetohydrodynamics* **24**, 139–141.)

Gvaramadze, V.V., Lominadze, J.G., Ruzmaikin, A.A., Sokoloff, D.D. & Shukurov, A.M. 1988*b* Turbulent generation of magnetic fields in astrophysical jets. *Astrophys. Space Sci.* **140**, 165–174.

Klapper, I. 1992 A study of fast dynamo action in chaotic cells. *J. Fluid Mech.* **239**, 359–381.

Klapper, I. 1994 Shadowing and the diffusionless limit in fast dynamo theory. *Nonlinearity* (in press).

Lau, Y.-T. & Finn, J.M. 1993 Fast dynamos with finite resistivity in steady flows with stagnation points. *Phys. Fluids B* **5**, 365–375.

Lortz, D. 1968 Exact solutions of the hydromagnetic dynamo problem. *Plasma Phys.* **10**, 967–972.

Moffatt, H.K. 1974 The mean electromotive force generated by turbulence in the limit of perfect conductivity. *J. Fluid Mech.* **65**, 1–10.

Moffatt, H.K. 1978 *Magnetic Field Generation in Electrically Conducting Fluids*. Cambridge University Press.

Moffatt, H.K. 1992 Relaxation under topological constraints. In *Topological Aspects of the Dynamics of Fluids and Plasmas* (ed. H.K. Moffatt, G.M. Zaslavsky, P. Comte & M. Tabor). NATO ASI Series E: Applied Sciences, vol. 218, pp.3–28. Kluwer.

Moffatt, H.K. & Kamkar, H. 1983 The time scale associated with flux expulsion. In *Stellar and Planetary Magnetism*, vol. 2 (ed. A.M. Soward) pp. 91–97. Gordon and Breach.

Moffatt, H.K. & Proctor, M.R.E. 1985 Topological constraints associated with fast dynamo action. *J. Fluid Mech.* **154**, 493–507.

Moffatt, H.K. & Tsinober, A. 1992 Helicity in laminar and turbulent flow. *Annu. Rev. Fluid Mech.* **24**, 281–312.

Otani, N.F. 1988 Computer simulation of fast kinematic dynamos. *Trans. Am. Geophys. Un.* **69**, 1366 (abstract no. SH51-15).

Otani, N.F. 1993 A fast magnetohydrodynamic dynamo in 2-d time-dependent flows. *J. Fluid Mech.* **253**, 327–340.

Ott, E., Du, Y, Sreenivasan, K.R., Juneja, A. & Suri, A.K. 1992 Sign-singular measures: fast magnetic dynamos and high Reynolds number fluid turbulence. *Phys. Rev. Lett.* **69**, 2654–2657.

Ottino, J.M. 1989 *The Kinematics of Mixing: Stretching, Chaos, and Transport*. Cambridge University Press.

Parker, E.N. 1979 *Cosmical Magnetic Fields, their Origin and their Activity*. Clarendon, Oxford.

Perkins, F.W. & Zweibel, E.G. 1987 A high magnetic Reynolds number dynamo. *Phys. Fluids* **30**, 1079–1084.

Ponomarenko, Yu B. 1973 On the theory of hydromagnetic dynamos. *Zh. Prikl. Mekh. & Tekh. Fiz., USSR* **6**, 47–51. (English transl.: *J. Appl. Mech. Tech. Phys.* **14**, 775–778.)

Ponty, Y., Pouquet, A., Rom-Kedar, V. & Sulem, P.L. 1993 Dynamo in nearly integrable chaotic flow. In *Solar and Planetary dynamos* (ed. M.R.E. Proctor, P.C. Matthews & A.M. Rucklidge), pp. 271–274. Cambridge University Press.

Reshetnyak, M., Sokoloff, D.D. & Sukurov, A.M. 1991 Evolution of a magnetic blob in a helical flow. *Astron. Nachr.* **312**, 33–39.

Roberts, G.O. 1972 Dynamo action of fluid motions with two-dimensional periodicity. *Phil. Trans. R. Soc. Lond.* A **271**, 411–454.

Roberts, P. H. & Soward, A.M. 1992 Dynamo theory. *Annu. Rev. Fluid Mech.* **24**, 459–512.

Ruzmaikin, A.A. Sokoloff, D.D. & Shukurov, A.M. 1988 A hydromagnetic screw dynamo. *J. Fluid Mech.* **197**, 39–56.

Shukurov, A.M. & Sokoloff, D.D. 1993 Evolution of magnetic fields in a swirling jet. In *Solar and Planetary dynamos* (ed. M.R.E. Proctor, P.C. Matthews & A.M. Rucklidge), pp. 271–274. Cambridge University Press.

Soward, A.M. 1972 A kinematic theory of large magnetic Reynolds number dynamos. *Phil. Trans. R. Soc. Lond. A* **272** 431–461.

Soward, A.M. 1987 Fast dynamo action in a steady flow. *J. Fluid Mech.* **180**, 267–295.

Soward, A.M. 1988 Fast dynamos with flux expulsion. In *Secular, Solar and Geomagnetic Variation in the Last 10,000 Years* (ed. F.R. Stephenson & A.W. Wolfendale). NATO ASI Series C: Mathematical and Physical Sciences, vol. 236, pp. 79–96. Kluwer.

Soward, A.M. 1989 On dynamo action in a steady flow at large magnetic Reynolds number. *Geophys. Astrophys. Fluid Dynam.* **49**, 3–22.

Soward, A.M. 1990 A unified approach to a class of slow dynamos. *Geophys. Astrophys. Fluid Dynam.* **53**, 81–107.

Soward, A.M. 1993a An asymptotic solution of a fast dynamo in a two-dimensional pulsed flow. *Geophys. Astrophys. Fluid Dynam.* **73**, 179–215.

Soward, A.M. 1993b Analytic fast dynamo solution for a two-dimensional pulsed flow. In *Solar and Planetary dynamos* (ed. M.R.E. Proctor, P.C. Matthews & A.M. Rucklidge), pp. 275–286. Cambridge University Press.

Soward, A.M. 1993c On the role of stagnation points and periodic particle paths in a two-dimensional pulsed flow fast dynamo model. *Physica D* (in press).

Soward, A.M. & Childress 1986 Analytic theory of dynamos. *Adv. Space Res.* **6**(8), 7–18.

Soward, A.M. & Childress, S. 1990 Large magnetic Reynolds number dynamo action in a spatially periodic flow with mean motion. *Phil. Trans. R. Soc. Lond. A* **331**, 649–733.

Stix, M. 1987 On the origin of stellar magnetism. In *Solar and Stellar Physics* (ed. E.-H. Schroter, M. Schussler). Lecture Notes in Physics, vol. 292, pp. 15–38. Springer.

Tabor, M. 1992 Stretching and alignment in general flow fields: classical trajectories from Reynolds number zero to infinity. In *Topological Aspects of the Dynamics of Fluids and Plasmas* (ed. H.K. Moffatt, G.M. Zaslavsky, P. Comte & M. Tabor). NATO ASI Series E: Applied Sciences, vol. 218, pp. 83–110. Kluwer.

Vainshtein, S.I. & Zel'dovich Ya.B. 1972 Origin of magnetic fields in astrophysics. *Sov. Phys. Usp.* **15**, 159–172.

Vincent, A. & Meneguzzi, M. 1991 The spatial structure and statistical properties of homogeneous turbulence. *J. Fluid Mech.* **225**, 1–20.

Vishik, M.M. 1988 On magnetic field generation by three-dimensional steady flow of a conducting fluid at high magnetic Reynolds number. *Izv. Acad. Sci. SSSR, Fiz. Zemli* no. 3, 3–12. (English transl.: *Izv., Acad. Sci. USSR, Phys. Solid Earth* **24**(3), 173–180.)

Vishik, M. 1989 Magnetic field generation by the motion of a highly conducting fluid. *Geophys. Astrophys. Fluid Dynam.* **48**, 151–167.

Weiss, N.O. 1966 The expulsion of magnetic flux by eddies. *Proc R. Soc. Lond. A* **293**, 310–328.

Woltjer, L. 1958 A theorem on force-free magnetic fields. *Proc. Nat. Acad. Sci.* **44**, 489–491.

Zaslavsky, G.M., Sagdeev, R.Z. & Chernikov, A.A. 1988 Stochastic nature of streamlines in steady state flows. *Zh. Eksp. Teor. Fiz. SSSR* **94**, 102–115. (English transl.: *Sov. Phys. JETP* **67**, 270–277.)

Zel'dovich, Ya.B., Ruzmaikin, A.A., Molchanov, S.A. & Sokoloff, D.D. 1984 Kinematic dynamo problem in a linear velocity field. *J. Fluid Mech.* **144**, 1–11.

Zheligovsky, V.A. 1993 Fast kinematic dynamo, sustained by a Beltrami flow in a sphere. *Geophys. Astrophys. Fluid Dynam.* **73**, 217–245.

CHAPTER 7

Nonlinear Planetary Dynamos

D. R. FEARN

Dept. of Mathematics
University of Glasgow
Glasgow, G12 8QW, UK

7.1 INTRODUCTION

The necessary fundamental theory which forms the background for this chapter is given in chapters 1 and 3. Here, we consider what features characterise planetary dynamos and the special problems these pose for finding solutions to the dynamo problem. This is best done by first looking at the governing equations. We shall use thermal buoyancy as the energy source driving the dynamo. While recognising that there are other energy sources (that are probably more important in the Earth and the terrestrial planets, see chapter 5), thermal buoyancy is the best understood and has been used as the basis for almost all dynamo models. Much of the discussion in this chapter relates to nonlinear processes which involve coupling between the core and the mantle. To make the discussion as straightforward as possible, we focus on viscous core-mantle coupling. Alternative coupling mechanisms (for example electromagnetic) may be more important but their role in the nonlinear dynamics discussed here is similar to that of viscous coupling. (There are some important differences though, see Fearn & Proctor 1992.)

The non-dimensionalised version of the equations governing the evolution of the magnetic field $\mathbf{B}$, fluid velocity $\mathbf{V}$ and temperature T are

$$Ro\left(\partial_t\mathbf{V} + (\mathbf{V}\cdot\nabla)\mathbf{V}\right) + \mathbf{1}_z\times\mathbf{V} =$$
$$-\nabla\Pi + (\nabla\times\mathbf{B})\times\mathbf{B} + q\widetilde{Ra}\,T\mathbf{r} + E\nabla^2\mathbf{V}\,, \qquad (7.1.1)$$

$$\partial_t\mathbf{B} = \nabla\times(\mathbf{V}\times\mathbf{B}) + \nabla^2\mathbf{B}\,, \qquad (7.1.2)$$

$$\partial_t T + \mathbf{V}\cdot\nabla T = q\nabla^2 T + \epsilon\,, \qquad (7.1.3)$$

$$\nabla\cdot\mathbf{V} = 0\,, \qquad \nabla\cdot\mathbf{B} = 0\,, \qquad (7.1.4,5)$$

where ϵ represents a uniform source of heat, and the gravitational acceleration has been taken to be $g\mathbf{r}$ where $\mathbf{r}$ is the non-dimensionalised radius vector. In (1.1–3) the dimensionless parameters, the Rossby number Ro, the Roberts

219

M.R.E. Proctor & A.D. Gilbert (eds.)
Lectures on Solar and Planetary Dynamos, 219–244
©1994 Cambridge University Press.

number q, the modified Rayleigh number $\widetilde{Ra}$ and the Ekman number E, are defined by

$$Ro = \frac{\eta}{2\Omega\mathcal{L}^2}, \quad q = \frac{\kappa}{\eta}, \quad \widetilde{Ra} = \frac{g\alpha\beta\mathcal{L}^2}{2\Omega\kappa}, \quad E = \frac{\nu}{2\Omega\mathcal{L}^2}, \quad (7.1.6)$$

where Ω, η, κ, α, β and ν have their usual meanings (see chapters 1 and 3), $\mathcal{L}$ is a typical length-scale (for example the radius of the Earth's core), times have been non-dimensionalised using the ohmic timescale $T_\eta = \mathcal{L}^2/\eta$, the magnetic field has been non-dimensionalised using $\mathcal{B} = (2\Omega\mu_0\rho\eta)^{1/2}$ as the magnetic field strength scale, and the flow speed has been non-dimensionalised using $\mathcal{V} = \eta/\mathcal{L}$ as the speed scale. A consequence of the choices for $\mathcal{B}$ and $\mathcal{V}$ is that the Elsasser number Λ defined in chapter 3 and the magnetic Reynolds number R defined in (1.2.37) do not appear in (1.1–3). This makes sense, since in a hydrodynamic dynamo, the field strength and flow speed (and hence Λ and R) emerge as part of the solution and so cannot be specified. Within the given non-dimensionalisation, Λ and R are given by the maximum magnitudes of $\mathbf{B}^2$ and $\mathbf{V}$ respectively. A further consequence of the choice of velocity scale is that the Rossby number Ro defined in (1.6) may not be a reliable measure of the strength of inertia, since the maximum flow speed may be considerably larger than $\eta/\mathcal{L}$.

What characterises planetary dynamos is the dominant effect of rotation and the corresponding smallness of the Rossby and Ekman numbers. The Rossby number is a measure of the ratio of the rotation period to the ohmic timescale. For the Earth, the former is a day while the latter (based on the radius of the core) is $O(10^5)$ years, so $Ro = O(10^{-8})$. Hence the inertial term in (1.1) is small (but not perhaps as small as the estimate of Ro suggests, see above). The viscosity of the core is very poorly determined but most estimates are very much smaller than the magnetic diffusivity. A typical value is $\nu = 3 \times 10^{-6}$ m^2/s (at the core-mantle boundary, Poirier 1988), giving $E = O(10^{-15})$. Consequently the viscous term in the momentum equation (1.1) is very small compared with the leading order magnetostrophic balance between Coriolis, pressure, Lorentz (and buoyancy) forces. The major problem for planetary dynamos is how to deal with the inertial and viscous forces, particularly the latter.

The inertial term is the less problematic of the two. If it is retained, then in any numerical model, we must be able to resolve the inertial timescale while integrating over at least several ohmic decay times. The complexity of dynamo models means that excessive computation time would be required if realistic values of the Rossby number are to be used. It is therefore desirable to neglect the inertial terms in (1.1). The justification for doing this is the smallness of Ro and, in general, it is thought possible to set $Ro = 0$ without causing any difficulties, provided the magnetic field is strong ($\Lambda \geq O(1)$). (It should be said though that the consequences of setting $Ro = 0$ have received much less attention than those associated with small E.) The case of a weak field is discussed below. Physically, neglecting inertial terms has the effect of filtering out inertial waves. Mathematically, it turns the character of the momentum equation from predictive to diagnostic.

The viscous term in (1.1) is the source of more fundamental problems. If it is retained, then any numerical code has to resolve very thin Ekman boundary layers, and perhaps internal Stewartson layers (see for example, Roberts & Soward 1978, Hollerbach & Proctor 1993). Numerical resolution therefore limits how small E can be made. Values of $O(10^{-3})$ have been attained (see for example Zhang & Busse 1988, 1989, 1990). It seems likely that this is not small enough to be characteristic of planetary interiors. The effect of Ekman boundary layers can be dealt with through a boundary-layer analysis and the derivation of appropriate boundary conditions for an inviscid interior flow (see section 7.2.1 for some details). This has been the approach adopted for example by Braginsky (see chapter 9). This eliminates the problem of numerical resolution of viscous boundary layers but there remains the problem of possible internal layers and the short length-scale of convection when the magnetic field is weak, see later. If viscous effects are completely neglected (in addition to neglecting inertia) then Taylor (1963) has shown that solutions of (1.1) only exist if the magnetic field **B** satisfies the constraint

$$\int_{C(s)} (\mathbf{J} \times \mathbf{B})_\phi \, dS = 0, \quad \forall s, \tag{7.1.7}$$

where **J** is the current density, $dS = s \, d\phi dz$, (s, ϕ, z) are cylindrical polar coordinates and $C(s)$ is the cylinder coaxial with the rotation axis, of radius s and contained within the outer core. This condition is modified if viscous boundary-layer (i.e., Ekman suction) effects are included. (Note that (1.7) must hold if viscous effects are absent, even if other core-mantle coupling effects are present, see for example Fearn & Proctor (1992).) The Taylor constraint has been the subject of a great deal of attention and is the source of most fundamental difficulties associated with solving the planetary dynamo problem. One example is the question of the existence of solutions satisfying (1.7). Another is the numerical problem of integrating over the cylinder $C(s)$ while the natural coordinates for the problem are spherical polars (r, θ, ϕ). The Taylor constraint gives the planetary dynamo problem its unique character. We discuss it in detail in section 7.2.

Much of the above discussion assumes that the Lorentz force is comparable with the Coriolis force, so applies to numerical calculations which start with a strong field ($\Lambda \geq O(1)$) and maintain a strong field. Some calculations, for example those of Zhang & Busse (1988, 1989), start from zero field and investigate the evolution of the system as it bifurcates from a non-magnetic non-convective state to a convective non-magnetic state and then to a convecting magnetic state. In such an approach it is not in general possible to neglect viscous or inertial forces, though Zhang & Busse (1990) look at the limit of infinite Prandtl number in which inertial forces are absent. In addition to having to resolve Ekman boundary layers (whose width is $O(E^{1/2})$), models that start from zero magnetic field are limited in how small the Ekman number can be made by the small length-scale ($O(E^{1/3})$) of convection at small E in a rotating non-magnetic system (see for example Roberts 1968, Busse 1970). Though $E^{1/2} \ll E^{1/3}$, this is an additional problem. The Ekman boundary layer requires good resolution in r at the boundary. This could be accommodated by a stretched grid, for example. Alternatively, viscous boundary-layer effects can be made less important if stress-free boundary conditions are used, see for example Zhang & Busse (1988, 1989, 1990). The convection has short length-scales in ϕ and s and is not at an *a priori* known location, so is much more difficult to deal with. It is the short length-scale of the convection that limits how small E can be made. The best that Zhang & Busse (1988, 1989, 1990) were able to achieve was $E = O(10^{-3})$. When a strong field is present (see for example Fearn & Proctor 1983a), such a problem is not met since convection is not localised and length-scales are typically of the same order as the dimensions of the container.

To summarise then, ideally we would like to solve (1.1–5) numerically as they stand to give a planetary dynamo model. In practice, computational limitations mean that it is not possible to use planetary values of Ro and E (which are both very small). We can use much larger values but it is not clear if the resulting solutions will be characteristic of planetary dynamos. Alternatively we can neglect inertia and possibly also viscous effects. In the latter case we have to deal with the Taylor constraint. The way forward is probably through attempts at all possible approaches and comparing the results. In section 7.4 we give some details of progress to date. Further discussion of the problems of formulating a realistic theory of the geodynamo can be found in Braginsky (1991) and in chapter 9.

7.2 TAYLOR'S CONSTRAINT AND ITS VARIATIONS

7.2.1 Taylor's Constraint

Setting $Ro = 0$, $E = 0$ in (1.1) (the magnetostrophic approximation), gives

$$1_z \times \mathbf{V} = -\nabla \Pi + (\nabla \times \mathbf{B}) \times \mathbf{B} + q\widetilde{Ra}\,T\mathbf{r}. \qquad (7.2.1)$$

Taking the ϕ-component gives

$$V_s = -\partial \Pi / \partial \phi + [(\nabla \times \mathbf{B}) \times \mathbf{B}]_\phi. \qquad (7.2.2)$$

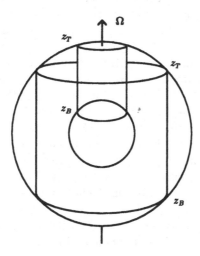

Figure 7.1 The Taylor cylinder $C(s)$, illustrated for the cases (a) where the cylinder intersects the inner core and (b) where s exceeds the inner-core radius r_i. The cylinder extends from $z = z_T = \sqrt{1 - s^2}$ to $z = z_B$ where (a) $z_B = \sqrt{r_i^2 - s^2}$ and (b) $z_B = -z_T$.

Integrating this over the cylinder $C(s)$ (see figure 7.1) gives

$$\int_{C(s)} V_s \, dS \; = \; \int_{C(s)} [(\nabla \times \mathbf{B}) \times \mathbf{B}]_\phi \, dS, \quad \forall \, s. \qquad (7.2.3)$$

The term on the left-hand side is the net flow of fluid out of the curved surface of the cylinder.

If viscosity is totally neglected, (2.2) applies throughout the core and the cylinder extends to the boundaries of the outer core. There can therefore be no flow of fluid into or out of the ends of the cylinder. Consequently, for an incompressible fluid, the left-hand side of (2.3) must vanish, giving

$$\int_{C(s)} [(\nabla \times \mathbf{B}) \times \mathbf{B}]_\phi \, dS \; = \; 0, \quad \forall \, s, \qquad (7.2.4)$$

which is just (1.7) with $\mathbf{J} = \nabla \times \mathbf{B}$ in our non-dimensionalised system. This condition was first derived by Taylor (1963) and is referred to as 'Taylor's condition' or 'Taylor's constraint'.

If viscous effects are retained in the problem (but are only important in thin Ekman layers at the boundaries of the outer core), then (2.2) is valid throughout the core except for the Ekman layers, and the cylinder $C(s)$ must be considered as extending, not to the boundaries of the outer core, but to the outer edges of the Ekman layers. The North-South flow in the Ekman layers leads to a net flow of fluid into the ends of the cylinder. This must be balanced by a net flow out of the curved surface of the cylinder, so the left-hand side of (2.3) is in general non-zero. To evaluate $\int_{C(s)} V_s \, dS$ we must calculate the flow in the Ekman layers.

The problem is analysed by splitting the core into three regions, a thin spherical shell that extends inward a short distance from the core-mantle boundary, a similar shell adjacent to the boundary with the inner core, and the *interior*, which is the remainder (and the bulk of) the outer core. In the interior, viscous effects are negligible, and (2.2) holds. In the two boundary regions, viscous effects are important. The short length-scale in the radial direction permits a simplification to the governing equations and an analytical solution. This must then be matched to the solution in the interior. The spherical geometry is unimportant in the boundary layers and can locally be approximated by a plane layer.

It can be shown that a uniform flow U in the x-direction close to a plane rigid boundary in an x-y plane results in an Ekman flux in the y-direction of

$$(U/2) \, (\nu/\Omega_3)^{1/2} \qquad (7.2.5)$$

per unit width in the x-direction, where Ω_3 is the component of the angular velocity Ω normal to the layer, see for example equation (4.4.17) of Batchelor (1967). Identifying (locally at the core-mantle boundary) x with ϕ, z with r, and y with $-\theta$, and integrating over ϕ, this gives the total flow into the top end of the cylinder due to the flow in the Ekman layer to be (in dimensional form)

$$\pi s \left(\frac{\nu}{\Omega \cos \theta} \right)^{1/2} \overline{V}_\phi \bigg|_{z_T} , \qquad (7.2.6)$$

where $\overline{V}_\phi$ is the azimuthally averaged (or *mean* or *axisymmetric*) part of V_ϕ. In general, we define, for any $f = f(r, \theta, \phi)$

$$\overline{f}(r, \theta) \equiv \langle f \rangle \equiv \frac{1}{2\pi} \int_0^{2\pi} f \, d\phi . \qquad (7.2.7)$$

By conservation of mass, the total flow of fluid into any cylinder $C(s)$ must be zero. If there is a net flow into the ends, then this must be balanced by a flow through the curved surface, so in general the left-hand side of (2.3) is non-zero. For example, in the case where $C(s)$ does not intersect the inner core and $\overline{V}(s, z_B) = \overline{V}(s, z_T)$ the Ekman flow into the bottom of the cylinder is the same as that into the top. Then the left-hand side of (2.3) is

$$2\pi s \left(\frac{2E}{\cos \theta} \right)^{1/2} \overline{V}_\phi \bigg|_{z_T} , \qquad (7.2.8)$$

(now in non-dimensional form). Then (2.3) gives

$$\left. V_\phi \right|_{z_T} = E^{-1/2} \left(\frac{\cos\theta}{2} \right)^{1/2} \int_{z_B}^{z_T} \langle (\nabla \times B) \times B \rangle_\phi \, dz \, . \tag{7.2.9}$$

This replaces (2.4) when the effects of Ekman boundary layers are included in the problem. We note here that (2.4) and (2.9) are very different in character. The former is a constraint on B while the latter is a means of determining $\left. V_\phi \right|_{z_T}$.

7.2.2 The 'Arbitrary' Geostrophic Flow $V_G(s)$

Taking the curl of (2.1) and using (1.4) gives

$$-\frac{\partial V}{\partial z} = -\nabla \times \big((\nabla \times B) \times B \big) + q\widetilde{Ra}\,(\nabla T \times r) \, . \tag{7.2.10}$$

Taking the axisymmetric part and integrating this with respect to z gives

$$V = \int_z^{z_T} \Big\langle \nabla \times \big((\nabla \times B) \times B \big) \Big\rangle dz' \; - \; q\widetilde{Ra} \int_z^{z_T} \langle \nabla T \times r \rangle \, dz' \; + \; C(s), \tag{7.2.11}$$

where $C(s)$ is an arbitrary function of integration. The flow V must satisfy the boundary condition that (to leading order) there is no normal flow at the top and bottom boundaries. These two boundary conditions give two expressions relating V_s and V_z, which then determine C_s and C_z. In a non-axisymmetric system, the third component V_ϕ of the flow would be determined from $\nabla \cdot V = 0$. However, in this axisymmetric system V_ϕ is independent of ϕ and so does not appear in $\nabla \cdot V$. Consequently C_ϕ is undetermined and we call it the *'arbitrary' geostrophic flow* V_G. We then have

$$V_\phi = V_M + V_T + V_G \, , \tag{7.2.12}$$

where

$$V_M = \int_z^{z_T} \Big\langle \nabla \times \big((\nabla \times B) \times B \big) \Big\rangle_\phi dz' \tag{7.2.13}$$

is the *magnetic wind*, and

$$V_T = -q\widetilde{Ra} \int_z^{z_T} \langle \nabla T \times r \rangle_\phi \, dz' \tag{7.2.14}$$

is the *thermal wind* familiar in the meteorological literature (see for example Roberts & Soward 1978). Note that with our choice of the limits of integration in (2.13, 14), $V_M = V_T = 0$ at $z = z_T$. Consequently the geostrophic flow $V_G = \left. V_\phi \right|_{z_T}$.

The apparent arbitrariness of the geostrophic flow V_G is a consequence of considering (2.1) in isolation, i.e., of considering the forcing terms on the right-hand side as prescribed, rather than as determined through (1.2–3). In practice V_G is not arbitrary. The manner in which it is determined depends on the importance of the Ekman suction.

7.2.3 Determination of V_G — Ekman States, Taylor States and Model-Z

If Taylor's condition (2.4) is satisfied, then (2.1) can be solved for $\mathbf{V}$ up to the unknown geostrophic flow. As Fearn & Proctor (1992) point out, it is the very existence of a 'homogeneous solution' of the form $V_G(s)\mathbf{1}_\phi$ that makes a 'solvability condition' of the form (2.4) necessary. Of course, $V_G(s)$ can only be considered as arbitrary in the context of solving (2.1) for a given right-hand side when (2.4) is satisfied. In practice, V_G is determined in one of two ways. Either Ekman suction is important, so the left-hand side of (2.3) is non-zero and Taylor's condition does not apply. Then we know $\overline{V}_\phi$ at the boundary through an expression like (2.9). This extra piece of information determines V_G explicitly. Alternatively, Taylor's condition does apply. The system (1.1–5) then must adjust the magnetic field so that Taylor's condition is satisfied. The mechanism used to achieve this is to adjust the differential rotation (the ω-effect discussed in chapter 1) by varying V_G. The differential rotation stretches out poloidal field to generate toroidal field. By varying V_G, B_ϕ can be adjusted and perhaps (2.4) satisfied. This mechanism determines V_G in a very complicated implicit manner.

There is no guarantee that (2.4) can be satisfied. Fearn & Proctor (1987a) tackled the problem of the determination of V_G through this mechanism, by choosing V_G to minimise the absolute value of the left-hand side of (2.4) for a given poloidal magnetic field and a flow that is prescribed (apart from the geostrophic flow). Their method was very successful for certain choices of field and flow, but gave poor results for other choices.

Axisymmetric kinematic α^2- and $\alpha\omega$-dynamo models and non-axisymmetric magnetoconvection models have been adapted to include a geostrophic flow determined by an expression similar to (2.9). Without the feedback due to the geostrophic flow, both problems are linear and would show exponential growth of their solutions for a sufficiently large force. For forcing just above critical, the systems typically find themselves in an 'Ekman state' where equilibration of the amplitude of the solution is achieved through the action of the geostrophic flow. In this respect, it is the condition (2.3) that provides the most important nonlinear effect, since it becomes important at much smaller amplitude of solution than all other nonlinear effects. The reason for this is the small value of the Ekman number, giving an equilibrated solution amplitude of $O(E^{1/4})$ (see section 7.3). As the driving force is increased, the system usually evolves to a state where (2.4) is satisfied (a 'Taylor state') and it is the other nonlinear effects that are responsible for equilibration, this time at higher ($O(1)$) amplitude; viscous effects no longer have a major influence on the solution.

This picture of the nonlinear evolution of a dynamo has come to be known as the Malkus–Proctor scenario, see Malkus & Proctor (1975). It is not the only possible manner in which a dynamo can evolve. An (or the) alternative is where the Taylor state is replaced by a state in which the solution amplitude is $O(1)$ but where viscous effects remain important, even in the limit $E \rightarrow 0$. This is Braginsky's (1975) model-Z. Its fundamental difference from a Taylor state is the manner in which Taylor's constraint is satisfied or almost satisfied. In a Taylor state, with $|\mathbf{B}|$ of $O(1)$, (2.4) is satisfied with $\tau \equiv [(\nabla \times \mathbf{B}) \times \mathbf{B}]_\phi$ of $O(1)$ everywhere and regions of positive τ cancelling with regions of negative τ when the integral over $C(s)$ is taken. (This cancellation effect is illustrated well in Fearn & Proctor (1987a).) There is strong coupling between adjacent cylinders (since τ is $O(1)$), providing the mechanism for Taylor's condition (2.4) to be maintained as the system evolves. By contrast, in model-Z (2.4) is almost satisfied by τ being small everywhere. This is achieved by having B_s close to zero, while B_ϕ, B_z are $O(1)$; see (2.22). The meridional field is then almost aligned with the z-axis, hence the name of the model. Since B_s is small, there is only small coupling between adjacent cylinders $C(s)$ and the system is unable to satisfy Taylor's condition exactly. Consequently, the geostrophic flow remains dependent on the strength of core-mantle coupling. Here, we have concentrated on viscous core-mantle coupling, so V_G remains dependent on E, and very large geostrophic flows are found in the limit of small E; see, for example, Braginsky & Roberts (1987).

Note that in the above discussion we have used the description '$O(1)$' rather loosely. This has been to avoid too detailed a discussion of the appropriate scalings and to focus on the important distinction between the small amplitude Ekman state and the high amplitude Taylor and model-Z states. Model-Z is discussed in detail in chapter 9, and the relationship between model-Z and Taylor states in Roberts (1989). We discuss the nonlinear role of the geostrophic flow on kinematic dynamo and magnetoconvection models in detail in section 7.3.

7.2.4 Alternative Forms of Taylor's Constraint
When considering magnetic fields that are axisymmetric and contained in a mantle that is a perfect electrical insulator, it has been found useful to rewrite the right-hand side of (2.3). We have, in cylindrical polar coordinates,

$$[(\nabla \times \mathbf{B}) \times \mathbf{B}]_\phi = \frac{1}{s}\left[B_s\frac{\partial}{\partial s}(sB_\phi) + sB_z\frac{\partial B_\phi}{\partial z} - \frac{1}{2}\frac{\partial}{\partial \phi}(B_s^2 + B_z^2)\right]. \quad (7.2.15)$$

We know $\nabla \cdot \mathbf{B} = 0$ so we can add $B_\phi \nabla \cdot \mathbf{B}$ to the right-hand side of (2.15) to get

$$\frac{1}{s}\left[\frac{1}{s}\frac{\partial}{\partial s}(s^2 B_\phi B_s) + s\frac{\partial}{\partial z}(B_\phi B_z) - \frac{1}{2}\frac{\partial}{\partial \phi}\left(B_s^2 + B_z^2 - B_\phi^2\right)\right], \quad (7.2.16)$$

which is the ϕ component of the divergence (1.2.46) of the Maxwell stress (1.2.47). Then

$$\int_{z_B}^{z_T} \langle (\nabla \times \mathbf{B}) \times \mathbf{B} \rangle_\phi \, dz = \frac{1}{s^2} \int_{z_B}^{z_T} \frac{\partial}{\partial s} \left(s^2 \langle B_\phi B_s \rangle \right) dz + \left[\langle B_\phi B_z \rangle \right]_{z_B}^{z_T}$$

$$= \frac{1}{s^2} \frac{d}{ds} \left(s^2 \int_{z_B}^{z_T} \langle B_\phi B_s \rangle \, dz \right) + \left[\langle B_\phi B_z \rangle \right]_{z_B}^{z_T} \quad (7.2.17)$$

$$- \left(\langle B_\phi B_s \rangle_{z_T} \frac{dz_T}{ds} - \langle B_\phi B_s \rangle_{z_B} \frac{dz_B}{ds} \right).$$

The last (bracketed) term in (2.17) (which arises from taking the s derivative outside the integral) may be written in the form

$$s \left[\frac{1}{z} \langle B_\phi B_s \rangle \right]_{z_B}^{z_T}. \quad (7.2.18)$$

This combines compactly with the $\left[\langle B_\phi B_z \rangle \right]_{z_B}^{z_T}$ term and (2.17) becomes

$$\int_{z_B}^{z_T} \langle (\nabla \times \mathbf{B}) \times \mathbf{B} \rangle_\phi \, dz = \frac{1}{s^2} \frac{d}{ds} \left(s^2 \int_{z_B}^{z_T} \langle B_\phi B_s \rangle \, dz \right) + \left[\frac{r}{z} \langle B_\phi B_r \rangle \right]_{z_B}^{z_T} \quad (7.2.19)$$

(see also Soward 1991, equation (4.6)), where B_r is the radial component of **B**. Note that it has been possible to write (2.18, 19) in forms that apply to both cases (*a*) and (*b*) of figure 7.1. We shall call the second term on the right-hand side of (2.19) the *boundary term*.

When the field is axisymmetric *and* the mantle is a perfect insulator *and* we are dealing with case (*b*) of figure 7.1, then $B_\phi = 0$ at $z = z_B, z_T$, so the boundary term vanishes and (2.19) becomes

$$\int_{z_B}^{z_T} \langle (\nabla \times \mathbf{B}) \times \mathbf{B} \rangle_\phi \, dz = \frac{1}{s^2} \frac{d}{ds} \left(s^2 \int_{z_B}^{z_T} B_\phi B_s \, dz \right) \quad (7.2.20)$$

(where we can dispense with the ϕ average since the field is axisymmetric). Then Taylor's condition (2.4) can be expressed as

$$\frac{1}{s^2} \frac{d}{ds} \left(s^2 \int_{z_B}^{z_T} B_\phi B_s \, dz \right) = 0. \quad (7.2.21)$$

Hence, since the integral must not be singular for $s = 0$, (2.4) is equivalent to

$$T \equiv \int_{z_B}^{z_T} B_\phi B_s \, dz = 0. \quad (7.2.22)$$

This result goes back to Childress (1969) (see also for example Roberts & Gubbins 1987). Braginsky (1975) used the result (2.20) for the case where (2.4) does not hold to express (2.9) in the form

$$V_\phi \big|_{z_T} = E^{-1/2} \left(\frac{\cos \theta}{2} \right)^{1/2} \frac{1}{s^2} \frac{d}{ds} \left(s^2 T \right). \quad (7.2.23)$$

This is equation (1.6) of Braginsky (1975). An alternative derivation, based on considering magnetic stresses, is given in chapter 9. Equation (2.23) has been reproduced in a variety of forms in many publications. For comparison purposes, note that the definition of T often involves integration from 0 to z_T rather than the limits used in (2.22). The two definitions differ by a factor of two.

The boundary term in (2.19) does not in general vanish. It is non-zero for an axisymmetric field with a conducting mantle or where the cylinder intersects the (conducting) inner core, and for a non-axisymmetric field irrespective of the conductivity of the boundaries (the latter because the identification of toroidal with zonal fields is not applicable in the non-axisymmetric case). In general, the rewriting of the Lorentz force term given in (2.17) gives a more complicated expression rather than a simplification, and it is probably best to use the original form. In some special cases though it is possible to evaluate the boundary term explicitly and to formulate a modified Taylor constraint. Examples are the effect of a finitely conducting inner core (Hollerbach & Jones 1993) and a weakly conducting mantle (see below).

Braginsky (1975) found the form (2.20) useful in two ways. First, it emphasizes how a poloidal field that is almost parallel to the z-axis (so B_s is close to zero) helps to satisfy Taylor's condition. This is the basis of Braginsky's model-Z dynamo. Secondly, for a weakly conducting mantle he was able to solve for the magnetic field in the mantle in the case where the conducting layer in the mantle is thin. (Fearn & Proctor (1992) do the same, but do not restrict themselves to a thin layer, nor to axisymmetric fields or dominant zonal flows as Braginsky (1975) did.) The solution for the mantle field then permitted a boundary condition to be determined for the core field, and eventually led to the boundary term in (2.19) being expressed in terms of a function proportional to $V_\phi|_{z_T}$ and proportional to the mantle conductivity. Thus, an expression analogous to (2.23) was derived for the case of electromagnetic core-mantle coupling (see Braginsky (1975) equation (2.5a) and chapter 9). This led Braginsky and later workers on model-Z to treat the modification to (2.4) due to mantle conductivity as being on the same footing as the Ekman layer correction (see Cupal 1985, Anufriev & Cupal 1987, Braginsky 1988). However care needs to be exercised here, since (2.4) remains necessary if $E = 0$, even with mantle conductivity, and in fact (2.4) will be violated by general initial conditions, whereas when $E \neq 0$ the appropriate modification (2.9) to (2.4) guarantees a solution for arbitrary **B**. Also, in going from the case of an insulating to a weakly conducting mantle (for $E = 0$) (2.4) must still be satisfied, so the (now) non-zero boundary term in (2.17) is balanced by a modified volume term $\int_{C(s)} B_\phi B_s \, dz d\phi$. The latter changes because **B** must change in response to the modified boundary condition at the core-mantle boundary.

7.3 EQUILIBRATION THROUGH THE GEOSTROPHIC FLOW

The complexity of the full hydrodynamic problem has prompted the study of simpler problems that focus on part of the whole problem. The two main classes are kinematic mean-field dynamos (see chapter 1) and convection in the presence of a prescribed magnetic field (*magnetoconvection*) (see chapter 3). In their simplest forms both are linear. Here, we look in turn at both classes of problem where the (nonlinear) effect of a geostrophic flow determined by Ekman suction through a condition of the form (2.9) is included. We shall refer to the mean-field dynamo problem as 'post-kinematic' since the term 'kinematic' implies that the velocity is completely prescribed whereas here, only part of the flow is prescribed, with the remainder (the geostrophic flow) being determined by the dynamics of the problem.

7.3.1 Post-kinematic Axisymmetric Dynamos — Energetics

The justification for, and the equations describing axisymmetric α^2- and $\alpha\omega$-dynamos are given in chapter 1. The equations describing the evolution of the mean field $\overline{\mathbf{B}} = B\mathbf{1}_\phi + \mathbf{B}_M$ are

$$\partial_t B + s\mathbf{V}_M \cdot \nabla \left(\frac{B}{s}\right) - \left(\nabla^2 - \frac{1}{s^2}\right) B = s\mathbf{B}_M \cdot \nabla \left(\frac{V}{s}\right) + (\nabla \times \overline{\mathcal{E}})_\phi, \quad (7.3.1)$$

$$\partial_t A + \frac{1}{s}\mathbf{V}_M \cdot \nabla (sA) - \left(\nabla^2 - \frac{1}{s^2}\right) A = \overline{\mathcal{E}}_\phi, \quad (7.3.2)$$

where $\mathbf{V} = V\mathbf{1}_\phi + \mathbf{V}_M$,

$$\mathcal{E} = \mathbf{V}' \times \mathbf{B}' \quad (7.3.3)$$

is taken to be equal to $\alpha\mathbf{B}$, and $\mathbf{B}_M \equiv \nabla \times A\mathbf{1}_\phi$. The source for the poloidal field is a prescribed α-effect. The toroidal field may be generated either by an α-effect (α^2-dynamo), an ω-effect ($\alpha\omega$-dynamo) or more generally by both ($\alpha^2\omega$-dynamo). With α and ω prescribed, the problem is linear in the magnetic field and if the combined strength of the α- and ω-effects is large enough, then the field strength will grow exponentially.

The differential rotation V in (3.1) is given by $V = \overline{V}_\phi$ in (2.12). In kinematic problems the thermal wind V_T is prescribed rather than being determined by (2.14). It can be shown that V_T is the only component of V that can lead to an increase in the toroidal field energy, see Childress (1969) and below. The magnetic wind V_M (see (2.13)) has no net effect on the toroidal field energy and the effect of the geostrophic flow depends on the role of Ekman suction.

If we take the case where the mantle is an insulator and consider case (*b*) *of figure* 7.1, then equations (2.22, 23) may be used in place of (2.4, 9) respectively. To determine the energetics of the toroidal flow, we multiply

(3.1) by B and integrate over the volume of the core. The left-hand side gives

$$\frac{\partial}{\partial t} \int_v \left(\frac{B^2}{2}\right) dV,$$

which is the rate of change of toroidal field energy. If we consider the term $s\mathbf{B}_M \cdot \nabla(V_G/s)$ in (3.1), since $V_G = V_G(s)$, this becomes $sB_s d\omega_G/ds$ where $\omega_G = V_G/s$. Multiplying by B and integrating over the core gives

$$E_G \equiv \int_0^{2\pi} \int_0^1 \int_{z_B}^{z_T} sBB_s \frac{d\omega_G}{ds} s\, dz\, ds\, d\phi$$

$$= 2\pi \int_0^1 s^2 \frac{d\omega_G}{ds} T\, ds.$$

(7.3.4)

When Taylor's constraint is satisfied this vanishes by (2.22), and so the geostrophic flow has no net effect on the generation of toroidal field.

When $T \neq 0$, integrating the right-hand side of (3.4) by parts gives

$$E_G = 2\pi \left[\omega_G s^2 T\right]_0^1 - 2\pi \int_0^1 \omega_G \frac{d}{ds}(s^2 T)\, ds.$$

(7.3.5)

The first term on the right-hand side vanishes since $T = 0$ at $s = 1$. Substituting for $\omega_G = s^{-1}\nabla_\phi\big|_{z_T}$ from (2.23) then gives

$$E_G = -2\pi E^{-1/2} \int_0^1 \left(\frac{\cos\theta}{2}\right)^{1/2} \frac{1}{s^3} \left(\frac{d}{ds}(s^2 T)\right)^2 ds,$$

(7.3.6)

(see Braginsky 1975, equation (6a)). This term is negative definite so when it is determined by Ekman suction, the geostrophic flow acts to reduce the toroidal field energy. The energy loss is due to viscous effects and, since they are only important in the Ekman boundary layers, it is viscous damping in the boundary layer that acts to equilibrate the system (3.1–2) when V_G is determined by (2.23). This equilibration occurs when $B^2 = O(E_G) = O(E^{-1/2}B^4)$, i.e., when $B = O(E^{1/4})$. At this amplitude, other nonlinear effects are unimportant.

The process of multiplying through (3.1) by B and integrating over the core also determines the roles of the other terms. For axisymmetric fields with $|B| \gg |\mathbf{B}_M|$, the magnetic wind (2.13) can be shown to be

$$V_M = B^2/s.$$

(7.3.7)

Then, the advective term $s\mathbf{V}_M \cdot \nabla(B/s)$ combines with the magnetic wind term $s\mathbf{B}_M \cdot \nabla(V_M/s)$ to give, after some manipulation, multiplication by B and integration over the core

$$\int_V \left(\nabla \cdot (\mathbf{V}_M B^2/2) - \nabla \cdot (\mathbf{B}_M B^3/s)\right) dV.$$

(7.3.8)

Applying the divergence theorem, this vanishes when $B = 0$ on the boundary, see Braginsky (1975). The diffusive term $(\nabla^2 - s^{-2})B$ gives

$$\int_V \left(\nabla \cdot (B\nabla B) - (\nabla B)^2 - B^2/s^2 \right) dV . \qquad (7.3.9)$$

Application of the divergence theorem and using $B = 0$ on the boundary removes the first term, leaving (as one would expect) a negative definite term, which may be expressed in the form

$$- \int_V \left(\nabla(sB)/s \right)^2 dV , \qquad (7.3.10)$$

see Braginsky (1975). Note that the equivalence of (3.9) and (3.10) depends on $B = 0$ on the boundary.

Thus both the magnetic diffusion and the ω-effect associated with the geostrophic flow are sinks of toroidal field energy. The advective and magnetic wind terms have no net effect and the only possible sources are the thermal wind and the α-effect (i.e., the electromotive force $\overline{\mathcal{E}}$). A similar analysis for (3.2) shows that the only term that can lead to the generation of meridional field is the term on the right-hand side of (3.2). Absence of this term (which arises from the mean effect of the nonlinear interaction of the non-axisymmetric components of the problem) means that A must decay and the dynamo fail. This result is Cowling's theorem (Cowling 1934).

Two points should be noted about the above. First, an alternative treatment involves multiplication through by B/s^2 and integrating over the core; see Childress (1969) and for example Roberts & Gubbins (1987). Then the advective and magnetic wind contributions vanish independently. Second, the above results have depended heavily on $B = 0$ on the boundary. As we have seen in the previous section, this applies only if the field is axisymmetric and the mantle is a perfect insulator. Otherwise the above manipulations lead to much more complicated expressions that are not so easy to interpret.

7.3.2 Post-kinematic Axisymmetric Dynamos with Geostrophic Flow

Here, we consider the effect of adding a geostrophic flow determined by (2.23) to what is otherwise a kinematic dynamo. The geostrophic flow will be included whether the original linear problem included an ω-effect or not. If we still wish to refer to α^2- or $\alpha\omega$-dynamos then these terms must be thought of as referring to the original linear problem and the ω-effect referred to must be thought of as the prescribed thermal wind.

The strengths of the α- and ω-effects are measured by the magnetic Reynolds numbers R_α and R_ω defined in section 1.4.3. Their product is called the dynamo number D; see (1.4.53). This is the appropriate control parameter for $\alpha\omega$-dynamos. For α^2-dynamos it is R_α. For the purposes of the following discussion, which deals with both α^2- and $\alpha\omega$-dynamos, it is convenient to talk in terms of a single control parameter, so in what follows, D should be understood to stand for R_α when applied to α^2-dynamos.

We expect that, for dynamo numbers D just above the linear critical value, the system will be in an Ekman state (see section 7.2.3). As D is increased, the amplitude of B will increase as greater driving can counteract greater viscous dissipation. However, if the spatial structure of BB_s remains unchanged then B must remain $O(E^{1/4})$ in the limit of small E. Alternatively, the spatial structure of BB_s may change in such a way that the positive and negative contributions almost cancel (see for example Fearn & Proctor 1987a). Then the amplitude of the field can grow while V_G remains $O(1)$. If this happens, then the system evolves to a Taylor state in which $B = O(1)$, its magnitude determined by nonlinear effects other than (2.23).

The Ekman state, the Taylor state and the transition between the two have been investigated in a variety of models. Much has been done in simpler geometries, where considerable simplifications to the equations are possible. The plane layer has been investigated for the α^2-problem by Soward & Jones (1983) and for the $\alpha\omega$-problem by Abdel-Aziz & Jones (1988), and a duct model by Jones & Wallace (1992) for the $\alpha\omega$-problem. The $\alpha\omega$-problem in the plane layer is a little simpler than the α^2-problem; see Jones (1991). There is no magnetic wind and poloidal flow can be included. It is the nonlinear effect of the poloidal flow that produces equilibration when Taylor's condition is satisfied. The solutions found show the behaviour sketched in figure 7.2. At D_c, the linear problem is neutrally stable. For $D \gtrsim D_c$ the solution is in an Ekman state. As D is increased, the system evolves and B, A change until Taylor's condition is satisfied at $D = D_T$ where the solution amplitude increases from $O(E^{1/4})$ to $O(1)$. This happens for each linear eigenmode. The Taylor state solutions are independent of E in the manner envisaged by Taylor (1963) and Malkus & Proctor (1975).

For the α^2-problem (Soward & Jones 1983), other manners of connection of the Ekman and Taylor states were found, see figure 7.3; subcritical instability was found as well as supercritical and it was possible for the solution branches for Taylor states to be disconnected from those for the Ekman states.

Bounding the (infinite) plane layer to form a duct gives a more realistic model, removing the unsatisfactory feature that in the plane-layer model dynamo waves propagate for ever (Jones 1991). Jones & Wallace (1992) have investigated the duct model. This has to be done using time-stepping (rather then the eigenvalue approach used for the plane-layer calculations) providing a very useful preliminary to calculations in a sphere. The (nonlinear) dissipative nature of the geostrophic flow turns out to have very important numerical consequences. Since it is this that provides the equilibration of the

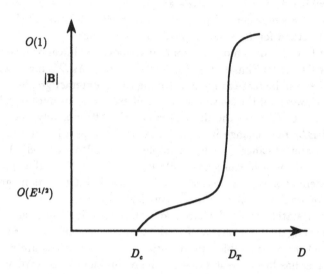

Figure 7.2 Sketch of the nonlinear behaviour of solutions to the plane-layer $\alpha\omega$-dynamo model (see Abdel-Aziz & Jones 1988).

nonlinear solution it is essential that the negative-definiteness of E_G in (3.6) be preserved in any numerical scheme (Jones 1991).

In general $\alpha\omega$-dynamos are oscillatory and the only robust method of making them steady is to include the effect of a meridional circulation (see Roberts 1972). Jones & Wallace (1992) include this in their model and, by varying the various parameters at their disposal, find a wide variety of solution behaviour; see also Jones (1991) who discusses this in much more detail.

Spherical models that include the Ekman-determined geostrophic flow include those based on Braginsky's model-Z idea. These form a special class of $\alpha\omega$-models; see for example Braginsky (1975, 1978, 1988) and Braginsky & Roberts (1987). We say no more about model-Z here as it is dealt with in detail in chapter 9.

Following on from the simple plane-layer models described above, there have been several calculations for α^2- and $\alpha^2\omega$-dynamos in spherical geometries; see Hollerbach & Ierley (1991), Barenghi & Jones (1991), Barenghi (1992, 1993) and Hollerbach, Barenghi & Jones (1992). As for the simple models, these use an Ekman suction condition similar to (2.9) to determine the geostrophic flow. Numerically this causes a significant problem since the natural coordinates for calculation are spherical polars while the integration involved is over cylinders $C(s)$. Many detailed results have come from these

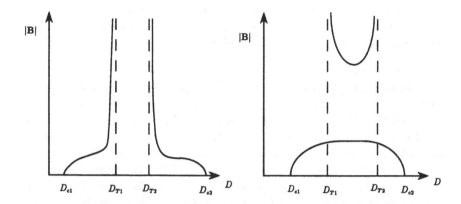

Figure 7.3 Sketches of two types of nonlinear behaviour of solutions to the plane-layer α^2-dynamo model (see Soward & Jones 1983).

models, but the main message they give is that for α^2-models the system almost always evolves to a Taylor state as the dynamo number is increased. This is not true for $\alpha\omega$-models (a result also found in the Jones & Wallace duct model). For a general α, in a finite geometry, in an $\alpha^2\omega$-model, the smaller the importance of the α-effect in generating toroidal field (i.e., the closer the system is to an $\alpha\omega$-dynamo), the more difficult it is to find a Taylor state. The explanation for this appears to be that for α^2-models, there are typically no secondary bifurcations between the primary bifurcation to a steady dynamo (at $D = D_c$) and the transition to a Taylor state. For $\alpha\omega$-dynamos, there are typically bifurcations to more and more complicated time behaviour following the primary bifurcation. In some cases an oscillatory Taylor state is eventually established. In others the solution is at some times close to satisfying Taylor's constraint, but at intermediate times the geostrophic flow is very much determined by the Ekman suction. The solution appears unable to make up its mind whether it wants to be in a Taylor state or an Ekman state; see Jones & Wallace (1992), Barenghi & Jones (1991) and Hollerbach et al. (1992).

Another class of spherical models has been studied for α^2-dynamos by Malkus & Proctor (1975), Proctor (1977), Ierley (1985) and Hollerbach & Jones (1993). These do not make the magnetostrophic approximation. Instead of determining V_G through Ekman suction, the full mean momentum equation is solved. This requires resolution of the Ekman layers in the limit of small E and so the approach is limited by numerical resolution. Ierley (1985), though improving on the earlier work, still found some dependence of his solutions on E at the smallest values ($E = 3 \times 10^{-4}$) he could achieve, so he could not claim to have reached a true Taylor state (which should be independent of E in the limit $E \to 0$). However this may simply be because he was unable to go to sufficiently small E.

7.3.3 Magnetoconvection

Perhaps the earliest analysis of the nonlinear coupling of the geostrophic flow with convection is by Roberts & Soward (1972). They anticipate that the effect must be a stabilising one and succeed in demonstrating that a MAC wave (that in the absence of the geostrophic flow would grow in amplitude exponentially) equilibrates at a finite amplitude due to the effect of the geostrophic flow. Their analysis neglects the poloidal field (which so effectively couples the concentric cylinders $C(s)$), leading to a greater strength of geostrophic flow and a probable overestimate of its stabilising effect. Additionally it should be noted that their analysis is weakly nonlinear and is for a non-dissipative problem, so is different in character from the dissipative work we describe below.

The evolution of the non-axisymmetric part of the field, flow and temperature are (for $Ro, E = 0$) described by

$$1_z \times \mathbf{V}' = -\nabla \Pi' + \left((\nabla \times \overline{\mathbf{B}}) \times \mathbf{B}' + (\nabla \times \mathbf{B}') \times \overline{\mathbf{B}} \right) + qRT'\mathbf{r}$$
$$+ \left[(\nabla \times \mathbf{B}') \times \mathbf{B}' - \langle (\nabla \times \mathbf{B}') \times \mathbf{B}' \rangle \right], \qquad (7.3.11)$$

$$\partial_t \mathbf{B}' = \nabla \times (\mathbf{V}' \times \overline{\mathbf{B}}) + \nabla \times (\overline{\mathbf{V}} \times \mathbf{B}') + \nabla^2 \mathbf{B}'$$
$$+ \left[\nabla \times (\mathbf{V}' \times \mathbf{B}') - \langle \nabla \times (\mathbf{V}' \times \mathbf{B}') \rangle \right], \qquad (7.3.12)$$

$$\partial_t T' + \mathbf{V}' \cdot \nabla \overline{T} + \overline{\mathbf{V}} \cdot \nabla T' = q\nabla^2 T' - \left[\mathbf{V}' \cdot \nabla T' - \langle \mathbf{V}' \cdot \nabla T' \rangle \right], \qquad (7.3.13)$$

$$\nabla \cdot \mathbf{V}' = 0, \qquad \nabla \cdot \mathbf{B}' = 0. \qquad (7.3.14, 15)$$

The linear problem for the onset of non-axisymmetric convection in the presence of a prescribed mean field is governed by the above system without the terms (contained in square brackets) quadratic in the non-axisymmetric quantities. This is discussed in detail in chapter 3. Here we are interested in the non-linear problem. The (nonlinear) terms in square brackets will become important when the amplitude of the non-axisymmetric quantities becomes $O(1)$. At smaller amplitude they can be ignored. Then the only nonlinear interaction is through the (mean) geostrophic flow V_G. It is determined through a relation of the form (2.9), with

$$\left\langle (\nabla \times \mathbf{B}) \times \mathbf{B} \right\rangle_\phi = \left((\nabla \times \overline{\mathbf{B}}) \times \overline{\mathbf{B}} \right)_\phi + \left\langle (\nabla \times \mathbf{B}') \times \mathbf{B}' \right\rangle_\phi. \qquad (7.3.16)$$

From linear studies (see Braginsky 1980, Fearn & Proctor 1983a, b, Fearn 1989 and chapter 3) we know that differential rotation acts to inhibit convection.

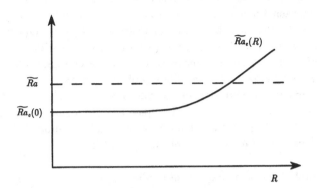

Figure 7.4 The mechanism of equilibration of magnetoconvection by the geostrophic flow.

The critical value $\widetilde{Ra}_c$ of the modified Rayleigh number increases with the magnetic Reynolds number R, with the differential rotation beginning to have a noticeable effect for $R \gtrsim q$ for $q \leq O(1)$. Consider now making a non-axisymmetric perturbation of a basic state $\overline{\mathbf{B}} = \mathbf{B}_0$, $\overline{\mathbf{V}} = \mathbf{0}$ and $\overline{T} = T_0$. Let $\widetilde{Ra}_c(0)$ be the critical value of $\widetilde{Ra}$ for this basic state. For any given mean flow $\overline{\mathbf{V}}$, we can solve the linear problem above to find the value $\widetilde{Ra}_c(\overline{\mathbf{V}})$ of $\widetilde{Ra}_c$. Typically $\widetilde{Ra}_c(\overline{\mathbf{V}}) > \widetilde{Ra}_c(0)$. In particular, for a given dynamically determined V_G, we can set $\overline{\mathbf{V}} = V_G \mathbf{1}_\phi$ and solve the linear problem to find $\widetilde{Ra}_c(V_G)$.

If we choose a value of $\widetilde{Ra} > \widetilde{Ra}_c(0)$, we expect that our non-axisymmetric solution will grow in a linear manner until $V_G = O(q)$. Then the shear (determined by (2.9)) will act back on the non-axisymmetric convection and inhibit growth. The amplitude will equilibrate when $\widetilde{Ra}_c(V_G) = \widetilde{Ra}$, i.e., there will be no further growth of the solution amplitude when the action of the geostrophic flow has been enough to raise the critical value of the modified Rayleigh number to the imposed value of $\widetilde{Ra}$, see figure 7.4. This is the manner in which the geostrophic flow acts to equilibrate the non-axisymmetric convection. As with the mean-field axisymmetric dynamos discussed in the previous section, this equilibration occurs when the solution amplitude is (for $q = O(1)$) $O(E^{1/4})$, see (2.9); $|\mathbf{B}'| = O(E^{1/4}) \Rightarrow \langle (\nabla \times \mathbf{B}') \times \mathbf{B}' \rangle_\phi = O(E^{1/2})$ and so $V_G = O(1)$.

We have to be more careful here however, since, as well as influencing the mean flow, the non-axisymmetric convection can act to generate mean magnetic field through the mean emf $\overline{\mathcal{E}}$ and so modify the first term on the right-hand side of (3.16). Even though $\mathbf{B}_0$ is chosen to make this term vanish, it will not necessarily be zero at later times as the system evolves. In particular, when the non-axisymmetric solution amplitude is $O(E^{1/4})$, $\overline{\mathcal{E}} = O(E^{1/2})$, and then the modification to $\overline{\mathbf{B}}$ is also $O(E^{1/2})$, making the size of both terms on the right-hand side of (3.16) comparable: $O(E^{1/2})$. This means that in general, we have to solve (3.11–15) together with (3.1–3) when we wish to investigate the nonlinear effect of the geostrophic flow on non-axisymmetric magnetoconvection. This problem has some very interesting aspects. There are two equilibration mechanisms acting. There is the viscous damping mechanism for the mean field (see the discussion in the previous section), and there is the action of shear on the non-axisymmetric convection discussed above. To date this complicated set of interactions has not been studied.

In the simple cases that have been studied, the model has been carefully chosen such that the term

$$\left((\nabla \times \overline{\mathbf{B}}) \times \overline{\mathbf{B}}\right)_\phi = -\frac{\partial A}{\partial z}\frac{1}{s}\frac{\partial}{\partial s}(sB) - \frac{\partial B}{\partial z}\frac{1}{s}\frac{\partial}{\partial s}(sA) \qquad (7.3.17)$$

is unimportant. For example, if $\mathbf{B}_0 = B_0(s)\mathbf{1}_\phi$ then the second term on the right-hand side of (3.17) is negligible. The first term can be integrated with respect to z and vanishes if the boundary conditions have been chosen such that A vanishes on the top and bottom boundaries (see for example Soward 1986). With the contribution from the mean field unimportant in (3.16), there is no need to solve the mean-field equations (3.1–3), and the problem becomes very much simpler.

The problem has been investigated in an infinite plane-layer model (Roberts & Stewartson 1974, 1975), in a duct (Soward 1986, Jones & Roberts 1990), and in a cylindrical annulus (Skinner & Soward 1988, 1990). In the Roberts and Stewartson model (which has the rotation and gravity vectors normal to the plane and a uniform horizontal magnetic field), for $\Lambda \leq \sqrt{3}/2$ the solution is a single convection roll perpendicular to the applied magnetic field. For higher values of Λ there are two oblique rolls. Each roll by itself satisfies Taylor's condition, but when both rolls coexist Taylor's constraint is not satisfied and there is a complicated interaction between the rolls and the geostrophic flow (see also Soward 1980). The infinite plane-layer problem is not typical of a confined geometry where typically a single mode that does not satisfy Taylor's condition is excited. This motivated Soward (1986) to investigate the effect of putting sidewall boundaries on the Roberts & Stewartson model to produce a duct model.

The duct model of Soward (1986) and the cylindrical annulus of Skinner & Soward (1988, 1990) both have anti-parallel gravity and rotation. In both cases, Taylor solutions were found for sufficiently high Rayleigh number. In contrast, Jones & Roberts (1990) showed that if the duct model was modified to have the rotation vector perpendicular to gravity and to the imposed magnetic field, then for $\Lambda \gtrsim 4$ and fixed wavenumber in the direction of the magnetic field, no Taylor state exists, no matter how high the Rayleigh number is made (see also Jones 1991).

In the models in which $\mathbf{g}$ and $\boldsymbol{\Omega}$ are anti-parallel, the limit $q \to 0$ was found to be problematical. The details of this difficulty were explored in most detail in Soward (1986). The problem arises out of the effect of shear on the convection. Since a balance is achieved between the action of shear and of diffusion, when $q \ll 1$ the temperature perturbation T' is affected to a much greater extent than $\mathbf{B}'$ since the former diffuses much more slowly than the latter (see Fearn & Proctor 1983a, b). Soward (1986) found that the duct divided into two distinct regions; in one V_G is $O(1)$, in the other it is $O(q^{-1})$, with a complicated internal layer separating the two regions. Finding (and resolving) solutions for small q is difficult. In the cylinder there was great difficulty in finding solutions for small q; the lowest value for which results are quoted is 10^{-2}. These calculations strike a cautionary note, since the geophysical value of q (based on molecular values of the diffusivities) is 10^{-6}.

7.4 HYDRODYNAMIC DYNAMOS

7.4.1 Discussion

In the preceding sections we have discussed the various features that characterise planetary dynamos and analysed their consequences. Our ultimate aim is to understand the planetary dynamo problem and one route to this is through numerical hydrodynamic models. Even with the latest developments in computational power, the stiffness of the problem means that simple brute force will not work. Progress is likely to be made through parallel studies of complementary problems, taking different approaches to the problems associated particularly with the very small value of the Ekman number E. These include resolution of the thin Ekman layers at the core boundaries, resolution of the short length-scale convection when the magnetic field is weak or absent, and the problem of possible discontinuities in the flow or its derivatives at the cylinder circumscribing the inner core (see Hollerbach & Proctor 1993).

Other problems identified include the role of inertia ($Ro \ll 1$), possible short length-scales and internal shear layers developing when $q \ll 1$ (see section 7.3.3), and the complex time behaviour found in $\alpha\omega$-dynamos (see section 7.3.2). In the following sections we review what progress has been made to date.

7.4.2 Models Including Viscosity and Inertia

Most progress has been made for models of this class. Two categories have been investigated. The main one is for $O(1)$ values of the Prandtl number $Pr = \nu/\kappa$ (Zhang & Busse 1988, 1989). The second is a special case of the first. In the infinite Prandtl number limit, inertial effects become unimportant (Zhang & Busse 1990).

Zhang and Busse adopt the approach of starting from a state of no motion and zero magnetic field. As the Rayleigh number is increased they find, first, a bifurcation to a convecting non-magnetic state, then at still higher $\widetilde{Ra}$, bifurcation from this convecting state to a convecting magnetic state (a dynamo). Their model uses stress-free boundary conditions to avoid having to resolve a narrow Ekman boundary layer. However, their approach of starting from a state of zero magnetic field means that they have to resolve the short length-scale convection that is the preferred mode in a rapidly rotating (non-magnetic) system. This is the principal restriction on how small they can make E. The best they were able to achieve was $E = 10^{-3}$. One might have hoped that once they had found a convecting state maintaining a strong magnetic field (and consequently with $O(1)$ length-scales), it would then be possible to reduce E. This was tried but, unfortunately, a dynamo solution did not persist at the lower values of E, the field decaying (Zhang, personal communication). Clearly a small-E limit had not been reached.

Zhang and Busse's solutions showed some interesting features. For $O(1)$ Prandtl number, they found solutions of $\alpha\omega$-type having a strong differential rotation and a mean toroidal field some 20 times larger than the mean poloidal field. For the infinite Prandtl number case, the solutions were more of α^2-type with poloidal and toroidal fields comparable in magnitude and smaller differential rotation. Dynamos were found for values of $\widetilde{Ra}$ exceeding some critical value, but fields were not maintained for all higher values of $\widetilde{Ra}$, there being a second critical value above which no dynamo existed (for a particular mode of solution).

More recently St. Pierre (1993) has investigated a plane-layer model.

7.4.3 Models Neglecting Inertia but Including Viscosity

Apart from the special case of the Zhang and Busse model discussed above there are so far no results from models of this class, but work is in progress by Hollerbach, whose spectral approach should be capable of achieving significantly lower values of the Ekman number than Zhang and Busse. He uses no-slip boundary conditions so must resolve the Ekman layers, but his approach will be to start from a finite magnetic field so should avoid the problem of the short length-scale convection found at zero field strength. Glatzmaier is also working on models of this type.

7.4.4 Models Neglecting Inertia and Viscosity

By 'neglecting viscosity' we mean neglecting viscosity in the main body of the core. Viscous effects are retained through the determination of the geostrophic flow by Ekman suction, involving performing integrals over the cylinders $C(s)$. Similar remarks apply here as to models of the class discussed in section 7.4.3; i.e., none have yet been completed. Progress has been made on the axisymmetric part of the problem (see Barenghi & Jones 1991, Barenghi 1992, 1993, Hollerbach *et al.* 1992), and work is in progress on the non-axisymmetric part and the merging of the axisymmetric and non-axisymmetric parts.

The above (time-stepping) work was preceded by an earlier attempt to solve the problem in an iterative manner. The idea was to solve the non-axisymmetric part of the problem in the presence of a prescribed mean field. The associated mean emf $\bar{\mathcal{E}}$ was then used as the source term in a mean-field dynamo and a mean field calculated. This process was repeated until the mean field generated was (to some prescribed tolerance) the same as that input to the non-axisymmetric magnetoconvection problem. Then the convection taking place in the presence of a given mean field is just what is required to generate exactly that field. A self-consistent dynamo solution has been found. Fearn & Proctor (1984, 1987*b*) adopted this approach with some success. Their simplest version was found to work. However, once more of the dynamics, for example the geostrophic flow being determined by that required to make the mean field satisfy Taylor's constraint (Fearn & Proctor 1987*a*), was included, converged solutions were not found.

Acknowledgements

This review was prepared while the author was working at the Isaac Newton Institute, Cambridge. Its support and provision of facilities are gratefully acknowledged. My knowledge and understanding of the material presented here has benefited from discussions with C.F. Barenghi, R. Hollerbach, C.A. Jones, M.R.E. Proctor, P.H. Roberts and A.M. Soward amongst others. I am grateful to S.I. Braginsky, R. Hollerbach, A.M. Soward and K. Zhang who have read a draft of the manuscript and have kindly supplied me with helpful suggestions for improvement and pointed out errors and omissions. Of course, all remaining errors and misunderstandings are my responsibility. My work on planetary dynamos is supported by the Science and Engineering Research Council of Great Britain under grant GR/H 03278.

REFERENCES

Abdel-Aziz, M.M. & Jones, C.A. 1988 $\alpha\omega$-dynamos and Taylor's constraint. *Geophys. Astrophys. Fluid Dynam.* **44**, 117–139.

Anufriev, A.P. & Cupal, I. 1987 Magnetic field at the core boundary in the nearly symmetric hydromagnetic dynamo Z. *Studia Geoph. Geod.* **31**, 37–42.

Barenghi, C.F. 1992 Nonlinear planetary dynamos in a rotating spherical shell II. The post Taylor equilibration for α^2-dynamos. *Geophys. Astrophys. Fluid Dynam.* **67**, 27–36.

Barenghi, C.F. 1993 Nonlinear planetary dynamos in a rotating spherical shell III. $\alpha^2\omega$-models and the geodynamo. *Geophys. Astrophys. Fluid Dynam.* **71**, 163–185.

Barenghi, C.F. & Jones, C.A. 1991 Nonlinear planetary dynamos in a rotating spherical shell I. Numerical methods. *Geophys. Astrophys. Fluid Dynam.* **60**, 211–243.

Batchelor, G.K. 1967 *An Introduction to Fluid Dynamics.* Cambridge University Press.

Braginsky, S.I. 1975 Nearly axially symmetric model of the hydromagnetic dynamo of the Earth. I. *Geomag. Aeron.* **15**, 122–128.

Braginsky, S.I. 1978 Nearly axially symmetric model of the hydrodynamic dynamo of the Earth. *Geomag. Aeron.* **18**, 225–231.

Braginsky, S.I. 1980 Magnetic waves in the core of the Earth II. *Geophys. Astrophys. Fluid Dynam.* **14**, 189–208.

Braginsky, S.I. 1988 The Z model of the geodynamo with magnetic friction. *Geomag. Aeron.* **28**, 407–412.

Braginsky, S.I. 1991 Towards a realistic theory of the geodynamo. *Geophys. Astrophys. Fluid Dynam.* **60**, 89–134.

Braginsky, S.I. & Roberts, P.H. 1987 A model-Z geodynamo. *Geophys. Astrophys. Fluid Dynam.* **38**, 327–349.

Busse, F.H. 1970 Thermal instabilities in rapidly rotating systems. *J. Fluid Mech.* **44**, 441–460.

Childress, S. 1969 A class of solutions of the magnetohydrodynamic dynamo problem. In *The Application of Modern Physics to the Earth and Planetary Interiors* (ed. S.K. Runcorn), pp. 629–648. Wiley.

Cowling, T.G. 1934 The magnetic field of sunspots. *Mon. Not. R. Astr. Soc.* **94**, 39–48.

Cupal, I. 1985 Z-model of the nearly symmetric hydromagnetic dynamo with electromagnetic core-mantle coupling. *Studia Geoph. Geod.* **29**, 339–350.

Fearn, D.R. 1989 Differential rotation and thermal convection in a rapidly rotating hydromagnetic system. *Geophys. Astrophys. Fluid Dynam.* **49**, 173–193.

Fearn, D.R. & Proctor, M.R.E. 1983*a* Hydromagnetic waves in a differentially rotating sphere. *J. Fluid Mech.* **128**, 1–20.

Fearn, D.R. & Proctor, M.R.E. 1983*b* The stabilising role of differential rotation on hydromagnetic waves. *J. Fluid Mech.* **128**, 21–36.

Fearn, D.R. & Proctor, M.R.E. 1984 Self-consistent dynamo models driven by hydromagnetic instabilities. *Phys. Earth Planet. Inter.* **36**, 78–84.

Fearn, D.R. & Proctor, M.R.E. 1987*a* Dynamically consistent magnetic fields produced by differential rotation. *J. Fluid Mech.* **178**, 521–534.

Fearn, D.R. & Proctor, M.R.E. 1987*b* On the computation of steady, self-consistent spherical dynamos. *Geophys. Astrophys. Fluid Dynam.* **38**, 293–325.

Fearn, D.R. & Proctor, M.R.E. 1992 Magnetostrophic balance in non-axisymmetric, non-standard dynamo models. *Geophys. Astrophys. Fluid Dynam.* **67**, 117–128.

Hollerbach, R. & Ierley, G.R. 1991 A modal α^2-dynamo in the limit of asymptotically small viscosity. *Geophys. Astrophys. Fluid Dynam.* **60**, 133–158.

Hollerbach, R. & Jones, C.A. 1993 A geodynamo model incorporating a finitely conducting inner core. *Phys. Earth Planet. Inter.* **75**, 317–327.

Hollerbach, R. & Proctor, M.R.E. 1993 Non-axisymmetric shear layers in a rotating spherical shell. In *Solar and Planetary dynamos* (ed. M.R.E. Proctor, P.C. Matthews & A.M. Rucklidge), pp. 145–152. Cambridge University Press.

Hollerbach, R., Barenghi, C.F. & Jones, C.A, 1992 Taylor's constraint in a spherical $\alpha\omega$-dynamo. *Geophys. Astrophys. Fluid Dynam.* **67**, 37–64.

Ierley, G.R. 1985 Macrodynamics of α^2-dynamos. *Geophys. Astrophys. Fluid Dynam.* **34**, 143–173.

Jones, C.A. 1991 Dynamo models and Taylor's constraint. In *Advances in Solar System Magnetohydrodynamics* (ed. E.R. Priest & A.W. Hood), pp. 25–50. Cambridge University Press.

Jones, C.A. & Roberts, P.H. 1990 Magnetoconvection in rapidly rotating Boussinesq and compressible fluids. *Geophys. Astrophys. Fluid Dynam.* **55**, 263–308.

Jones, C.A. & Wallace, S.G. 1992 Periodic, chaotic and steady solutions in $\alpha\omega$-dynamos. *Geophys. Astrophys. Fluid Dynam.* **67**, 37–64.

Malkus, W.V.R. & Proctor, M.R.E. 1975 The macrodynamics of α-effect dynamos in rotating fluids. *J. Fluid Mech.* **67**, 417–443.

Poirier, J.P. 1988 Transport properties of liquid metals and viscosity of the Earth's core. *Geophys. J.* **92**, 99–105.

Proctor, M.R.E. 1977 Numerical solution of the nonlinear α-effect dynamo equations. *J. Fluid Mech.* **80**, 769–784.

Roberts, P.H. 1968 On the thermal instabilities of a fluid sphere containing heat sources. *Phil. Trans. R. Soc. Lond. A* **263**, 93–117.

Roberts, P.H. 1972 Kinematic dynamo models. *Phil. Trans. R. Soc. Lond. A* **272**, 663–703.

Roberts, P.H. 1989 From Taylor state to model-Z? *Geophys. Astrophys. Fluid Dynam.* **49**, 143–160.

Roberts, P.H. & Gubbins, D.G. 1978 Origin of the main field: Kinematics. In *Geomagnetism*, vol. 2 (ed. J.A. Jacobs), pp. 185–249. Academic Press.

Roberts, P.H. & Soward, A.M. 1972 Magnetohydrodynamics of the Earth's core. *Annu. Rev. Fluid Mech.* **4**, 117–154.

Roberts, P.H. & Soward, A.M. 1978 *Rotating Fluids in Geophysics.* Academic Press.

Roberts, P.H. & Stewartson, K. 1974 On finite amplitude convection in a rotating magnetic system. *Phil. Trans. R. Soc. Lond. A* **277**, 93–117.

Roberts, P.H. & Stewartson, K. 1975 Double roll convection in a rotating magnetic system. *J. Fluid Mech.* **68**, 447–466.

St. Pierre, M.G. 1993 The strong field branch of the Childress–Soward dynamo. In *Solar and Planetary dynamos* (ed. M.R.E. Proctor, P.C. Matthews & A.M. Rucklidge), pp. 295–302. Cambridge University Press.

Skinner, P.H. & Soward, A.M. 1988 Convection in a rotating magnetic system and Taylor's constraint. *Geophys. Astrophys. Fluid Dynam.* **44**, 91–116.

Skinner, P.H. & Soward, A.M. 1990 Convection in a rotating magnetic system and Taylor's constraint II. *Geophys. Astrophys. Fluid Dynam.* **60**, 335–356.

Soward, A.M. 1980 Finite amplitude thermal convection and geostrophic flow in a rotating magnetic system. *J. Fluid Mech.* **98**, 449–471.

Soward, A.M. 1986 Non-linear marginal convection in a rotating magnetic system. *Geophys. Astrophys. Fluid Dynam.* **35**, 329–371.

Soward, A.M. 1991 The Earth's dynamo. *Geophys. Astrophys. Fluid Dynam.* **62**, 191–209.

Soward, A.M. & Jones, C.A. 1983 α^2-dynamos and Taylor's constraint *Geophys. Astrophys. Fluid Dynam.* **27**, 87–122.

Taylor, J.B. 1963 The magnetohydrodynamics of a rotating fluid and the Earth's dynamo problem. *Proc R. Soc. Lond. A* **274**, 274–283.

Zhang, K. & Busse, F.H. 1988 Finite amplitude convection and magnetic field generation in a rotating spherical shell. *Geophys. Astrophys. Fluid Dynam.* **44**, 33–54.

Zhang, K. & Busse, F.H. 1989 Convection driven magnetohydrodynamic dynamos in rotating spherical shells. *Geophys. Astrophys. Fluid Dynam.* **49**, 97–116.

Zhang, K. & Busse, F.H. 1990 Generation of magnetic fields by convection in a rotating spherical fluid shell of infinite Prandtl number. *Phys. Earth Planet. Inter.* **59**, 208–222.

CHAPTER 8

The Chaotic Solar Cycle

E. A. SPIEGEL

Astronomy Department
Columbia University
New York NY 10027, U.S.A.

8.1 PREFACE

Almost 400 hundred years ago Galileo noticed that the period of a pendulum is the same for all small amplitudes. Not long afterwards, Galileo and his contemporaries (see figure 8.1) proved that sunspots really were on the sun. So the same person was involved in discovering the paradigm of periodicity and establishing an exemplar of irregularity. But just how irregularly do sunspots behave? In modern terms, this question comes down to asking how many degrees of freedom are involved in the phenomenon. If the mechanism I am going to describe here, *on/off intermittency*, is operative, this question cannot be answered soon (Platt, Spiegel & Tresser 1993a).

That I should begin this discussion by mentioning aperiodicity is a sign of where we are in the long saga of sunspot studies. Shortly after Galileo's discoveries, serious work on sunspots got under way. This was somewhat disappointing for a time because sunspots had become quite scarce, with only a few per year being detected. This intermission in solar activity lasted approximately throughout the life of Newton, being most extreme when he was in his prime and ending about a decade before his death (Eddy 1978). So the question of the changing level of solar activity must have been much on astronomers' minds at that time. By the time this puzzle was fading from memory, a new issue was raised in the middle of the nineteenth century, when it was noticed that the level of solar activity (as judged mainly by sunspots) was found to vary with some regularity. The variation was taken to be periodic with a ten-year period on (at first) insubstantial evidence, perhaps because the assumption of periodicity came naturally to those indoctrinated with the behaviour of the pendulum. This variation was supported by ancient Chinese observations (Fritz 1882) and it may have been one motivation for the careful recording of sunspot (or Wolf) numbers in Zurich for the last hundred years. (A history of the sunspot number is given by Izenman (1985).) In

M.R.E. Proctor & A.D. Gilbert (eds.)
Lectures on Solar and Planetary Dynamos, 245–265
©1994 Cambridge University Press.

MACVLÆ   SOLARES
ex felectis obfernationibus Petri Saxonis Holsati
Altorfii in Academia Norica factis
AD
MAGNIFICVM SENATVM INCLITÆ REIPVBLICÆ NORINBERGENSIS

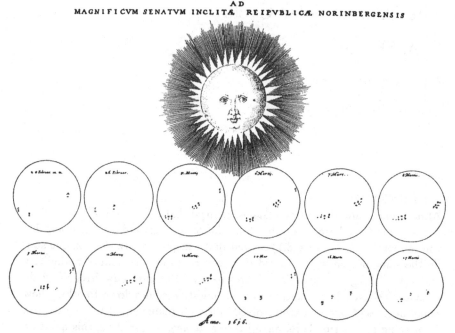

Figure 8.1 Sunspots in the seventeenth century from Heß (1911), courtesy of E.L. Schucking.

any case, it quickly must have become clear that the sunspot number was not varying periodically but, as someone wisely put it, cyclically, with a time scale of eleven years.

The variation of the annual sunspot number with time over the past two centuries is shown in figure 8.2. It is natural to look for an oscillator driving this phenomenon and to ask how many degrees of freedom are represented in the mechanism. I shall argue here that we can model the process as a relatively simple dynamical system that has both the desired cyclic character and the strong intermittency revealed in the so-called Maunder minimum that occurred in Newton's time. Perhaps, from such a mathematical model, we can attempt to read something of the physical nature of the process itself. This is not the usual direction of astrophysical research, which begins by trying to isolate the physical mechanisms behind observed processes. In trying to proceed in terms of a generic mathematical description, I am illustrating the approach of what I call *astromathematics*. However, both approaches have been used in getting to the model described here so that a certain amount of physical background is given. The end product is a set of equations whose

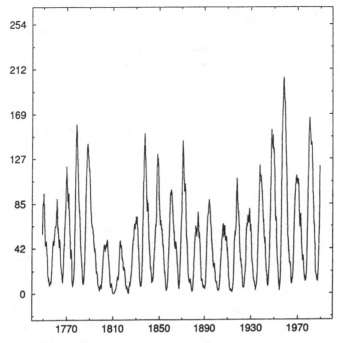

Figure 8.2 The yearly mean sunspot number as a function of time with the period of the Maunder minimum (ca. 1650–1720) not shown.

output looks like the observed variation of the sunspot number. Nevertheless, I have had to forgo calling this 'an astromathematician's apology' since I have not strictly followed the rules of the game, as well as for more obvious reasons.

For those who like to read only introductions, let me say here that the proposed mathematical model has two essential ingredients. First, it contains a simple oscillator. And secondly, the model exhibits extended periods with little activity. This behavior is like the intermittency seen in turbulent fluids and has been called on/off intermittency. It is built into the mathematics by arranging for the equations to admit an invariant manifold within which the system does not exhibit the behavior that will be called activity. The manifold has both stable directions along which the system is occasionally drawn into its neighborhood for extended periods and unstable directions in which it flies out again to resume the large oscillations that here represent the solar cycle. One can make several versions of this process, differing in detail, but what I am after here is the isolation of specific mathematical mechanisms that may be incorporated in such models so as to capture the main temporal features of the global solar cycle (Platt, Spiegel & Tresser 1993b). Such models can also be made spatio-temporal, and this is being done.

8.2 THE SOLAR TACHOCLINE

At any stage of the solar cycle, sunspots are concentrated in a particular band of latitude whose location drifts toward the equator as the cycle progresses, beginning at ±40° and decreasing to ±5° in the course of eleven years. By the time a given cycle is ending at ±5°, the next one has already begun to appear. Of course, it is not the individual spots themselves that move toward the equator, for spots rarely last more than a month or so. This progression in latitude gives the impression that there are *solitary* waves of solar activity whose propagation time is eleven years. The nature of such waves and their fate when they meet at the equator raise questions that I will address presently. Why should the spots not appear all over the place, given that they are appearing at all? The confinement in latitude is a hint that the activity might originate in a physically distinct layer and the wave-like motion of the locale is suggestive of the influence of a wave guide whose thickness reflects the width of the activity zone. One such layer might be the convection zone itself, whose thickness, $R_\odot/3$, is not all that much greater than the width of the band of sunspot activity. However, the strong spots have fields of several kilogauss. At fields well below this, magnetic tubes will have lowered density inside them and be buoyed up to the surface. Unless they are in the deep convection zone, or below this, they will emerge before they have time to develop the field strength seen in many spots. But in the deep convection zone ten kilogauss fields do not lead to significant evacuation; so buoyancy forces are less important. Brandenburg & Tuominen (1991) report that there is sufficiently strong downwelling in compressible convection to overcome the effects of magnetic buoyancy. Hence the lower convection zone might serve as the seat of solar activity (DeLuca 1986). Other possible sites for the origin of solar activity have been considered as well (for example, Layzer, Rosner & Doyle 1979).

Another distinct layer with the right properties is the one that mediates the transition between the differential rotation of the solar convection zone and the rotation of the bulk of the sun, or radiative interior. Such a layer was discussed twenty years ago (Spiegel 1972) but its existence became a reality when helioseismologists were able to infer the distribution of the solar differential rotation well into the sun (Brown et al. 1989, Goode et al. 1991). According to them, the variation of rotational velocity with latitude that is seen on the solar surface continues with little change through the solar convection zone. Throughout the convection zone, the equator turns faster than the poles with a velocity that is constant on cones. I will take this motion in the convection zone as specified, much as oceanographers take the wind stress on the surface of sea as prescribed, though I am sure we are both somewhat in error. Of course, oceanographers allow for time dependence of the wind stress, but helioseismology is not old enough to give us accurately the corresponding variability for the large-scale flow of the solar convection zone. Even the picture of constancy on cones remains tentative (Gough *et al.* 1992).

The inner sun turns rigidly, at least down to depths at which acoustic sounding works. And between these two regimes there is an unresolved transition that is reminiscent of the thermal transition layer between the earth's atmosphere and the deep ocean. In fact, it is perhaps even more like the layers in planetary atmospheres that produce lively activity and are called weather layers. At least, I am claiming that this transition layer produces the magnetic weather in the sun called solar activity. This analogy to geophysical layers like the oceanic thermocline led to the name *tachocline* for the solar rotational transition layer (Spiegel & Zahn 1992).

Even now that the tachocline's existence has been confirmed, it is not clear why it is there. We may reasonably assume that the stresses exerted by the differential rotation of the convection zone on the interior will produce effects in the stable layers. But the implied turbulent spindown process will tend to spread the effects well into the interior and not leave a well-defined layer. However, strongly anisotropic turbulent stresses that are produced by horizontal shear in a vertically stabilized layer (Zahn 1975) could short-circuit this spreading, as could strong horizontal magnetic stresses. To show how this works, in the case of turbulent viscosity, let me give an equationless summary of our estimate of the tachocline thickness on the assumption of a steady tachocline.

If we take the large-scale rotational flow in the convection zone as given, its mismatch to the interior rotation will cause a large-scale convective pumping process that drives a vertical velocity, w, just below the convection zone (Bretherton & Spiegel 1968). This will generate a meridional current with north–south component

$$u \sim \frac{R_\odot}{\ell} w \tag{8.2.1}$$

with vertical extent ℓ. Strictly, a density gradient term is needed, but this is not important as long as ℓ is less than the density scale height, which is rather large in the tachocline.

We need to balance the Coriolis force caused by the north–south motion. If we do this with eddy viscosity operating on the azimuthal flow, v, we have

$$\Omega u \sim \left(\frac{\nu_H}{R_\odot^2} \right) v, \tag{8.2.2}$$

where ν_H is the eddy viscosity of the horizontal turbulence. We have included only horizontal turbulent stresses since they may be expected to dominate in a medium with strongly stable vertical stratification (Zahn 1975).

The azimuthal flow will also produce a Coriolis force and we need a north-south pressure head to balance it:

$$\rho v \Omega \sim \frac{\Delta p}{R_\odot}. \tag{8.2.3}$$

We are assuming that the scale of variation in latitude is of the order of the solar radius. The pressure perturbation has a vertical derivative which is hydrostatically balanced:

$$\frac{\Delta p}{\ell} \sim g \Delta \rho \sim \frac{g \rho \Delta T}{T}, \tag{8.2.4}$$

where signs are ignored.

Once we bring in the temperature perturbation, we need to worry about maintaining it and that requires advection of heat to balance radiative diffusion:

$$w \frac{d\Theta}{dz} \sim \frac{\kappa}{\ell^2} \Delta T, \tag{8.2.5}$$

where κ is the thermal diffusivity and $d\Theta/dz$ is the vertical, unperturbed potential temperature gradient (that is, the entropy gradient in some units). Since we are in the radiative zone, κ is the radiative diffusivity, and the contribution by turbulence is small.

The condition that these balances should be mutually compatible is

$$\ell \sim R_\odot \left(\frac{\tau_H}{t_{ES}} \right)^{1/4}, \tag{8.2.6}$$

where the horizontal eddy time, τ_H, is $R_\odot^2/\nu_H$ and $\tau_{ES} = (NR_\odot)^2/(\kappa\Omega^2)$ is the Eddington–Sweet time. If the theory is carried out with an isotropic turbulent stress tensor, spin-down spreads the effects vertically and the tachocline thickens inexorably. But as long as the stable vertical stratification favors a strong horizontal turbulence, we can maintain a thin tachocline, though there will be some vertical spreading from the initial mismatch, or from any time-dependent forcing.

A thin tachocline with horizontal turbulence will engender the coherent structures — vortices and flux tubes — that are between the lines of this discussion. On the other hand, (2.6) does not stand by itself as we know neither ℓ nor ν_H, but the observations ought to tell us the former before long. For now, we may note that the value of ℓ does not depend sensitively on the details of the flow in the convection layer and requires only that there be a mismatch between that flow and that of the deep interior. Then $\ell \sim 20,000(\kappa/\nu_H)^{1/4}$ km. Although a similar story might be made with magnetic stresses, the eddy viscosity approach leads to qualitative agreement between the empirical isorotation curves (Paternò 1991, Morrow 1988) and the theoretical ones (Spiegel & Zahn 1992) as shown in figure 8.3.

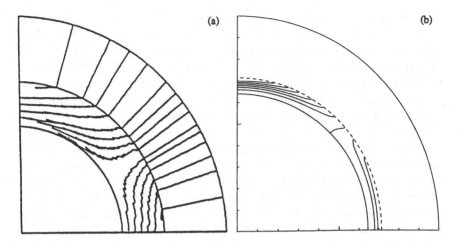

Figure 8.3 The structure of the solar tachocline from (a) the observed viewpoint (figure reconstructed by Paternò (1991) from reductions given by Morrow (1988)), and (b) the theoretical viewpoint (Spiegel & Zahn 1992). The tachocline thickness in (b) is arbitrary.

8.3 THE SOLAR OSCILLATOR

In the analogy between the solar tachocline and the oceanic thermocline, the solar convection zone is like the earth's atmosphere and the solar interior is the abyssal ocean. Instead of rain we have plumes twisting downward through the convection zone, dragging down (and perhaps enhancing) magnetic fields. Such thermals are known in experimental convection and in the earth's atmosphere. Simulations of highly stratified convection shows that the descending plumes are frequent, but there are no comparable rising plumes. In the simulations reported by Brandenburg & Tuominen (1991) downwelling brings magnetic fields to the depths of the convection zone with a vigour that may overcome the opposing tendencies of magnetic buoyancy.

The descending matter, with its trapped magnetic field, will be entrained by the turbulent motions in the tachocline where it is sheared out to build up a toroidal component over long times. How extensive this reservoir is, or how it is structured, are questions that have troubled me for a long time. Other issues like the structure at high latitudes and the effects of the meridional circulations are also worrisome, since they may bear on observational details. To get to the mathematical model we do not need to answer these questions, but they must be faced some day. For now I will simply assume that the toroidal field is there in the tachocline in describing the scenario that S. Meacham and I have been trying to develop over the past few summers in Woods Hole for feeding this field into the convection zone to maintain some kind of balance and, incidentally, to produce spots in the process.

If the tachocline is like an atmospheric weather layer, such as the oceanic thermocline, we must expect it to develop vortices, as does every such layer we can observe well (Dowling & Spiegel 1990). These vortices will have more or less vertical axes and, when a toroidal magnetic filament impinges on one, it will wind the field up. If the process were confined to the tachocline, we might expect flux expulsion from the vortex (Parker 1979, chapter 16). But the local strengthening of the field produces magnetic buoyancy that will lift the field-containing region up into the convection zone. In this way, a rising magnetic tube will be extruded from the tachocline like the output of a cotton-candy machine. Such rising helical tubes return the field to the convection zone in a process that is the surrogate of evaporation in the magnetic weather cycle. A buoyant tube will ultimately protrude through the solar surface to form a single spot or a strong tube may go beyond the surface before falling back to produce a second, more diffuse region of magnetic disturbance.

Whatever the details of such a cycle, the general picture is that the tachocline has a source of field from above to which it may return the field by this and other processes (Spiegel & Weiss 1980). If there is field stretching, much of it occurs as the helix is twisted out of the tachocline. One form of such a process is in Cattaneo, Chiueh & Hughes 1990. (For another vision of the role of vortices, see Parker 1992.)

I mention these images to motivate the construction of (what engineers call) a lumped model of the solar cycle. At that coarse level, we ignore all the spatial detail implied by the magnetic meteorology and simply introduce a parameter, β say, that measures the degree of instability of the magnetic field in the tachocline. When $\beta > 0$, the convection zone is feeding the process abundantly and the magnetic buoyancy is able to extrude strong, ordered fields. This could work in several ways.

There could simply be overstable magnetoconvection giving rise to oscillatory instability and β would measure something like the difference between a magnetic Rayleigh number and its critical value (Childress & Spiegel 1981). Or there could be a dynamo process, such as an α–ω dynamo and β could be related to the dynamo number. In the lumped model, we simply need a potentially unstable oscillator and may assume that its operation is described by the normal form for the appropriate bifurcation, either a Hopf bifurcation or a BLT bifurcation (Bogdanov–Lyapunov–Takens).

In the former, for fixed β, the complex amplitude of the oscillation is given by the normal form for a Hopf bifurcation

$$\dot{A} = (\beta + i\omega)A - |A|^2 A. \tag{8.3.1}$$

I have presumed for definiteness that the bifurcation is supercritical and have scaled the coefficient in the nonlinear term equal to unity. If we were starting from first principles, we should be able to relate the parameters to the physical properties of the model. For now, I simply assume that $\pi/\omega \approx 11$ yrs and leave β free. If this is the oscillator that describes the solar cycle at some level, some property of A should be the measure of the toroidal field that is somehow forced to poke out of the sun and produce spots.

Alternatively, we might favor the more subtle BLT bifurcation (as, for some years, I did). In a simplified version with linear friction, the real amplitude of the oscillation is governed by

$$\ddot{A} = \beta A - \gamma \dot{A} - A^3. \tag{8.3.2}$$

As they stand, neither of these oscillators will by itself adequately describe the complications of the solar cycle. To make the oscillations aperiodic and intermittent — in a word, chaotic — we allow β to vary slowly.

8.4 ON/OFF INTERMITTENCY

An oscillator like (3.2) becomes chaotic when its parameters are made to vary suitably in time. We may impose this time dependence, or it may come about through a feedback of the oscillation on the mechanism that determines the value of the parameter. For example, suppose that instead of having β constant in (3.2), we let it vary according to

$$\dot{\beta} = -c[\beta + a(A^2 - 1)], \tag{8.4.1}$$

where a and c are specified parameters. A simple transformation turns (3.2) and (4.1) into the Lorenz equations, originally devised in the study of thermal convection. So there is little doubt that this is a system capable of producing aperiodic behavior for appropriate values of the parameters.

This way of producing chaotic systems, by letting simple oscillators feed back on their parameters (Marzec & Spiegel 1980) may be used to generate equations for excitable media, so perhaps in a case like this, we ought to refer to hysterical media. But I would prefer to reserve this usage for the case of intermittency, for an example of which, suppose that in (3.2) $\beta = Z - 2Y$ and that for Y and Z we have the equations

$$\ddot{Y} = -Y^3 + ZY - \gamma \dot{Y} - A^2, \tag{8.4.2}$$

$$\dot{Z} = -\epsilon[Z + a(Y^2 + A^2 - 1)]. \tag{8.4.3}$$

When $A = 0$, equations (4.2) and (4.3) constitute the form of the Lorenz equations that I just mentioned. So $A = 0$ is an invariant manifold of the fifth-order system that combines these two equations with (3.2). Figure 8.4 (from Spiegel 1981) shows $A(t)$ for $\epsilon = 0.1$, $a = 6.5$ and $\gamma = 0.4125$. This example of intermittent behavior with episodes of inactivity in A recalls the inactive sun of Newton's time.

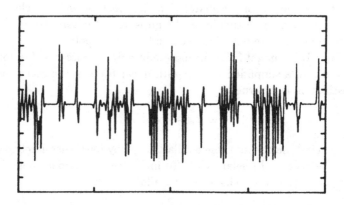

Figure 8.4 On/off intermittency from equations (3.2), (4.2) and (4.3) (after Spiegel 1981). Even when the oscillator is inactive, there is chaos in the invariant manifold.

The term intermittency has been used in dynamical systems theory to describe alternation between two modes of activity, as in the Pomeau & Manneville (1980) theory (see also Manneville 1990). To restore the meaning of the word as used by fluid dynamicists, the term *on/off intermittency* has been proposed (Platt *et al.* 1993*a*) to connote alternation between activity of a certain kind and inactivity, as in figure 8.4. The present interest of the model is that there is continuous chaos in the invariant manifold, but the behavior of $\mathcal{A}$ alone shows on/off intermittency. In this metaphor for the solar cycle, chaos in the Lorenz system represents convection and $\mathcal{A}$ the solar activity. The merit of the model is that it captures the kind of intermittency that the cycle manifests, but otherwise figure 8.4 does not look very much like figure 8.2. One reason is that the effect of the solar activity ($\mathcal{A}$) on the convection (Y, Z), is pronounced and this makes for great irregularity. There must really be such coupling, but it is likely to be weaker than in this model. We turn to a model which better captures the nature of the solar cycle. In this one there is no feedback of the oscillator on the chaotic driver.

In on/off intermittency, the intermittent behaviour is organized by an unstable invariant manifold with stable and unstable manifolds coming into and out of it (Platt *et al.* 1993*a*). When the system moves away from the manifold, it bursts into activity until it is brought back very close to the manifold along a stable manifold to hover inactively before being sent out again. This may be seen as a chaotic relaxation oscillation, or a higher dimensional version of homoclinic chaos, or as what is called bursting in neurophysiology (Hindmarsh & Rose 1984). The general idea is to make a potentially unstable oscillator whose stability parameter is the variable of an associated chaotic system. There are many ways to set this up, so what we are isolating is not a particular model but a particular mechanism, on/off intermittency. Whether the oscillation really is generated by an instability of the tachocline is a separate issue that is not central to the mathematical description. We do not even need the tachocline for the mathematical model to work, though it is useful to think in such explicit terms. An interesting analysis of on/off intermittency has recently been given by Heagy, Platt & Hammel (1993) and there are by now several discussions of this kind of process (Yamada & Fujisaka 1986, Fujisaka et al. 1986, Fujisaka & Yamada 1987, Hughes & Proctor 1990, Pikovsky & Grassberger 1991). One key result is that, if this process is going on in the solar cycle, we have no real hope of determining the dimension of the solar attractor by any of the presently known means. It is not just that the data are inadequate for the purpose, as has already been objected (Spiegel & Wolf 1987), but that the on/off process imposes a sort of indeterminism on dimension determination (Platt et al. 1993a).

Here is a mathematical model for the solar cycle (Platt et al. 1993b) that has the features I have outlined. We take the standard form (3.1) for the oscillator, which we couple to a chaotic system by letting $\beta = \beta_0(\mathcal{U} - \mathcal{U}_0)$ where β_0 and $\mathcal{U}_0$ are fixed parameters. So (3.1) becomes

$$\dot{A} = [\beta_0(\mathcal{U} - \mathcal{U}_0) + i\omega]A - |A|^2A. \tag{8.4.4}$$

This says that the instability is strongly affected by $\mathcal{U}$, which is determined by something else in the system. In particular we generate $\mathcal{U}$ with this third order system:

$$\ddot{\mathcal{U}} = r\mathcal{U} - \mathcal{U}^3 - q\dot{\mathcal{U}} - \mathcal{V}, \tag{8.4.5}$$

$$\dot{\mathcal{V}} = \delta[\mathcal{V} - p\mathcal{U}(\mathcal{U}^2 - 1)], \tag{8.4.6}$$

where (r, q, p) are more parameters. Like (4.2), (4.5) is a modification of (3.2).

This time, the chaotic driver is a particular case of a model that was constructed to clarify the physics of doubly diffusive convection (Moore & Spiegel 1966). The system (4.4–6) makes a fair model of the solar cycle, at least in the coarse-grained sense. We do have to make some decision about what to compare to the sunspot number, though this appears not to be crucial. In figure 8.5 we see a plot of the square of Re A vs. time showing several intermissions in activity. Within a long period of activity, the cycle will be chaotic as we see clearly in figure 8.6, a portion of figure 8.5 with an enlarged time scale. These results are robust and we do not need a lot of fine tuning of all these parameters to get this behaviour.

In fact, the observed sunspot number variation is much more ragged than this model predicts, as we see from figure 8.2. So there is evidence that more

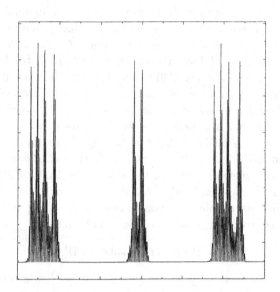

Figure 8.5 The activity predicted by (4.4-6) for $r = 0.7$, $q = 0$, $p = -0.5$, $\beta_0 = 1$ and $U_0 = -0.15$, $\delta = 0.03$.

is happening than just an intermittent oscillation such as is shown here. If the cycle does come from a deep layer, we are seeing it through the convection zone, which will add its own direct input while distorting the 'true' signal. That can be modeled too (Platt *et al.* 1993*b*) and, when such effects are included, the qualitative agreement seems (to us) very good. But here I omit such fine points of the cycle in the belief that they are incidental.

8.5 SOLAR ACTIVITY WAVES

A plot showing the latitudes of vigorous sunspot activity vs. time looks like a row of butterflies (see Weiss' discussion in chapter 2). This Maunder butterfly diagram is a space-time plot showing the propagation of solar activity. Lines along the activity maxima are world lines of motion toward the equator. But what is moving? The most likely prospect is that we are seeing some kind of wave motion and, in one version, these are dynamo waves (Parker 1979).

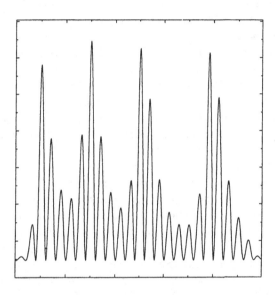

Figure 8.6 A blowup of a portion of figure 8.5.

The idea I wish to describe next is that the the butterfly diagram represents the propagation of solitary waves (Proctor & Spiegel 1991).

If an oscillation arises in a thin layer like the thermocline, we might expect to see simple waves produced. Since the layer is thin, there should be a dense spectrum of allowed wavenumbers. If the wave numbers are closely spaced, there is effectively a continuum of them. A packet of such waves could have a solitary wave as envelope that would make a nice descriptor of the activity band in latitude. The generic form of the propagation equation should be the same for any simple overstability. The astromathematical approach allows us to discuss the butterfly diagram in a general way, even though the precise instability mechanism has not yet been isolated. There are in fact several possible instabilities, including magneto-convective overstability, instability caused by the vertical shear of the tachocline or instability of a dynamo in the tachocline. Since they would have a common mathematical description in the present coarse-grained discussion, we are not hampered on that account.

The amplitude equation for the Hopf bifurcation is based on a model in which one mode has, in linear theory, a time dependence like $\exp(\beta t + i\omega t)$ with $|\beta|$ small, and all the other modes are rapidly damped. If there is just one mode with small $|\beta|$, its complex amplitude, A, evolves according to (3.1). If the seat of the instability is a thin layer like the tachocline, there can be a band of modes with small β. But now β is a function of the wavenumber along the channel, k, and such modes can propagate.

To describe the nonlinear development of the instability, we construct a packet of waves, in which the amplitude of each component, A, depends on k. If the system is axisymmetric in the large, we need consider only a one-dimensional case. We factor out the carrier frequency and wavenumber, defined as those of the most unstable mode, and we characterize the packet by $\Psi(x, t)$, the Fourier transform of A, where x is latitude. The amplitudes measure the size of all disturbance quantities but further details will differ from case to case depending on things like the relation of packet width to the wavelength of the carrier wave. That in turn depends on subtleties like interface conditions which we are here blurring over.

On general grounds, we expect the equation for Ψ to be the complex Ginzburg–Landau (G–L) equation, which is like (3.1), but with spatial derivatives as well. Strictly speaking, the governing equations are two coupled Ginzburg–Landau equations, one for each direction. Though we know how to write these down (Bretherton & Spiegel 1983), we do not as yet have solutions relevant to the solar case, so I shall discuss only the single G–L equation here (Manneville 1990). The reason for the limited progress is that there is a more serious complication that has to be dealt with first, one that Proctor and I (1991) have so far treated in a phenomenological way. This is the variation of underlying conditions, such as local stability, with latitude in the sun.

In the phenomenological view, the magnetic rain probably varies with latitude, and certainly the shear in the tachocline does. This inhomogeneity should induce a drift mode into the problem in addition to the one we are already omitting. However, we have so far left out this extra mode and have attempted to make amends by putting a positional dependence into the coefficients in the G–L equation. As the correct positional dependences are as yet unknowable, we have used simple forms for it. This parameterization will have to serve until we have a better understanding of the underlying variations in tachocline structure.

In the wave packet, frequencies and growth rates depend on the wave number in linear theory. We treat only situations where the width of the packet is small, as measured by some small parameter, ϵ. Linear theory provides a group velocity c_0 that we use to provide a basic reference frame. The peak of the packet is nearly stationary in the frame with coordinate $\xi = x - c_0 t$.

The form of the equation, when we choose units to minimize the number of parameters, is

$$\partial_t \Psi - c(\xi)\partial_x \Psi - (\epsilon + i)\partial_x^2 \Psi + (\nu - i)|\Psi|^2 \Psi = [\beta(\xi) + i\omega(\xi)]\Psi. \qquad (8.5.1)$$

Here we have allowed for a dependence of the stability parameter, β, and of the linear frequency, ω, on the location of the solitary wave. The parameter $c(\xi)$ is a local drift speed with respect to the preferred frame.

We assume that the instability is weak and write $\beta(\xi) = \epsilon\mu(\xi)$. When $\epsilon \to 0$, (5.1) reduces to the cubic Schrödinger equation, which admits a soliton solution that we may write as

$$\Psi(x,t) = \mathcal{R}\, e^{i\Theta(x,t)}, \qquad (8.5.2)$$

where

$$\mathcal{R}(x,t) = \sqrt{2}R\,\mathrm{sech}[R(x - x_0)], \qquad (8.5.3)$$

and

$$\Theta(x,t) = U(x - x_0) + \int (U^2 + R^2)\,dt. \qquad (8.5.4)$$

This soliton contains two arbitrary parameters, R and U, with $x_0 = 2Ut$. The presence of arbitrary parameters is related to symmetry groups of the nonlinear Schrödinger equation.

The soliton, for all its remarkable stability, is a rather dull object when left to itself. When we introduce dissipation and instability into the system, a richer behavior arises. The arbitrariness of the parameters permits us to accommodate the dissipation and instability terms that come in when $\epsilon \neq 0$. For small ϵ, we let both R and U be functions of ϵt. Then the methods of singular perturbation theory lead to equations of motion for the parameters. These equations form a dynamical system that control the behaviour of the solitary wave, much as a mind does for a person. In this way, the otherwise mindless soliton is provided with a rather simple mind in the case of the standard complex G–L equation that goes right to a fixed point. However, that situation is enriched when the domain is large enough to allow instabilities that produce other solitary waves (Bretherton & Spiegel 1983).

In the solar case, when the parameters depend on position, even a single activity wave shows a certain amount of interesting behaviour. The theory for $\epsilon \neq 0$ shows that when the amplitude and position of the solitary waves depend on ϵt, (5.2–4) represent a solution of (5.1) provided that these equations are satisfied:

$$\dot{R} = 2R[\mu(\xi) - \kappa(\xi) - U^2] - \frac{2}{3}(1 + 4\nu)R^2, \qquad (8.5.5)$$

$$\dot{U} = U[2\kappa(\xi) - \frac{4}{3}R^2] + \lambda(\xi), \qquad (8.5.6)$$

$$\dot{\xi} = 2U - c_0, \qquad (8.5.7)$$

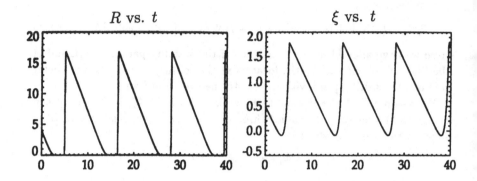

Figure 8.7 The dynamics of a single-winged butterfly according to (5.5-7) (after Proctor & Spiegel 1991).

where $\kappa = dc/d\xi$ and $\lambda = d\omega/d\xi$ (Proctor & Spiegel 1991).

   In modeling the dependences of the given quantities on latitude, we need to look at the structure of the tachocline. The helioseismological studies suggest that the rotation in the solar interior is the same as the surface rotation at somewhere around 35° latitude. The model (Spiegel & Zahn 1992) agrees with this and predicts that the vertical shear has a minimum at this latitude. Since we expect the shear to drive instability in a dynamo, either directly or indirectly, we represent this either as a quadratic dependence over a whole hemisphere or, more crudely, as a linear dependence over the zone of sunspot activity. In either case the qualitative results are similar and, for the linear case, with μ proportional to ξ, we get results like those in figure 8.7 (Proctor & Spiegel 1991). The results are for a single hemisphere and thus represent a series of one-winged butterflies. To this extent, the model is satisfactory. It suggests that at the end of a cycle the activity wave survives and returns rapidly to midlatitudes maintaining very small amplitude, there to begin another trip to the equator.

   The observations reveal that a new cycle begins in midlatitudes before the previous cycle ends near the equator. This is not seen in figure 8.7. On the other hand, that picture is based on the solitary wave being a rigid object regarded as a small particle. In fact, the real reflection process is a more complicated affair lasting about the time it takes the wave to travel its own width. This seems about right for the overlap period of the two cycles. Moreover, we ought to see some of the spots associated with the return trip of the activity wave to the midlatitudes, so part of the overlap may be on that account.

Another feature of figure 8.7 that is not in good agreement with the facts is that the maximum of activity occurs virtually at the beginning of the cycle. This may be a result of the form adopted for the latitude dependence of the parameters. If this model turns out to be on the right track, the phase of maximum activity may ultimately permit us to study the latitude variation of the tachocline structure.

The cycle shown in figure 8.7 is periodic, but this is not surprising at this stage of the story. In the next level of development, when we include two solitary waves in the description, one for each hemisphere, we obtain a coupled pair of sets of equations like (5.6-7). This leads to chaos and north–south asymmetry, more in accord with observation. However, Proctor and I are not yet sure about coupling terms in this description of both hemispheres, so I do not give details here. In fact the major cause of aperiodicity is likely to come from input variations from the convection zone, expressed once again by a chaotic origin of β in (5.1). This will produce spatio-temporal on/off intermittency of the kind we see in the sun and the next step should be to include this mechanism in the theory.

8.6 FINAL REMARKS

Since we do not have a theory of turbulence, it is not possible to make a fully deductive theory of the solar cycle on account of the involvement of the solar convection zone. Nevertheless, we can hope to make phenomenological models of increasing precision. In the work described here, there are two parallel developments along those lines, one physical and one mathematical. Both are frankly qualitative, but in the mathematical case, this may be a desirable feature.

The mathematical models discussed here are aimed at showing how the apparently complicated spatio-temporal behavior of the solar cycle can be reasonably well reproduced with relatively simple equations. This encourages us to attack the physical model in a more detailed way, despite our inability to cope with the turbulence problem. The equations describe a simple oscillatory instability fed by an aperiodic process. The sun provides the necessary ingredients for all the processes that can be read from the model equations.

The solar tachocline, the rotational transition layer between the convection zone and the deep interior, offers a natural site in which to unfold our scenario. Fed from above by plunging plumes it can entrain fluid carrying tangled magnetic field and stretch the field out into some more orderly configuration only to expel it in discrete structures. We have several promising mechanisms to choose from before setting out to follow one through to a quantitative model. But before embarking on such daunting calculations, we need to see a way through some of the unsolved problems. The main ones seem to me to be concerned with time dependence.

The solar differential rotation appears to vary on the time scale of the activity cycle (Howard & LaBonte 1980). We do not know whether this is incidental or fundamental. In the picture I am describing, the dynamical coupling between the tachocline and the convection zone is strengthened when the spot fields link them. This time-dependent interaction could modify the structure of the tachocline and produce large-scale motions like azimuthal rolls. Whether such effects are fundamental or just secondary is not yet clear. Similarly, I do not know whether the polarity reversals that occur with the solar cycle point to some deep process or represent some superficial feature of the cycle. The true physical nature of this behavior is not reliably known and the corresponding mathematical explanation in generic instability models has not been isolated.

Another question that has to be faced at some stage is the quantitative determination of properties of the cycle, such as the eleven-year time scale. Eleven years is very long compared to the travel time of any of the obvious waves across the tachocline, which is about a quarter of an hour for sound waves. On the other hand, eleven years is quite short compared to the conventional Kelvin–Helmholtz, or thermal, time of the tachocline of a million years, or so. The changes that the solar cycle must work in the tachocline would seem to encounter a sort of fluid-dynamical impedance mismatch between the driving frequency and these response times of the tachocline. However, the hydrostatic adjustment time of the tachocline is approximately the geometric mean of the acoustic travel time and the thermal time (Spiegel 1987), which is of the order of years. So it may be that the period of the cycle is less of a clue to the actual process than it is to the structure of the tachocline. If this is true, we have another means of estimating its thickness. I offer this as an example of a feature of the cycle that might be fundamental but might just as well be secondary.

There are many places to seek further clues to the processes discussed such as other solar type stars (Belvedere 1991) and turbulent disks that might show solar type processes. Indeed there are hot stars that seem to show activity resembling that of the sun (Casinelli 1985). It difficult to know which phenomena are central to the sunspot cycle and the decision is usually subjective. The models I have described are rooted in elementary mathematical processes that seem robust. They suggest a vision of the solar activity process that differs from the conventional solar dynamo and avoid some of the difficulties solar dynamo theory faces. I am sure that the present approach will also face numerous problems as it is elaborated and I look forward to learning what these will be.

Finally, I would like to stress that the models described here are not unique. They are constructed to show that many of the complicated looking features of the solar activity cycle may have simple and mathematically natural causes. The main aim of this work is to try and identify these causes.

Acknowledgments
The work described here was supported by the Air Force Office of Scientific Research under contract number F49620-92-J-0061 to Columbia University. Nigel Weiss kindly acted as a 'referee' and I am grateful for his comments; his own views appear in chapter 2. Gaetano Belvedere kindly helped trace the source of figure 8.3(*a*).

REFERENCES
Belvedere, G. 1991 Stellar activity belts as potential indicators of internal rotation and angular momentum distribution. In *Angular Momentum Evolution of Young Stars* (ed. S. Catalano & J.R. Stauffer), pp. 281–288. Reidel, Utrecht.

Brandenburg, A. & Tuominen, I. 1991 The solar dynamo. In *The Sun and Cool Stars: Activity, Magnetism, Dynamos* (ed. I. Tuominen, D. Moss & G. Rüdiger), pp. 223–233, IAU Coll. 130, Lecture Notes in Physics, vol. 380. Springer–Verlag.

Bretherton, F.P. & Spiegel, E.A. 1968 The effect of the convection zone on solar spin-down. *Astrophys. J.* **153**, 277.

Bretherton, C.S. & Spiegel, E.A. 1983 Intermittency through modulational instability. *Phys. Lett. A* **96**, 152–156.

Brown, T.M., Christensen-Dalsgaard, J., Dziembowski, W.A., Goode, P., Gough, D.O. & Morrow, C.A. 1989 Inferring the sun's internal angular velocity from observed P-mode frequency splittings. *Astrophys. J.* **343**, 526–546.

Casinelli, J.P. 1985 Evidence for non-radiative activity in hot stars. In *The Origin of Nonradiative Heating/Momentum in Hot Stars* (ed. A.B. Underhill & A.G. Michalitsianos), pp. 2–23. NASA 2358.

Cattaneo, F., Chiueh, T. & Hughes, D.W. 1990 Buoyancy-driven instabilities and the nonlinear breakup of a sheared magnetic layer. *J. Fluid Mech.* **219**, 1–23.

Childress, S. & Spiegel, E.A. 1981 A prospectus for a theory of variable variability. In *Variations of the Solar Constant* (ed. S. Sofia), pp. 273–292, NASA Conf. Publ. 2191, Goddard Space Flight Center, Greenbelt, MD.

DeLuca, E.E. 1986 Dynamo theory for the interface between the convection zone and the radiative interior of a star. Doctoral thesis. University of Colorado, NCARCT–104.

Dowling, T.E. & Spiegel, E.A. 1990 Stellar and jovian vortices. In *Fifth Florida Workshop on Nonlinear Astrophysics* (ed. S. Gottesman & J.R. Buchler), pp. 190–216. Ann. N.Y. Acad. Sci., vol. 617.

Eddy, J.A. 1978 Historical and arboreal evidence for a changing sun, In *The New Solar Physics* (ed. J.A. Eddy), pp. 11–34, AAAS Selected Symposium, vol. 17. Westview Press, Coulder, Co.

Fritz, G. 1882 Zur Bestimmung der älteren Sonnenflecken-Perioden. *Sirius: Zeit. Pop. Astron.* **15**, 227–231.

Fujisaka, H. & Yamada, T. 1987 Intermittency caused by chaotic modulation. 3. Self-similarity and higher-order correlation functions. *Prog. Th. Phys.* **77**, 1045–1056.

Fujisaka, H., Ishii, H., Inoue, M. & Yamada, T. 1986 Intermittency caused by chaotic modulation. 2. Lyapunov exponent, fractal structure and power spectrum. *Prog. Th. Phys.* **76**, 1198–1209.

Goode, P.R., Dziembowski, W.A., Korzennik, S.G.& Rhodes, E.J. 1991 What we know about the sun's internal rotation from solar oscillations. *Astrophys. J.* **367**, 649–657.

Gough, D.O., Kosovichev, A.G., Sekii, T., Libbrecht, K.G. & Woodard, M.F. 1992 The form of the angular velocity in the solar convection zone. In *Seismic Investigation of the Sun and Stars* (ed. T.M. Brown), pp. 213–216. Conf. Ser. Astron. Soc. Pacific, vol. 42.

Heagy, J.F., Platt, N. & Hammel, S.M. 1993 Characterization of on–off intermittency. *Phys. Rev. E* (in press).

Heß, W. 1911 *Himmels- und Naturerscheinungen in Einblattdrucken des XV. bis XVIII. Jahrhunderts*. Druck u. Verlag v. W. Drugulin, Leipzig.

Hindmarsh, J.L. & Rose, R.M. 1984 A model of neuronal bursting using three coupled first order differential equations. *Proc. Roy. Soc. Lond. B* **221**, 87–102.

Howard, R. & LaBonte, B.J. 1980 The Sun is observed to be a torsional oscillator with a period of 11 years. *Astrophys. J.* **239**, L33–36.

Hughes, D.W. & Proctor, M.R.E. 1990 Chaos and the effect of noise in a model of 3-wave mode coupling. *Physica D* **46**, 163–176.

Izenman, A.J. 1985 J.R. Wolf and the Zurich sunspot relative numbers. *Mathematical Intelligencer* **7**, 27–33.

Layzer, D., Rosner, R. & Doyle, H.T. 1979 On the origin of solar magnetic fields. *Astrophys. J.* **229**, 1126–1137.

Manneville, P. 1990 *Dissipative Structures and Weak Turbulence*. Academic Press, New York.

Marzec, C.J. & Spiegel, E.A. 1980 Ordinary differential equations with strange attractors. *SIAM J. App. Math.* **38**, 403–421.

Moore, D.W. & Spiegel, E.A. 1966 A thermally excited nonlinear oscillator. *Astrophys. J.* **143**, 871–887.

Morrow, C.A. 1988. In *Seismology of the Sun and Sun-like Stars* (ed. E.J. Rolfe), pp. 91–98. ESA, vol. S.P.–286. ESTEC, Noordwijk.

Parker, E.N. 1979 *Cosmical Magnetic Fields*. Clarendon Press, Oxford.

Parker, E.N. 1992 Vortex attraction and the formation of sunspots. *Astrophys. J.* **390**, 290–296.

Paternò, L. 1991. The solar internal rotation and its implications. In *The Sun and Cool Stars: Activity, Magnetism, Dynamos* (ed. I. Tuominen, D. Moss, G. Rüdiger), pp. 182–186. Proc. IAU Coll. 130., Lecture Notes in Physics, vol. 380. Springer–Verlag.

Pikovsky, A.S. & Grassberger, P. 1991 Symmetry breaking bifurcation for coupled chaotic attractors. *J. Phys. A: Math. Gen.* **24**, 4587–4597.

Platt, N., Spiegel, E.A. & Tresser, C. 1993a On-off Intermittency: a mechanism for bursting. *Phys. Rev. Lett.* **70**, 279–282.

Platt, N., Spiegel, E.A. & Tresser, C. 1993b The intermittent solar cycle. *Geophys. Astrophys. Fluid Dynam.* **73**, 147–161.

Pomeau, Y. & Manneville, P. 1980 Intermittent transition to turbulence in dissipative dynamical systems. *Commun. Math. Phys.* **74**, 187–195.

Proctor, M.R.E. & Spiegel, E.A. 1991 Waves of solar activity. In *The Sun and the Cool Stars* (ed. D. Moss, G. Rüdiger & I. Tuominen), pp. 117–128. Springer–Verlag.

Spiegel, E.A. 1972 A history of solar rotation. In *Physics of the Solar System* (ed. S.I. Rasool), pp. 61, NASA SP-300. Sci. & Tech. Information Office, NASA, Washington, D.C.,

Spiegel, E.A. 1981 A class of ordinary differential equations with strange attractors. In *Nonlinear Dynamics* (ed. R.H.G. Helleman), pp. 305–312. Ann. N. Y. Acad. Sci., vol. 357.

Spiegel, E.A. 1987 Hydrostatic adjustment time of the solar subconvective layer. In *The Internal Solar Angular Velocity* (ed. B.R. Durney & S. Sophia), pp. 321–328.

Spiegel, E.A. & Weiss, N.O. 1980 Magnetic activity and variations in solar luminosity. *Nature* **287**, 616–617.

Spiegel, E.A. & Wolf, A.N. 1987 Chaos and the solar cycle. In *Chaotic Phenomena in Astrophysics* (ed. J.R. Buchler & H. Eichhorn), pp. 55–60. Ann. N.Y. Acad. Sci., vol. 497.

Spiegel, E.A. & Zahn, J.-P. 1992 The solar tachocline. *Astron. Astrophys.* **265**, 106–114.

Yamada, T. & Fujisaka, H. 1986 Intermittency caused by chaotic modulation. I. Analysis with a multiplicative noise model. *Prog. Th. Phys.* **76**, 582–591.

Zahn, J.-P. 1975 Differential rotation and turbulence in stars. *Mém. Soc. Roy. Sci. Liège*, 6e série, **8**, 31–34.

CHAPTER 9

The Nonlinear Dynamo and Model-Z

S. I. BRAGINSKY

Institute of Geophysics and Planetary Physics
University of California at Los Angeles
Los Angeles CA 90024-1567, USA.

9.1 INTRODUCTION

Many astrophysical bodies possess magnetic fields that arise from dynamo action. The case of the Earth is a unique one because the observational data available are much more detailed for the Earth than for any other astrophysical body, making possible a rather detailed comparison of geodynamo theory with observations. To meet this unique opportunity we therefore need a geodynamo theory that is very detailed. To develop the fully-fledged theory of such a complicated system as the geodynamo, even with the help of modern computers, it is however necessary to possess a qualitative understanding of its structure. This can be achieved by preliminary 'scouting' calculations of some artificially simplified models that are much simpler than the full geodynamo model but nevertheless help to understand it. A kinematic dynamo theory is the first step towards this goal. Kinematic models provide us with an understanding of its electrodynamics (the magnetic field generation process). The next necessary step is an understanding of its mechanics. The model-Z geodynamo emerges as a result of this step of scouting calculations. It may be considered as a specific case of a more general model that we call the nonlinear (pseudo-) axisymmetric dynamo model. This is a natural generalisation of the linear, nearly axisymmetric, kinematic dynamo model (Braginsky 1964a, b, c, d), and it is 'intermediate' between the kinematic and the complete theories of the geodynamo.

The nonlinear axisymmetric dynamo model aims at understanding the specific features of the main convective flow and the production of axisymmetric field in the core while the field generation due to the non-axisymmetric motion (α-effect) is considered as given. Another direction for an essential 'intermediate' investigation is to explore non-axisymmetric magnetoconvection. This magnetoconvection develops in the form of nonlinear waves, and the basic state on which the waves are running is the axisymmetric magnetoconvection

M.R.E. Proctor & A.D. Gilbert (eds.)
Lectures on Solar and Planetary Dynamos, 267–304
©1994 Cambridge University Press.

pattern. This research needs, therefore, information about the features of this pattern that are to be revealed by the axisymmetric model.

The fundamentals necessary for this chapter are given by Roberts in chapter 1, and by Fearn in chapter 7, where approaches different from the one used in this chapter are also presented (e.g., Malkus & Proctor 1975, Proctor 1977) and related references are given. We start here nevertheless from the basic equations of the Earth's dynamo (see, e.g., Braginsky 1991), and proceed quickly to the specific focus of this chapter. An effort is made to explain geodynamo physics in a simple way, using rather compact formulae, so that the reader is able to grasp most of the mathematics directly by eye, without paper and pencil. Careful attention was paid to convenience of notation. Some often-used notation is listed below:

(a) t is time. (r, θ, ϕ) and (z, s, ϕ) are spherical and cylindrical polar coordinates in the frame rotating around the z-axis with the angular velocity of the Earth, $\Omega = 0.729 \times 10^{-4}$ s^{-1}; θ is the colatitude, ϕ is the longitude, s is the distance from the z-axis, with $z = r \cos \theta = r\mu$, $s = r \sin \theta = r\mu_s$, $\mu = \cos \theta$ and $\mu_s = \sin \theta$.

(b) ∂_t, ∂_r, ∂_θ, ... are partial derivatives, $\partial/\partial t$, $\partial/\partial r$, $\partial/\partial \theta$, ...

(c) 1_r, 1_z, 1_θ, ... are unit vectors in the directions of increasing r, z, θ, ...

(d) $\sim$ is an order of magnitude sign and $\propto$ is a proportionality sign.

(e) $R_0 = 6371$ km, $R_1 = 3480$ km, $R_2 = 1222$ km are the radii of the Earth, the core mantle boundary (CMB) and the solid inner core boundary (ICB), according to Dziewonski *et al.* (1981). The subscript 1 marks the CMB; here $r = R_1$, $z = \pm z_1$ where $z_1 = (R_1^2 - s^2)^{1/2}$. The subscript 2 marks the ICB, $r_2 = R_2/R_1 = 0.351$.

The good rule 'denote different values by different symbols and the same value by the same symbol' is mostly obeyed, but with the following two exceptions. First both the dimensional variables A_{dim} and corresponding nondimensional A_{nd} variables, connected by the equations $A_{dim} = A_I A_{nd}$, are denoted by the same symbol A, here A_I is the appropriate unit of A. Second the overbars and some superscripts are often omitted (after due warning) for brevity.

9.2 NONLINEAR AXISYMMETRIC GEODYNAMO MODELS

9.2.1 The Main Equations of the Geodynamo

The system of equations governing the complete three-dimensional model of the geodynamo must describe the convection of an electrically conducting fluid interacting with the magnetic field generated by the same fluid motion. This convection is driven by the buoyancy that is created by the density excess due to the compositional and thermal inhomogeneities of the core. The simplest set of governing equations can be written down in the Boussinesq

approximation with only one driving agent (the compositional one is preferred here). In the Boussinesq approximation the real distribution of density
is everywhere replaced by a constant ρ_0 except in terms in which the density variation is crucial. For example the density excess is defined by the
expression $C = (\rho - \rho_a)/\rho_0$, where ρ is the true density of the fluid and $\rho_a(r)$
is the adiabatic equilibrium distribution of density. The system of dynamo
equations has the following form (see, e.g., Braginsky 1991)

$$d_t \mathbf{V} = -\nabla P + \sum \mathbf{F}^K, \tag{9.2.1}$$

$$\partial_t \mathbf{B} = \nabla \times (\mathbf{V} \times \mathbf{B}) + \eta \nabla^2 \mathbf{B}, \tag{9.2.2}$$

$$\partial_t C + \nabla \cdot (\mathbf{V}C + \mathbf{I}) = G^C, \tag{9.2.3}$$

$$\nabla \cdot \mathbf{V} = 0, \qquad \nabla \cdot \mathbf{B} = 0. \tag{9.2.4,5}$$

Here $\mathbf{V}$ is the fluid velocity, $\mathbf{B}$ is the magnetic field, $d_t = \partial_t + \mathbf{V} \cdot \nabla$ is
the Lagrangian time derivative, P is the effective pressure divided by ρ_0,
$\eta = 1/\mu_0\sigma$ is the magnetic diffusivity, and σ is the electrical conductivity of
the core (μ_0 is the permeability of vacuum). $\mathbf{F}^K$ are the forces acting upon a
unit mass of fluid:

$$\mathbf{F}^\Omega = 2\mathbf{V} \times \mathbf{\Omega}, \tag{9.2.6a}$$

$$\mathbf{F}^B = (\nabla \times \mathbf{B}) \times \mathbf{B}/\mu_0\rho_0, \tag{9.2.6b}$$

$$\mathbf{F}^C = \mathbf{g}C, \tag{9.2.6c}$$

$$\mathbf{F}^\nu = \nu\nabla^2\mathbf{V}. \tag{9.2.6d}$$

Here $\mathbf{F}^\Omega$ is the Coriolis force, $\mathbf{F}^B$ is the Lorentz (magnetic) force, $\mathbf{F}^C$ is the
Archimedean (buoyancy) force, $\mathbf{F}^\nu$ is the force of Newtonian viscosity, $\mathbf{g}$ is
the acceleration due to gravity, and $\mathbf{I}$ is the diffusional flux of the density
excess, C, produced mainly by turbulent mixing.

The equation of motion (2.1) and the induction equation (2.2) describe
the evolution of the fluid velocity $\mathbf{V}$ and the magnetic field $\mathbf{B}$. They obey
well-known boundary conditions: the no-slip condition of no relative motion
between the fluid and solid surfaces and continuity of the magnetic field.
Equation (2.3) describes the creation and transport of density excess. Its
boundary conditions, and also the expressions for the flux $\mathbf{I}$ and the source
G^C, are determined by conditions and processes in the core that are not
completely known. They can only be modelled with some uncertainty; see,
e.g., Braginsky (1991) for details concerning compositional convection.

It is worth specially noting that the second term in the expression $\rho = \rho_a + \rho_0 C$ is very small. This can be seen from $gC \sim 2\Omega V$; for example, if
$V \sim 10^{-1}$ cm/s, then $C \sim 10^{-8}$. The Archimedean force due to this small
density inhomogeneity drives the geodynamo. The constant ρ_0 is used in all
coefficients of the equations in the Boussinesq approximation instead of the

adiabatic equilibrium density ρ_a despite the fact that the relative change of ρ_a in the core is of the order of $10^{-1} \gg C \sim 10^{-8}$. There is no contradiction here because $\mathbf{F}^C = \mathbf{g}C$ appears directly in (2.1) whereas $\nabla \rho_a$ would manifest itself only as a slight inhomogeneity in the coefficients of the equations.

A quantity with the dimension of magnetic field can be composed from the coefficients of equations (2.1) and (2.2), which represents a characteristic magnetic field magnitude

$$B_* = (\mu_0 \rho_0 2\Omega \eta)^{1/2}. \tag{9.2.7}$$

We take $\rho_0 = 10.9$ g/cm^3 and $\eta = 2$ m^2/s ($\sigma = 4 \times 10^5$ S/m); this gives $B_* = 20$ G.

From now on we suppose that the magnetic field, $\mathbf{B}$, is divided by $(\mu_0 \rho_0)^{1/2}$, and therefore coincides with the corresponding Alfvén velocity, 1 cm/s being equivalent to 11.7 G for $\rho_0 = 10.9$ g/cm^3. The magnetic force (2.6b) is now expressed as $\mathbf{F}^B = (\nabla \times \mathbf{B}) \times \mathbf{B} = (\mathbf{B} \cdot \nabla)\mathbf{B} - \nabla B^2/2$, and the characteristic magnitude of magnetic field is $B_* = (2\Omega \eta)^{1/2}$ corresponding to 1.71 cm/s for $\eta = 2$ m^2/s.

9.2.2 Nearly Axisymmetric Kinematic Dynamo Model

The above complete system of equations is very complicated. To begin with it is natural to investigate with the help of the induction equation (2.2) magnetic field generation by a prescribed fluid velocity. This is the task of 'kinematic' dynamo theory. Cowling's (1934) theorem forbids an axially symmetric dynamo. The 'working' dynamo must therefore be three-dimensional, and hence rather complicated. These complications can be greatly simplified in the case when the deviation from axisymmetry is small and the fluid conductivity is high, more precisely, the magnetic Reynolds number, $\mathcal{R}_m = LV/\eta$, should be large enough. Here L is the characteristic length scale of the system and V is the characteristic velocity. It was shown by Braginsky (1964a, b, c) that the self-excitation of a magnetic field is possible in this case of a nearly axisymmetric system.

All fields in the theory of nearly axisymmetric dynamos are subdivided into two parts. One part is the average over the longitude; it is axisymmetric about the z-axis and is marked by an overbar. Another part has zero ϕ-average and is marked by a prime: $\mathbf{V} = \overline{\mathbf{V}} + \mathbf{V}'$, $\mathbf{B} = \overline{\mathbf{B}} + \mathbf{B}'$ and $C = \overline{C} + C'$. The averaged components can be written in forms automatically satisfying the continuity equation

$$\overline{\mathbf{V}} = \mathbf{1}_\phi \overline{V}_\phi + \overline{\mathbf{V}}_p, \qquad \overline{\mathbf{V}}_p = \nabla \chi \times \mathbf{1}_\phi s^{-1}, \tag{9.2.8a}$$

$$\overline{\mathbf{B}} = \mathbf{1}_\phi \overline{B}_\phi + \overline{\mathbf{B}}_p, \qquad \overline{\mathbf{B}}_p = \nabla \psi \times \mathbf{1}_\phi s^{-1}. \tag{9.2.8b}$$

Here $\psi = s\overline{A}_\phi$ where A is the magnetic vector potential, and χ is the flux function of the poloidal (meridional) motion. In such a nearly axisymmetric dynamo, the axisymmetric parts of the toroidal (azimuthal) velocity and the magnetic field, $\overline{V}_\phi$ and $\overline{B}_\phi$, are much greater than the asymmetrical components, $\mathbf{V}'$ and $\mathbf{B}'$. The large toroidal velocity is used in the definition of the magnetic Reynolds number, $\mathcal{R}_m \sim L\overline{V}_\phi/\eta$. By expansion of equation (2.2) in the small parameter $\mathcal{R}_m^{-1/2}$ the 'equations of generation' for the averaged quantities were obtained in two spatial variables, e.g., (z, s) or (r, θ):

$$\partial_t \overline{A}_\phi + s^{-1}\mathbf{V}_p \cdot \nabla(s\overline{A}_\phi) - \eta\Delta^{(1)}\overline{A}_\phi = \eta\Gamma_\alpha\overline{B}_\phi, \qquad (9.2.9a)$$

$$\partial_t \overline{B}_\phi + s\mathbf{V}_p \cdot \nabla(s^{-1}\overline{B}_\phi) - \eta\Delta^{(1)}\overline{B}_\phi = s\overline{\mathbf{B}}_p \cdot \nabla\zeta. \qquad (9.2.9b)$$

Here $\Delta^{(1)} = \nabla^2 - s^{-2}$, $\zeta = \overline{V}_\phi/s$. The function $\Gamma_\alpha(r, \theta)$ was expressed explicitly as the ϕ-average of some quadratic combinations of the components of the asymmetrical velocity, $\mathbf{V}'$. The following relations required by self-consistency of equations (2.9a, b) are assumed in their derivation:

$$\partial_t \sim \eta/L^2, \qquad \overline{V}_p/\overline{V}_\phi \sim \overline{B}_p/\overline{B}_\phi \sim \mathcal{R}_m^{-1},$$

$$V'/\overline{V}_\phi \sim B'/\overline{B}_\phi \sim \mathcal{R}_m^{-1/2}, \qquad \Gamma_\alpha \sim L^{-1}\mathcal{R}_m^{-1}.$$

The stretching of the field lines of 'poloidal' field $\overline{B}_p$ by the fast non-uniform rotation, ζ, produces a large 'toroidal' field $\overline{B}_\phi \sim \mathcal{R}_m\overline{B}_p$. The action of the source term, $\eta\Gamma_\alpha\overline{B}_\phi$, often called the α-effect, then creates the poloidal field $\overline{B}_p$ from $\overline{B}_\phi$, thus completing the generation loop. The toroidal field is generated by the axisymmetric velocity, $\overline{V}_\phi = s\zeta$ (this is sometimes called the ω-effect), while the asymmetrical velocity, $\mathbf{V}'$, is essential for poloidal field generation. If $\mathbf{V}' = 0$, (2.9a) has no source term, and therefore $\overline{A}_\phi \to 0$ and $\overline{B}_p \to 0$ as $t \to \infty$. Then the source of $\overline{B}_\phi$ in (2.9b) vanishes and hence $\overline{B}_\phi \to 0$ as $t \to \infty$. This is Cowling's theorem. It should be noted that the large toroidal field, $\overline{B}_\phi \sim \mathcal{R}_m\overline{B}_p$, is invisible; only the poloidal field, $\overline{B}_p$, reaches the Earth's surface $r = R_0$. Such a dynamo is often referred to as an '$\alpha\omega$-dynamo' or as a 'strong-field dynamo', because its main field, $\overline{B}_\phi$ is much greater than the observed one.

Equations (2.9a, b) have broader applicability than might be suggested from their derivation. The creation of an α-effect by a motion in which $\mathbf{V}'$ has a short characteristic length-scale is also possible; see, e.g., Moffatt (1978) and Krause & Rädler (1980). Two-dimensional equations similar to (2.9a, b) were obtained for the mean magnetic field in this case also. The name α-effect was introduced by Steenbeck, Krause & Rädler (1966) for the generation of a mean electric current parallel to the mean magnetic field by the term $\nabla \times (\alpha\overline{\mathbf{B}})$, obtained after averaging (2.2) over the small length-scale. The quantity α may be a tensor in the case of anisotropy. In the $\alpha\omega$-dynamo we have $\alpha\overline{\mathbf{B}} = \mathbf{1}_\phi\eta\Gamma_\alpha\overline{B}_\phi$, but the concept of an α-effect is applicable also for the case $\overline{B}_\phi \sim \overline{B}_p$, in which a significant α-effect term appears also in the equation for $\overline{B}_\phi$. The dynamo with $\overline{B}_\phi \sim \overline{B}_p$, where both $\overline{B}_p$ and $\overline{B}_\phi$ are generated by an α-effect while rotation ζ is insignificant, is referred to as an 'α^2-dynamo' or a 'weak-field dynamo'. The self-explanatory label $\alpha^2\omega$-dynamo is also used.

The kinematic theory is now broadly developed (see chapter 1). Many different models of both $\alpha\omega$ and α^2 types have been obtained by solving numerically either the two-dimensional generation equations (see, e.g., Braginsky 1964c, Roberts 1972) or the much more complicated three-dimensional induction equation (2.2) directly. We are now certain that dynamo generation by some 'plausible' velocities is possible that gives magnetic fields similar to that of the Earth. The models that take into account significant forces and sources of energy in the core are necessary to resolve the indeterminacy. The complete geodynamo theory is very complicated and therefore some intermediate theory should be the reasonable next step.

For the geodynamo in the Earth's core one can roughly estimate the fluid velocity from the observed westward drift of the magnetic field as $V \sim 3 \times 10^{-2}$ cm/s; with the characteristic length-scale of the system, $L \sim 10^3$ km, and $\eta \sim 2$ m^2/s, this gives a rather large (but not extremely large) value of $\mathcal{R}_m$, namely $\mathcal{R}_m \sim 10^2$. This means that the nearly axially symmetric model with large toroidal magnetic field is fairly well applicable to the Earth's dynamo. It was shown by Braginsky (1964d) that in order of magnitude the main forces also can be consistently included in the nearly axisymmetric scheme. The equations of the nearly axisymmetric model including the mechanics were developed by Tough & Roberts (1968) and by Soward (1972).

We consider below an axisymmetric dynamo model that aims primarily at an investigation of the mechanics of the axisymmetric motion and the production of buoyancy; the process of poloidal magnetic field generation (α-effect) is taken to be known from kinematic theory. It is shown that the mechanics of the axisymmetric motion seriously restrict the dynamo configuration. This should be taken into account in every geodynamo model. This consideration shows also that the nearly axisymmetric strong field structure of the geodynamo can be considered as the result of the adjustment of the non-linear geodynamo to the rather intensive source of energy of convection in the Earth's core.

9.2.3 Nonlinear Axisymmetric Dynamo Model

After the rather delicate problem of field generation is reduced to equations (2.9a, b) we may concentrate on the mechanical part of the problem. To do this we may simply consider the axisymmetric system but with a prescribed α-effect term on the right hand side of equation (2.9a). This geodynamo is not really axisymmetric because the non-axisymmetric motion and field, $\mathbf{V}'$ and $\mathbf{B}'$, are also present and these are responsible for the assumed α-effect, but they are neglected in all other respects. The model could be named more accurately by the perhaps too cumbersome word 'pseudo-axisymmetric'. A theory of such pseudo-axisymmetric dynamo models is the natural next step after nearly axisymmetric kinematic dynamo theory.

The inertia term $d_t \mathbf{V}$ in (2.1) is much smaller than the Coriolis one, $\mathbf{F}^\Omega$, and may be neglected. This reduces (2.1) to an equation of equilibrium from which the fluid velocity at any moment can be expressed in terms of the forces at the same instant. The viscous force inside the main body of the core, $\mathbf{F}^\nu \sim \nu V/L^2$, is also much smaller than $\mathbf{F}^\Omega$, for it is certainly true that $\nu \ll \Omega L^2$ cm^2/s.

The natural way to develop two-dimensional dynamo theory is by adding the axisymmetric components of equations (2.1) and (2.3) to the system (2.9a, b). The axisymmetric velocity, $\overline{V}$, and density excess, $\overline{C}$, are then to be found from these equations instead of being prescribed. The problem is now a nonlinear one. A characteristic magnetic field strength (2.7), $B_* = 20$ G, appears, that is greater than $\overline{B}_p$ but smaller than $\overline{B}_\phi$.

To make the equations nondimensional we use units based on the quantities R_1, η and B_*. The radius of the core is used as a unit of length. The thermal and compositional diffusivities are very small, $\kappa^T \sim 10^{-5}\eta$ and $\kappa^C \sim 10^{-10}\eta$, and for problems of global scale they should be replaced by some turbulent diffusivities. We expect these to be (very roughly) of the order of η, and just this value is used for scaling. There are a few different characteristic toroidal velocities and length scales in the axisymmetric dynamo model considered below, and that is why we use an 'effective' magnetic Reynolds number, $\mathcal{R} = B_{\phi I}/B_{pI}$, instead of the Reynolds number based on a specific 'characteristic' velocity and length. The expressions for the units (marked by the subscript I) are:

$$t_I = R_1^2/\eta, \quad V_{pI} = R_1/t_I = \eta/R_1, \quad V_{\phi I} = R_1 \zeta_I = V_{pI}\mathcal{R}, \quad (9.2.10a, b, c)$$

$$B_{pI} = B_* \mathcal{R}^{-1/2}, \quad B_{\phi I} = B_* \mathcal{R}^{1/2}, \quad (9.2.10d, e)$$

$$\Gamma_{\alpha I} = R_1^{-1}\mathcal{R}^{-1}, \quad C_I = (B_*^2/g_1 R_1)\mathcal{R} = (2\Omega/g_1)V_{\phi I}, \quad (9.2.10f, g)$$

where $g_1 = g(R_1)$ is the acceleration due to gravity on the CMB. The 'effective' magnetic Reynolds number, $\mathcal{R}$, defined here, measures a ratio of field components that we assume is large: $B_{\phi I}/B_{pI} = \mathcal{R} \gg 1$. The units (2.10) make the non-dimensional field 'of order unity'. Here the quotation marks indicate that these values are also functions of other model parameters (e.g., the small parameter ϵ_ν; see below); therefore they may vary significantly. Formally, the quantity $\mathcal{R} = B_{\phi I}^2/B_*^2$ coincides with the Elsasser number (compare chapter 7) but B_ϕ cannot be prescribed when the complete hydromagnetic dynamo problem is considered. The assumption $\overline{B}_p \ll \overline{B}_\phi$ simplifies greatly the expression for the magnetic force.

From now on only axisymmetric quantities will be used (except when specially noted) and we will omit the overbar indicating the averaging over the longitude.

The ϕ-projections of (2.1) and of its curl give us expressions for V_s and $\partial_z V_\phi$ in terms of the ϕ-components of the magnetic force, F_ϕ^B, and its curl, $(\nabla \times \mathbf{F}^B)_\phi = \partial_z(s^{-1}B_\phi)^2$; in the latter expression a very small term of order $\mathcal{R}^{-2}$, due to $\mathbf{B}_p$, is neglected. We assume that $\mathbf{g} = -1_r g_1 r/R_1$, then

$$2\Omega V_s = F_\phi^B = s^{-1}\nabla \cdot (sB_\phi \mathbf{B}_p),$$

$$2\Omega \partial_z V_\phi = s^{-1}\partial_z B_\phi^2 - (g_1/R_1)\partial_\theta C(r, \theta).$$

Integrating this equation and $\partial_z \chi = -sV_s$ over dz we obtain $V_\phi = s\zeta$, χ and $\mathbf{V}_p$, thus expressing velocities in terms of the magnetic field and density excess. The axisymmetric part of the system (2.1-5) can be written in the nondimensional form:

$$\partial_t A_\phi + s^{-1}\mathbf{V}_p \cdot \nabla(sA_\phi) - \Delta^{(1)}A_\phi = \Gamma_\alpha B_\phi, \qquad (9.2.11)$$

$$\partial_t B_\phi + s\mathbf{V}_p \cdot \nabla(s^{-1}B_\phi) - \Delta^{(1)}B_\phi = s\mathbf{B}_p \cdot \nabla\zeta, \qquad (9.2.12)$$

$$\partial_t C + \mathbf{V}_p \cdot \nabla C + \nabla \cdot \mathbf{I}_p = G^C, \qquad (9.2.13)$$

where $\Delta^{(1)} = \nabla^2 - s^{-2}$. In the insulating mantle $\Delta^{(1)}A_\phi = 0$ replaces equation (2.11). The nondimensional expressions for velocities are

$$V_s = s^{-1}\nabla \cdot (sB_\phi \mathbf{B}_p), \qquad (9.2.14)$$

$$\zeta = \zeta_C + \zeta_B + \zeta_G(s), \qquad (9.2.15)$$

where $\zeta_B = s^{-2}B_\phi^2$. The term ζ_C in (2.15) can be directly expressed by integration of $\partial_\theta C$ over z:

$$\zeta_C = s^{-1}\int_z^{z_1} \partial_\theta C \, dz; \qquad (9.2.16)$$

here $z_1 = (1 - s^2)^{1/2}$, the subscript 1 corresponds to the CMB where it has been assumed that $\zeta_C = 0$. The counterpart of $s\zeta_C$ in dynamical meteorology is called 'the thermal wind', similarly $s\zeta_B$ is sometimes called 'the magnetic wind'. The term $s\zeta_G(s)$ is called the 'geostrophic velocity'; so far it is an arbitrary function of s, resulting from the integration of (2.1b) over z. It is related to the velocity on the CMB, $\zeta_1 = \zeta(z_1, s)$ by $\zeta_1(s) = \zeta_{B1}(s) + \zeta_G(s)$ because $\zeta_{C1} = 0$ by definition. In the case of an insulating mantle $\zeta_{B1} = 0$ and $\zeta_1(s) = \zeta_G(s)$.

To use (2.13) we need expressions for G^C and $\mathbf{I}$. Our understanding of the creation and transport of the density excess in the core is unfortunately far from being complete. We assume that the compositional inhomogeneity dominates. The fluid core is composed of iron and some light constituent(s). During the crystallisation of the inner core the light admixture is released at the inner core boundary ($r = R_2$) lowering the fluid density there, and creating a flux of the density excess, $I_r = -I^N$. The lighter fluid redistributes itself throughout the core, thus creating an effectively homogeneous source of G^C in the volume of the core. It was suggested by Braginsky (1963, 1964d), see also Braginsky (1991), that this process is the principal cause of the density excess in the Earth's core. It can be imagined also that the creation of a density excess is due to a flux, I^M, of heavy constituents from the mantle to the core. In the simplest case, when injection of the flux I^N is the sole mechanism of creation of the density excess, the light fluid balance implies the relation $G^C = I^N S_2/V_{12}$ where S_2 is the inner core surface area and V_{12} is the volume of the fluid part of the core.

It is obvious that the diffusional transport of the admixtures cannot be performed by their molecular diffusivity, which is extremely small. The diffusional flux $\mathbf{I}$ is created by a turbulent mixing that is very efficient; e.g., for the rather small mixing length $l^t \sim 10$ km and turbulent velocity $v^t \sim 10^{-2}$ cm/s we have the turbulent diffusivity $\kappa^t = l^t v^t \sim 1$ m^2/s $\sim \eta$. Although a detailed quantitative theory of the turbulence in the core is lacking at this time, some heuristic considerations of Braginsky & Meytlis (1990) show that small-scale turbulence in the core should develop which would mix the core efficiently, so that κ^t is comparable with η. A simple diffusion-like expression for the flux may be, therefore, a reasonable choice: $\mathbf{I} = -\kappa^t \cdot \nabla C$. The strong magnetic field and the fast rotation makes the turbulence highly anisotropic and therefore the turbulent diffusivity, κ^t, is a tensor. Writing it as $\kappa^t = \eta \mathbf{D}$ we obtain in equation (2.13) the diffusional term $\partial_i I_i$ where

$$I_i = -D_{ij}\partial_j C. \tag{9.2.17}$$

Here D_{ij} is a nondimensional diffusivity tensor containing parameters that should be found by comparing geodynamo theory and the theory of turbulence with observations.

The value of I^N is proportional to the rate of inner core freezing which is determined by the slow processes of Earth's cooling. In geodynamo theory, this value may be considered as given. A simple calculation based on the assumption that the inner core growth rate is of order of $\partial_t V_2 \sim V_2/t_2$, where $t_2 \approx 4 \times 10^9$ yr, gives an estimate of I^N and G^C (Braginsky 1964d, 1991). In nondimensional form these values are:

$$G^C \approx 3I^N r_2^2 \sim 10^4/\mathcal{R}. \tag{9.2.18}$$

Here $r_2 = R_2/R_1$, the unit (2.10g) is used for C_I; correspondingly G^C and $\mathbf{I}$ are measured in the units C_I/t_I and $C_I R_1/t_I$. The value of G^C is proportional to the density jump $\Delta\rho_0$ on the ICB which is assumed to be $\Delta\rho_0 = 0.5$ g/cm^3 in the estimate (2.18). To make the nondimensional values of G^C and $\mathbf{I}$ 'of order unity' (not too big) we should adopt a rather large magnetic Reynolds number $\mathcal{R}$. In other words the hydromagnetic dynamo of the Earth operates far beyond the threshold of self-excitation because the source of buoyancy is very powerful. The geodynamo adapts itself to this large fuelling, through nonlinear effects, by developing a large magnetic field and a high dissipation rate. That is why the geodynamo is nearly axisymmetric, and operates in the strong-field regime. Large numerical coefficients (even as large as 10^2) arise in the problem, mainly because the relevant characteristic length, L, is significantly smaller than the 'natural' $R_1/3$. In practice $\mathcal{R} \sim 100$ or even smaller should be chosen.

It should be noted that the 'diffusional' parametrisation of the transport of density excess is not the only one conceivable. Other possibilities were also considered by Frank (1982) and Moffatt (1988), and there is still much room for discussion.

9.2.4 Geostrophic Velocity and Core-Mantle Interaction

To complete the description of the axisymmetric model it is necessary to show how the geostrophic velocity can be obtained.

The fluid velocity on the CMB, ζ_1, is determined by the interaction of the core fluid with the mantle. The wall subjects the fluid to tangential stresses which may be expressed as the sum of coupling forces of different origin. The three main components considered are viscous, magnetic and topographic couplings (see, e.g., Roberts 1989b). The magnitudes of all three kinds of coupling are weak in comparison with magnetic interactions within the core, where the fluid has a large electrical conductivity. Simple expressions for the viscous and magnetic couplings can be obtained, while there is no way at present of reliably estimating the topographic coupling; the latter will not be considered here.

It is well known (Batchelor 1967) that so-called 'Ekman layers' develop on the boundaries of a rotating vessel where the fluid velocity adjusts itself to the no-slip condition. The velocity of fluid relative to the mantle, $V_{\phi 1}$, drops to zero there, and its gradient is very large: $\partial_r V_{\phi 1} = V_{\phi 1}/\delta_{\nu r}$, where the thickness of the layer is $\delta_{\nu r} = (\nu/\Omega_r)^{1/2} = \delta_\nu \mu^{1/2}$. Here ν is the fluid viscosity, $\delta_\nu = (\nu/\Omega)^{1/2}$, and Ω_r is the component of Ω normal to the boundary, which is relevant for the dynamics of the layer ($\Omega_r = \Omega\cos\theta = \Omega\mu$). The corresponding viscous stress is

$$f_\phi^\nu = -f_\nu V_{\phi 1}, \qquad f_\nu = (\nu/\delta_\nu)\mu^{1/2} = (\nu\Omega\mu)^{1/2}. \qquad (9.2.19a,b)$$

Tough & Roberts (1968) obtained the boundary condition determining $V_{\phi 1}$ in the case of viscous core-mantle friction by utilising the properties of the

Ekman layer. An expression for $V_{\phi 1}$ can be obtained in a more universal way from the condition of equilibrium of the z-projection of the moment of the forces acting on a thin cylindrical shell within the core that has generators parallel to the z-axis. This shell occupies the space between s and $s + ds$; its 'top' and 'bottom' are situated at $z = \pm z_1$, but with the assumption of z-symmetry we will consider only its upper part, $z > 0$. The side surfaces of the shell experience magnetic stresses $B_\phi B_s$. The total moment acting on the side surfaces of the shell is proportional to the so-called 'Taylor integral'

$$T = \int_0^{z_1} B_\phi B_s \, dz. \qquad (9.2.20)$$

The z-moment of the Coriolis force is zero because it is proportional to the integral of V_s over dz, which is equal to a net mass flux through the surface of the shell (which is absent). The Archimedean force has no ϕ-component at all. Two large forces are, therefore, ineffective. The total ϕ-component of magnetic force acting per unit length (along the ϕ-coordinate) of the side surface of the shell is given by (2.20). If there were no direct core-mantle interaction ('friction') on the CMB then the equation of equilibrium for the shell could be written in the form

$$T(s) = 0. \qquad (9.2.21)$$

This is the so-called 'Taylor condition'. However in reality the viscous, magnetic and topographic couples act at $z = \pm z_1$. Let us consider the viscous coupling first. Equating the momenta acting on the side surfaces of the shell, $2\pi \partial_s(s^2 T) \, ds$, and on its 'top' surface, $2\pi s^2 f_\phi^\nu \, ds/\mu$, we obtain the shell equilibrium condition; in non-dimensional form (note that $z_1 = \mu$) it can be written as:

$$\zeta_1 = \epsilon_\nu^{-1}(2z_1^{1/2}s^{-3})\partial_s(s^2 T), \qquad (9.2.22)$$

where

$$\epsilon_\nu = (\delta_\nu/R_1)\mathcal{R} = (\nu/\Omega)^{1/2}R_1^{-1}\mathcal{R}. \qquad (9.2.23)$$

At the boundaries $r = 1$ and $r = r_2$, where $r_2 = R_2/R_1$, the conditions of impenetrability, $V_r = 0$, and the conditions of continuity for magnetic field $[\mathbf{B}] = 0$, should be fulfilled. Assuming the mantle is insulating we should match at $r = 1$ the internal field with the potential external field: $B_\phi = 0$, $\mathbf{B}_p = \mathbf{B}^{ex} = -\nabla\Psi$, where $\nabla^2\Psi = 0$ and $\Psi \to 0$ as $r \to \infty$. The flux, I_r, is prescribed at $r = 1$ and $r = r_2$. If inner core freezing is the cause of the production of density excess then I_r is zero at $r = 1$, and $I_r = -I^N$ at $r = r_2$, where I^N is the given source of density excess determining the geodynamo energetics.

The expression (2.22) for $\zeta_1(s)$ is valid only for $s > r_2$. To obtain a similar expression for the case $s < r_2$, when the shell intersects the inner core, the details of the interaction of the fluid with the inner core should be specified. The expression for $\zeta_1(s)$ for $s < r_2$ can be established using the condition of equilibrium of the inner core as a whole, and by specifying the details of fluid–inner core interaction. The jump on ζ_1 arises at $s = r_2$ because conditions on $\zeta_1(s)$ are completely different for $s > r_2$ and $s < r_2$.

Now let us take the magnetic coupling into account, by the simplified method of Braginsky (1975). The electrical conductivity of the mantle, σ_M, is small, and we assume that it is mainly concentrated in a thin layer (of thickness L_M) near the CMB, so that the magnetic coupling has a local character, like that of the viscous coupling. The electric current, J_θ, in the mantle creates magnetic field according to the relation $\partial_r B_\phi = -\mu_0 J_\theta = -\mu_0 \sigma_M E_\theta$. The electric field, E_θ, is continuous, and on the core side of the CMB it is determined by Ohm's law for a moving fluid, $\partial_r B_{\phi 1} = -\mu_0 \sigma(E_{\theta 1} + V_{\phi 1} B_{r 1})$, where $\partial_r B_\phi$, etc., are evaluated at the CMB and marked by the subscript 1. Integrating $\partial_r B_\phi$ in the mantle over the conducting layer we obtain the magnetic field, $B_{\phi 1}$, on the CMB (the subscript 1 on B_r is omitted):

$$B_{\phi 1} = -\mu_0 \sigma_M L_M (V_{\phi 1} B_r + \partial_r B_{\phi 1}). \tag{9.2.24}$$

The Maxwell stresses acting on the fluid are $B_r B_{\phi 1}$; they can be written as

$$f_\phi^M = -\mu_0 \sigma_M L_M (B_r^2 V_{\phi 1} + B_r \partial_r B_{\phi 1}). \tag{9.2.25}$$

The first term here gives magnetic friction of the form $-f_M V_{\phi 1}$ where

$$f_M = -\mu_0 \sigma_M L_M B_r^2 = (\tau_M / L_M) B_r^2.$$

Here $\tau_M = \mu_0 \sigma_M L_M^2$ is a characteristic time of screening of electromagnetic signals (here the geomagnetic secular variation) by the conducting layer in the mantle. The ratio of viscous friction to magnetic friction can be expressed as

$$f_\nu / f_M = (\nu / \nu_M)^{1/2}, \qquad \nu_M = (\mu_0 \sigma_M L_M)^2 (\Omega \mu)^{-1} B_r^4. \tag{9.2.26a, b}$$

A measurement of the electrical conductivity of the lower mantle material was made by Peyronneau & Poirier (1989); the total mantle conductivity can be estimated from their results as $\sigma_M L_M = \int \sigma_M \, dr \approx 3.5 \times 10^7$ S, and $L_M \approx 500$ km. With these parameters and $B_r \sim 5G$ on the CMB it follows that $\nu_M \sim 10^2$ cm^2/s and $\tau_M \approx 0.7$ yr (so that the mantle does not screen even decade variations).

The set of equations (2.11–23) together with the boundary conditions on the solid boundaries and the given source of density excess, describe the pseudo-axisymmetric dynamo model. They govern the evolution of three functions, A_ϕ, B_ϕ and C, through which all other interesting quantities can be evaluated. We will use for this model a short label 'ABC-model'.

To simplify the problem even more, the axisymmetric Archimedean force gC in (2.1), or equivalently the 'wind' ζ_C, may be prescribed, instead of I^N, thus avoiding the need to analyse the complicated processes of the production and transfer of the density excess. This model contains equations governing the evolution of only two functions, A_ϕ and B_ϕ, and we call it the 'AB-model'. This is much simpler than the more complete ABC-model; in particular, it is much easier to analyse the AB-geodynamo when the inner core is absent. All results presented in section 9.3 are obtained from an AB-model.

9.2.5 Some General Relations for the Axisymmetric Model

The equation of balance of the poloidal magnetic flux, $\psi = sA_\phi$, can be obtained by multiplying equation (2.11) for the core and its counterpart $-\Delta^{(1)}A_\phi = 0$ for the mantle, by $s^2A_\phi = s\psi$, integrating over the volume, adding, and integrating by parts (Braginsky 1964a). This gives

$$\partial_t \int \psi^2/2 \, dV = -\int (\nabla\psi)^2 \, dV + \int s\Gamma_\alpha \psi B_\phi \, dV. \qquad (9.2.27)$$

The integral of $(\nabla\psi)^2$ is taken over all space while the other two integrals are taken over the volume of the core. If $\Gamma_\alpha = 0$ then $\psi \to 0$ as $t \to \infty$ according to (2.27). This is Cowling's theorem.

The magnetic energy is principally that of the toroidal magnetic field because the poloidal field is relatively small. The magnetic energy balance can be obtained by multiplying equation (2.12) by B_ϕ, integrating over the volume of the core, and integrating by parts (Braginsky 1975). The result is

$$\partial_t \int B_\phi^2/2 \, dV = \mathcal{A} - Q_J - Q_f, \qquad (9.2.28)$$

where the rate of working of the Archimedean force, $\mathcal{A}$, and the total rates of heat production due to Joule heating and due to core-mantle (viscous) friction, Q_J and Q_f, can be expressed in alternative forms (here $g_r = -r$):

$$\mathcal{A} = \int B_\phi s \mathbf{B}_p \cdot \nabla\zeta_C \, dV = \int CV_r g_r \, dV = \int s^{-1}\chi\partial_\theta C \, dV, \qquad (9.2.29a,b,c)$$

$$Q_J = -\int B_\phi \Delta^{(1)} B_\phi \, dV = \int (s^{-1}\nabla s B_\phi)^2 \, dV, \qquad (9.2.29d,e)$$

$$Q_f = -\int B_\phi s B_s \partial_s \zeta_1 \, dV = -\int sT \partial_s \zeta_1 \, s \, ds \qquad (9.2.29f,g)$$

$$= \epsilon_\nu \int (2z_1^{1/2})^{-1} s^2 \zeta_1^2 \, s \, ds = \epsilon_\nu^{-1} \int (2z_1^{1/2})(s^{-2}\partial_s s^2 T)^2 \, s \, ds. \qquad (9.2.29h,i)$$

The integrations in (2.27) and (2.28) are performed only over the upper hemisphere because $z \to -z$ symmetry is assumed. The factor 2π is removed from dV in (2.29) so that here $dV = r^2 \, dr \, d\mu = s \, ds \, dz$.

The only positive term on the right hand side of equation (2.28) is $\mathcal{A}$ while $-Q_J$ and $-Q_f$ are negative definite. This means that the sole source of energy driving the dynamo model considered is the rate of working of the Archimedean force. In particular the rate of working of the magnetic force on the geostrophic velocity is negative; it is transformed into the heat, Q_J, of core-mantle friction ($2.29f$, g, h). The geodynamo cannot sustain magnetic field without energy input from the Archimedean force, see Childress (1969) and Braginsky (1972, 1975).

The system of equations considered has an interesting property of self-similarity: if a solution of these equations is known, one can obtain from it an entire family of solutions by the following scaling ($\propto$ is a proportionality sign):

$$A_\phi \propto B_p \propto \lambda^{-1}, \quad B_\phi \propto \lambda, \quad V_p \propto \lambda^0, \quad \zeta \propto C \propto I^N \propto \lambda^2, \quad \epsilon_\nu \propto \Gamma_\alpha \propto \lambda^{-2}.$$
$$(9.2.30)$$

This property can be seen directly from the fact that the units used for reducing the system to non-dimensional form contain an arbitrary parameter $\mathcal{R}$. By varying this parameter we would change the non-dimensional quantities proportional to some power of this parameter (without changing the physical situation); in this case $\lambda = \mathcal{R}^{-1/2}$. Another approach is to consider the non-dimensional solution as given, and A_ϕ, B_ϕ, etc., as in (2.30) as dimensional values having magnitudes proportional to their units; in this case $\lambda = \mathcal{R}^{1/2}$.

It can be seen from the self-similarity of the solution how the fields and velocities in the geodynamo vary as the intensity of convection during the geological evolution of the Earth changes. The unit, C_I, of density excess is proportional to the 'driving flux' I^N so that $\lambda \propto C_I^{1/2} \propto (I^N)^{1/2}$, and (2.30) directly shows the scaling dependencies. It is interesting to note that as the intensity of convection increases the magnitude of the poloidal magnetic field $\mathbf{B}_p$ decreases. Intensification of the fuelling of the geodynamo therefore increases C and B_ϕ^2 but tends to decrease the observed geomagnetic field. The above scaling law is, of course, of limited practical significance because for its strict applicability the non-dimensional solution should be independent of time (e.g., the turbulent diffusivity tensor D_{ij} should not change, and the core-mantle friction coefficient should not vary with the change of intensity of convection). One could hope, however, that the tendency it indicates is roughly correct.

9.3 MODEL-Z GEODYNAMO

9.3.1 Model AB Equations
We rewrite here the system of equations for the AB-model in a slightly different form (Braginsky 1978). It consists of equations of evolution of two functions, $\psi = sA_\phi$ and B_ϕ, together with subsidiary relations expressing all other quantities through these functions:

$$\partial_t \psi + \nabla \cdot (\psi \mathbf{V}_p) - \Delta^{(-)}\psi = \mathcal{R}_\alpha s\alpha B_\phi, \qquad (9.3.1)$$

$$\partial_t B_\phi + \mathcal{D}(B_\phi) - \Delta^{(1)} B_\phi = s B_s \partial_s \zeta_1 + \mathcal{R}_\omega s \mathbf{B}_p \cdot \nabla \zeta_\omega, \qquad (9.3.2)$$

where

$$\mathcal{D}(B_\phi) = \nabla \cdot (B_\phi \mathbf{V}_p) - s^{-1} B_\phi (3 \mathbf{B}_p \cdot \nabla B_\phi - s^{-1} B_\phi B_s),$$

$$\mathbf{V}_p = \nabla \times (\mathbf{1}_\phi \chi s^{-1}), \qquad \mathbf{B}_p = \nabla \times (\mathbf{1}_\phi \psi s^{-1}), \qquad (9.3.3a, b)$$

$$\chi = -s \left(B_\phi B_s + s^{-2} \partial_s \left(s^2 \int_0^z B_\phi B_s \, dz \right) \right), \qquad (9.3.4)$$

$$\zeta_1 = \epsilon_\nu^{-1} (2 z_1^{1/2} s^{-3}) \partial_s (s^2 \mathcal{T}), \qquad (9.3.5)$$

$$\mathcal{T} = \int_0^{z_1} B_\phi B_s \, dz. \qquad (9.3.6)$$

The inner core conductivity is supposed to be equal to the fluid core conductivity. Equations (3.1) and (3.2) for the regions of the inner core, $r < r_2$, and for the insulating mantle, $r > 1$ (the 'external region'), change into

$$\partial_t \psi - \Delta^{(-)} \psi = 0, \qquad \Delta^{(-)} \psi^e = 0, \qquad (9.3.1'a, b)$$

$$\partial_t B_\phi - \Delta^{(-)} B_\phi = 0, \qquad B_\phi^e = 0. \qquad (9.3.2'a, b)$$

Here $\Delta^{(-)} \psi = s \Delta^{(1)} (s^{-1} \psi)$ and $\Delta^{(1)} B_\phi = (\nabla^2 - s^{-2}) B_\phi$. Equation (2.14) in the form $V_s = s^{-1} \nabla \cdot (s B_\phi \mathbf{B}_p)$ follows from (3.3a) and (3.4). The substitution $\Gamma_\alpha = \mathcal{R}_\alpha \alpha$ and $\zeta_C = \mathcal{R}_\omega \zeta_\omega$ is made, where $\mathcal{R}_\alpha$ and $\mathcal{R}_\omega$ are 'amplitudes' while α and ζ_ω are 'forms' of the nondimensional source functions Γ_α and ζ_C. The dimensional source functions $\Gamma_{\alpha\text{dim}}$ and $\zeta_{C\text{dim}}$ are written in the form $\Gamma_{\alpha\text{dim}} = R_I^{-1} \mathcal{R}^{-1} \mathcal{R}_\alpha \alpha$ and $\zeta_{C\text{dim}} = t_I^{-1} \mathcal{R} \mathcal{R}_\omega \zeta_\omega$ where $\Gamma_{\alpha I} = R_I^{-1} \mathcal{R}^{-1}$ and $\zeta_1 = t_I^{-1} \mathcal{R}$ are the corresponding units according to (2.10c, f). The coefficients $R_I^{-1} \mathcal{R}^{-1} \mathcal{R}_\alpha$ and $t_I^{-1} \mathcal{R} \mathcal{R}_\omega$ give the dimensional amplitudes of $\Gamma_{\alpha\text{dim}}$ and $\zeta_{C\text{dim}}$. If the same unit is used for both B_p and B_ϕ then $\mathcal{R} = 1$, and the coefficients $\mathcal{R}_\alpha R_I^{-1}$ and $\mathcal{R}_\omega t_I^{-1}$ therefore give the dimensional amplitudes of $\Gamma_{\alpha\text{dim}}$ and $\zeta_{C\text{dim}}$. This choice ($\mathcal{R} = 1$) is usually made in kinematic theory when α^2, $\alpha^2 \omega$ and $\alpha \omega$ dynamo models are considered together.

Boundary conditions for ψ, B_ϕ may be written in the form

$$\psi = 0, \qquad B_\phi = 0, \qquad \theta = 0, \qquad (9.3.7a)$$

$$\partial_\theta \psi = 0, \qquad B_\phi = 0, \qquad \theta = \pi/2, \qquad (9.3.7b)$$

$$\psi = 0, \qquad B_\phi = 0, \qquad r = 0, \qquad (9.3.7c)$$

$$\partial_r \psi = \partial_r \psi^e - \chi_1 B_r, \qquad B_\phi = 0, \qquad r = 1; \qquad (9.3.7d)$$

$$\chi_1 = \epsilon_\nu (2 z_1^{1/2})^{-1} s^2 \zeta_1. \qquad (9.3.7e)$$

Here χ_1 is the velocity flux function at $r = 1$, just below the Ekman layer. The external field can be found by developing the solution of (3.1'b) into a series of terms $r^{-n} \mu_s P_n^1$, where P_n^1 are the associated Legendre functions, which satisfy the equation $\Delta^{(1)}P_n^1 = -n(n+1)P_n^1$; then $\psi = \psi^e$ and $\partial_r\psi = \partial_r\psi^e$ at $r = 1$ can be found:

$$\partial_r\psi^e = -\mu_s \sum n\psi_n P_n^1(\mu), \qquad n = 1, 3, 5, \ldots, \qquad (9.3.8a)$$

$$\psi_n = \int \psi(1, \theta)\mu_s^{-1} P_n^1 \, d\mu, \qquad \mu = 0, 1. \qquad (9.3.8b)$$

There is a large current density in the Ekman layer: $\partial_r B_\theta = -\eta^{-1}V_\theta B_r$. This leads to jump of $\partial_r\psi = -sB_\theta$ that is equal to the following integral over the layer $[\partial_r\psi] = -s\int \partial_r B_\theta \, dr$. In non-dimensional form the jump is $[\partial_r\psi] = B_r\chi_1$. It is rather small because of the smallness of δ_ν, even though the velocity in the Ekman layer, $V_\theta \sim s\zeta_1$, is large.

The system (3.1–8) contains two non-dimensional prescribed functions $\zeta_C = \mathcal{R}_\omega\zeta_\omega$ and $\Gamma_\alpha = \mathcal{R}_\alpha\alpha$, and also a parameter, ϵ_ν, characterising core-mantle friction. If the forms α and ζ_ω are given, the solution of the system is determined by three parameters, $\mathcal{R}_\alpha, \mathcal{R}_\omega$ and ϵ_ν. The system (3.1–8) is invariant under the scaling transformation (2.30) to which we add $\mathcal{R}_\alpha \propto \lambda^{-2}$ and $\mathcal{R}_\omega \propto \lambda^2$. There are, therefore, self-similar solutions occupying subspaces in the three-dimensional parameter space according to the scaling transformation. It may be convenient to choose the 'dynamo number' $\mathcal{R}_D = \mathcal{R}_\alpha\mathcal{R}_\omega$ as the characteristic parameter because it is an invariant of the scaling transformation.

9.3.2 Two Scenarios: Taylor's State and the Model-Z State

The system (3.1–8) contains a small parameter ϵ_ν. Its magnitude is not known precisely but it can be roughly estimated as $\epsilon_\nu \sim 10^{-3}$ for $\nu \sim 0.1 \text{ m}^2/\text{s}$. This value may be considered as a viscous equivalent of the joint action of all friction mechanisms. The presence of the small parameter ϵ_ν in the dynamo model invites the question 'is this parameter significant?'; in other words, can we take $\epsilon_\nu = 0$ in the leading approximation? Taylor (1963) considered the case $\epsilon_\nu = 0$. He suggested that $\zeta_1(s)$ will automatically adjust itself in such a way that the equilibrium condition (2.21) is fulfilled, so that the function $\zeta_1(s)$ can be found from the condition on another function: $T(s) = 0$. In this case there is no small parameter in the equations; therefore ζ_1 must be of order unity. For small but non-zero T this corresponds to $T \sim \epsilon_\nu$ and $\zeta_1 \sim 1$. One may, however, also anticipate another possibility: $T \sim \epsilon_\nu^n$, where $0 < n < 1$, in which case $\zeta_1 \sim T/\epsilon_\nu \sim \epsilon_\nu^{-(1-n)}$. The value of ζ_1 is very large in the second case. Which of these possibilities is adopted by the geodynamo? It depends on the nature of the mechanism that causes $T(s)$ to tend to zero and its interaction with the adjustment of ζ_1. If such a mechanism produces $T \sim \epsilon_\nu$ then Taylor's alternative is satisfied with $\zeta_1 \sim 1$. If it produces $T \ll 1$ but $T \gg \epsilon_\nu$ then we have $\zeta_1 \gg 1$, i.e., the second alternative.

The equation governing the evolution of T can be obtained by combining the equations for $\partial_t B_\phi$ and for $\partial_t B_s$; the latter is obtained by applying ∂_s to equation (3.1). After some rearrangements the result can be written in the following form (Braginsky 1975, 1976):

$$\partial_t(s^2 T) = \epsilon_\nu^{-1} \mathcal{D}_s s^3 \partial_s (2 z_1^{1/2} s^{-3} \partial_s (s^2 T)) + s^2 \mathcal{F}, \qquad (9.3.9a)$$

where

$$\mathcal{D}_s = \int B_s^2 \, dz$$

and

$$\mathcal{F} = - \int (s \mathbf{B}_p \cdot \nabla s^{-1} B_\phi)^2 \, dz + \int (B_s \Delta^{(1)} B_\phi + B_\phi \Delta^{(1)} B_s) \, dz$$

$$+ \int (B_s s \mathbf{B}_p \cdot \nabla \zeta_C + \Gamma_\alpha B_\phi \partial_z B_\phi) \, dz.$$

Here ζ_C and Γ_α are written in place of $\mathcal{R}_\omega \zeta_\omega$ and $\mathcal{R}_\alpha \alpha$ for brevity. The equation $(3.9a)$ is valid for $s > r_2$, and all integrals in $\mathcal{F}$ are taken from 0 to z_1. If we neglect the inner core the boundary conditions on $s^2 T$ are

$$s^2 T = 0 \qquad \text{at } s = 0, 1. \qquad (9.3.9b)$$

Equation $(3.9a)$ resembles a diffusion equation with $\epsilon_\nu^{-1} \mathcal{D}_s$ playing the role of diffusion coefficient. If $B_s \sim 1$ then $\mathcal{D}_s \sim 1$. In this case the diffusion is very fast and, because of the zero boundary conditions $(3.9b)$, this results in a quick decay of the quantity $s^2 T$, even if $T \sim 1$ at the initial moment.

Multiplying $(3.9a)$ by $\epsilon_\nu (s \mathcal{D}_s)^{-1} T$ and integrating from $s = 0$ to $s = 1$ it is possible to obtain, after integrating by parts,

$$\int \epsilon_\nu \mathcal{D}_s^{-1} \partial_t (T^2/2) \, s \, ds = - \int (s^{-2} \partial_s s^2 T)^2 2 z_1^{1/2} \, s \, ds + \int \epsilon_\nu \mathcal{D}_s^{-1} T \mathcal{F} \, s \, ds. \quad (9.3.10)$$

If $\mathcal{D}_s T \gg \epsilon_\nu$ then the first (negative) term dominates on the right-hand side of (3.10). This implies a rapid decrease of T^2; its value drops significantly in a time of order $\epsilon_\nu \mathcal{D}_s$. It is the value of B_ϕ that changes quickly when $T \sim 1$, because equation (3.2) contains the term $\partial_s \zeta_1 \sim \epsilon_\nu^{-1} T$, which is very large, while B_s and $\mathcal{D}_s$ change much more slowly. After some evolution a balance becomes established between the terms on the right-hand side of (3.10) in which $\mathcal{D}_s T \sim \epsilon_\nu$. In the stationary state, the 'equilibrium' T could be calculated from equation (3.9) using its Green's function. Equilibrium is achieved when either $\mathcal{D}_s \sim 1$ and $T \sim \epsilon_\nu$ or $\mathcal{D}_s \ll 1$ and $T \gg \epsilon_\nu$. In either case the equilibrium T is small. The crucial question is 'what is the order of magnitude of the equilibrium T'? If $B_s \sim \mathcal{D}_s \sim 1$ then $T \sim \epsilon_\nu$, according to (3.10), and $\zeta_1 \sim 1$, so that ζ_1 does not depend on ϵ_ν. This is the Taylor state. In this state $T \sim \epsilon_\nu$ because of the perfect cancellation of terms in the integrand of (2.20). If however B_s and $\mathcal{D}_s$ become very small then $T \ll B_\phi B_p L$ because $B_s \ll B_p$ (and possibly also because of partial cancellation of terms), but in this case $T \gg \epsilon_\nu$, so that ζ_1 is large and depends on ϵ_ν. For example if $T \sim \epsilon_\nu^n$, where $0 < n < 1$, then $\zeta_1 \sim \epsilon_\nu^{-(1-n)}$. This is the model-Z state. Although the viscosity is small, it is significant in such a state and the solution cannot be obtained by setting $\epsilon_\nu = 0$ as was done by Taylor (1963).

9.3.3 Form of the Model-Z Solution

Braginsky (1975) suggested a form of the solution for $T \ll 1$ but $\zeta_1 \gg 1$. In this solution the field lines of $\mathbf{B}_p$ are predominantly parallel to the z-axis because B_s is small, which is why it was named 'model-Z'. To clarify its pattern let us make the assumptions $B_s \ll B_z$, $\zeta_1 \gg 1$ and $\partial_s \zeta_1 \gg 1$, and consider their consequences. An immediate one is the existence of a sharp 'break' in the lines of force of $\mathbf{B}_p$ near the core-mantle boundary. This break is necessary to match the internal field, which has $B_s \ll B_z$, to the potential field in the mantle, where $B_s \sim B_z$. A narrow layer, of thickness $\delta_J \ll 1$, with high current density should develop at $r = 1$. The solution, therefore, consists of two terms: a smooth bulk solution and a sharp current layer solution; the latter depends on the stretched coordinate

$$\xi = (1 - r)/\delta_J.$$

We may write the solution in the form $\mathbf{V} = \mathbf{V}^0 + \mathbf{V}^\delta(\xi)$ and $\mathbf{B} = \mathbf{B}^0 + \mathbf{B}^\delta(\xi)$, where $\mathbf{V}^\delta(\xi)$ and $\mathbf{B}^\delta(\xi)$ drop to zero as $\xi \to \infty$.

In the current layer B_r^δ and V_r^δ can be neglected with sufficient accuracy because $B_r^\delta/B_\theta^\delta \sim V_r^\delta/V_\theta^\delta \sim \delta_J$ and, recalling that $B_r^0 \ll B_z^0$, we may adopt the following relations in the leading approximation:

$$\partial_r = -\delta_J^{-1}\partial_\xi, \quad \Delta^{(1)} = \partial_r^2, \quad B_r = B_r^0 = \mu B_z^0, \quad B_s^\delta = \mu B_\theta^\delta, \quad V_s^\delta = \mu V_\theta^\delta.$$

Equations (3.1, 2, 2.14) for the current layer take the form

$$\delta_J B_r V_\theta^\delta - \partial_\xi B_\theta^\delta = \delta_J \Gamma_\alpha^\delta B_\phi, \tag{9.3.11a}$$

$$\partial_\xi^2 B_\phi^\delta = -\delta_J^2 \mu B_\theta^\delta s \partial_s \zeta_1, \tag{9.3.11b}$$

$$\delta_J \mu V_\theta^\delta = -B_r \partial_\xi B_\phi^\delta. \tag{9.3.11c}$$

The thickness of the layer can be defined as

$$\delta_J = (B_r^2 s \partial_s \zeta_1)^{-1/2}. \tag{9.3.12}$$

A large (nondimensional) value $\partial_s \zeta_1 = 10^3$ is obtained below; this corresponds to $\delta_J \sim 3 \times 10^{-2}$ or, in dimensional form, $\delta_J \sim 10^2$ km.

The following equation for B_ϕ in the layer may be obtained (Braginsky 1975) from (3.11a, b, c):

$$\partial_\xi^3 B_\phi - \partial_\xi B_\phi = G^\delta(\xi) B_\phi. \tag{9.3.13}$$

An enhanced generation concentrated in the layer, Γ_α^δ, is assumed here. The scaled coefficient of concentrated generation is written as

$$G^\delta(\xi) = \delta_J \Gamma_\alpha^\delta(\xi)\mu/B_r^2.$$

Generation in the layer is significant when $\Gamma_\alpha^\delta \sim \delta_J^{-1}$. This opportunity looks a natural one according to Braginsky (1964a), and the author plans to investigate it in detail. In this paper, however, the full equation (3.13) is not solved. Only a simpler case $G^\delta(\xi) = 0$ is considered here. In this case (3.13) with the boundary condition $B_\phi(\xi = 0) = 0$ has a simple solution

$$B_\phi = B_{\phi 1} + B_\phi^\delta(\xi) = B_{\phi 1}(1 - e^{-\xi}), \tag{9.3.14}$$

where $B_{\phi 1} = B_\phi^0(r = 1)$, and (3.11a, c) give

$$V_\theta^\delta(\xi) = -\delta_J^{-1}\mu^{-1}B_r B_{\phi 1}e^{-\xi}, \qquad B_\theta^\delta(\xi) = \mu^{-1}B_r^2 B_{\phi 1}e^{-\xi}.$$

Note that according to (3.12) such a layer is possible only for $\partial_s \zeta_1 > 0$. This means that such a layer cannot cover all the CMB because the condition of mantle equilibrium implies that ζ_1 changes sign as does $\partial_s \zeta_1$. There is a 'jump' of χ across the layer equal to $[\chi] = -\chi^\delta(0)$. Using $sV_\theta^\delta = -\partial_r \chi^\delta = \delta_J^{-1}\partial_\xi \chi^\delta$ we obtain $[\chi] = \delta_J \int sV_\theta^\delta(\xi)\,d\xi = -\mu^{-1}sB_{\phi 1}B_r$ and the substitution $B_r = \mu B_z^0$ gives $[\chi] = -sB_z B_{\phi 1}$. The contribution of the layer to the Taylor integral (note that $B_s^\delta\,dz = B_\theta^\delta\,dr$ and $B_r = \mu B_z^0$) is

$$T^\delta = \int B_\phi B_s^\delta\,dz = \delta_J \mu B_z^2 B_{\phi 1}^2/2. \tag{9.3.15}$$

The superscript 0 is omitted here and it will usually be omitted below.

In the bulk of the core the leading term of the poloidal field is $\mathbf{B}_p = \mathbf{1}_z B_z(s)$. Then (3.4) and (3.3a) with the B_s component neglected give

$$\chi = -sB_z B_\phi, \qquad V_s = B_z \partial_z B_\phi, \qquad V_z = -s^{-1}\partial_s(sB_z B_\phi). \tag{9.3.16a,b,c}$$

The quantity $\chi^0(r = 1) = -sB_z B_{\phi 1}$ coincides with $-\chi^\delta(0)$; therefore the bulk velocity matches with the layer velocity in the assumed approximation, $B_s^0 \ll B_z^0$. Equation (3.1) turns into

$$B_z^2 \partial_z B_\phi - \Gamma_\alpha B_\phi = \partial_s B_z(s) - s^{-1}\partial_t \psi(s), \tag{9.3.17}$$

where $\Gamma_\alpha = \mathcal{R}_\alpha \alpha$. By comparing the first terms on the left- and on the right-hand sides of (3.17) we obtain $B_z B_\phi \sim 1$ and $V_p \sim B_z B_\phi \sim 1$. With these estimates (3.15) gives $T^\delta \sim \delta_J$.

The Joule dissipation rate in the layer is of order B_ϕ^2/δ_J; it is much greater than its integral over the bulk of the core which is of order of B_ϕ^2. Comparing the main terms in the energy balance equation (2.28), $\mathcal{A} \sim B_z B_\phi \zeta_C$ and

$Q_J \sim B_\phi^2/\delta_J$, we obtain $\zeta_C \sim B_\phi^2/\delta_J$. To compensate for large dissipation in the current layer, the Archimedean force should be large, so that the dominant terms in (3.2) are those on the right hand-side. Equation (3.2) for the bulk of the core turns into

$$B_s = -(B_z/\partial_s\zeta_1)\partial_s\zeta_C. \qquad (9.3.18)$$

The contribution of the bulk of the core to the Taylor integral is

$$T^0 = \int B_\phi B_s \, dz = -(B_z/\partial_s\zeta_1) \int B_\phi \partial_s \zeta_C \, dz, \qquad (9.3.19)$$

where integration is performed (from 0 to z_1) excluding the current layer. Using the estimates $B_\phi B_s \sim 1$, $\zeta_C \sim \delta_J^{-1}B_\phi^2$ and $\zeta_1 \sim \delta_J^{-2}B_z^{-2}$ from (3.12), we find from (3.18) that $B_s/B_z \sim \delta_J$. Comparison of (3.15) and (3.19) shows that both contributions to the Taylor integral are of the same order of magnitude, and $T \sim \delta_J$. Now equation (3.5) implies that $\zeta_1 \sim \epsilon_\nu^{-1}T \sim \epsilon_\nu^{-1}\delta_J$; this relation together with $\zeta_1 \sim \delta_J^{-2}B_z^{-2}$ gives $\delta_J^3 \sim \epsilon_\nu B_z^{-2}$. Gathering all estimates together, expressing all quantities through ϵ_ν and ζ_C, and replacing ζ_C by $\mathcal{R}_\omega$, we may write

$$B_\phi B_z \sim 1, \qquad B_s/B_z \sim T \sim \delta_J, \qquad \delta_J \sim \epsilon_\nu^{1/2}\mathcal{R}_\omega^{1/2}, \qquad (9.3.20a,b,c)$$

$$B_z \sim \epsilon_\nu^{-1/4}\mathcal{R}_\omega^{-3/4}, \qquad \zeta_1 \sim \epsilon_\nu^{-1/2}\mathcal{R}_\omega^{1/2}. \qquad (9.3.20d,e)$$

It should be stressed that the large Archimedean stirring, $\zeta_C \sim \mathcal{R}_\omega \sim \delta_J^{-1}B_\phi^2 \gg B_\phi^2$, is necessary for $\delta_J \ll 1$, that is, for the very existence of the model-Z solution, because of large Joule dissipation in the current layer. If we suppose that $\mathcal{R}_\omega \sim \epsilon_\nu^{-1/3}$ then (3.20) simplifies and takes the form

$$B_\phi \sim B_z \sim 1, \qquad B_s \sim T \sim \delta_J, \qquad \delta_J \sim \epsilon_\nu^{1/3}, \qquad \zeta_1 \sim \epsilon_\nu^{-2/3}. \qquad (9.3.21a,b,c,d)$$

The z-dependence of $B_\phi(z,s)$ can be found directly from (3.17). This is a first-order differential equation for $\partial_z B_\phi(z,s)$; its right-hand side does not depend on z. This equation can be solved, and the z-dependence of B_ϕ can be written down as

$$B_\phi(z,s)/B_\phi(z_1,s) = \exp(u(z,s) - u(z_1,s))U(z,s)/U(z_1,s), \qquad (9.3.22)$$

where

$$U(z,s) = \int_0^z \exp(-u(z',s) \, dz', \qquad u(z,s) = B_z^{-2} \int_0^z \Gamma_\alpha(z',s) \, dz'.$$

If $B_z \sim 1$, and $\Gamma_\alpha(z,s)$ is positive and significantly greater than unity then $B_\phi(z,s)$ decreases sharply with increasing distance, $z_1 - z$, from the CMB. For example, if we let $B_z^{-2}\Gamma_\alpha \approx a(s)z$ where $a \gg 1$, then $u \approx az^2/2$ and $U(z,s)/U(z_1,s) \approx 1$, but the first factor in (3.22) decreases exponentially as

$\exp(-a(z_1^2 - z^2)/2)$, with $z_1 - z$. This effect results in expulsion of toroidal field from the region where Γ_α is rather large, so that generation of poloidal field turns out to be impossible. Braginsky (1978) encountered this effect during his numerical integration of the axisymmetric dynamo model. The only remedy found in that work was to set $\Gamma_\alpha = 0$ in the greater part of the core where $\partial_s \zeta_1$ is large and positive, B_s is small, and (3.22) is valid. If $\Gamma_\alpha = 0$ then $\partial_z^2 B_\phi(z,s) = 0$ according to (3.17). This relation, or the limit $\Gamma_\alpha \to 0$ in (3.22), implies a simple form for the field B_ϕ in the bulk of the core, namely

$$B_\phi(z,s) = B_{\phi 1}(s)\, z/z_1.$$

For the generation of poloidal field an α-effect is necessary. With this demand in mind a rather large Γ_α was assumed in the comparatively narrow equatorial belt where $\partial_s \zeta_1$ is significantly smaller and $B_s \sim B_z$, so that (3.22) is not applicable.

The pattern of fields and motions in the equatorial belt is quite different from that in the main part of the core. It is complicated and significantly two-dimensional. The poloidal field is generated there by the zonal electric current produced by the α-effect, so that in order of magnitude we may write $B_z \sim \Gamma_\alpha B_\phi L_b$ where L_b is the equatorial belt thickness ($L_b \sim 0.1$). Toroidal field is also screened in the belt, and from equation (3.2) the estimate $B_\phi \sim \zeta_C B_z L_b^2$ can be obtained. From both these estimates it can be deduced that $\Gamma_\alpha \zeta_C L_b^3 > 1$ or $\mathcal{R}_D = \mathcal{R}_\alpha \mathcal{R}_\omega > L_b^{-3}$. The sign $\sim$ is here replaced by an inequality sign, thus anticipating the possibility that the screening is stronger when the fields decrease on a length-scale smaller than L_b. In order of magnitude we may write roughly $\mathcal{R}_D > \mathcal{R}_{D0}$ where $\mathcal{R}_{D0} \sim L_b^{-3} \sim 10^3$. In fact, numerical calculations of model-Z (Roberts 1989a, Hofflin & James 1992) gave $\mathcal{R}_{D0} \approx 5000$, which is natural because a coefficient of about $k \sim 3$–10 is usually present in estimates such as $\Delta^{(1)} B_\phi \sim k B_\phi L_b^{-2}$.

9.3.4 Model-Z Numerical Solutions

The direct, but insufficiently general, way to understand the form of the solution and to decide between the two possible AB-model scenarios is by integrating the system of equations (3.1–8) forward in time, starting from some arbitrary initial state. By making ϵ_ν small but finite, the system can approach either of the two possible states. This programme was realised by Braginsky (1978), who integrated the equations numerically by time-stepping the functions $\psi(t,r,\theta)$ and $B_\phi(t,r,\theta)$, starting from an arbitrary smooth initial state, $\psi(0,r,\theta)$ and $B_\phi(0,r,\theta)$, and obtained a stationary solution of model-Z type. Model-Z solutions were also obtained by Braginsky & Roberts (1987), Braginsky (1988, 1989), Roberts (1988, 1989a), Cupal & Hejda (1989) and Anufriev, Cupal & Hejda (1993).

This subsection is based on Braginsky (1978) and Braginsky & Roberts (1987). Both papers used nearly the same algorithm to obtain the numerical solution. For simplicity, only viscous friction was assumed in these works. This implies the boundary condition $B_\phi = 0$ at $r = 1$, so that $\zeta_G = \zeta_1$. Computational expediency dictated choices of ϵ_ν that were an order of magnitude greater than geophysically realistic values, but nevertheless a rather thin boundary current layer developed in the solution.

The solution was defined on a numerical grid (r_i, θ_j). For better resolution of the current layer, the radial grid points of the numerical mesh were chosen to crowd together as the core surface was approached. There are N_r intervals in the radial grid. These intervals, h_i, are unequal: they are constant from the centre until $r = 0.5$, then they decrease, being at $r = 1$ about an order of magnitude smaller than near the centre. The uniform θ-grid of N_θ intervals has a constant step, $h_\theta = \pi/2N_\theta$. Braginsky (1978) used $N_r = 30$ and $N_\theta = 24$ (h_i changes from 0.05 to 0.006). In Braginsky & Roberts (1987) two grids were used, a coarse grid with $N_r = 32$ and $N_\theta = 32$ (h_i changes from 0.05 to 0.00628 according to $h_{i+1} = qh_i$ where $q = 0.9014706286$), and a fine grid with $N_r = 64$ and $N_\theta = 64$ (h_i changes from 0.025 to 0.00314 with $q = 0.9476702642$).

The numerical grid (s, y), where $y = z/z_1(s)$, was introduced in order to carry out the z-integrations in (3.4, 6) and to evaluate ζ_1 by (3.5). The grid in s was equally spaced with step $h_s = 1/N_r$. The grid in $y = z/z_1$ was equi-spaced in Braginsky (1978) with step $h_y = 1/30$. In Braginsky & Roberts (1987) the y-grid (z-grid) was stretched like the r-grid, the intervals between the y_i were precisely the same as the h_i used for r_i. The interpolations between (r, θ) and (s, y) grids were linear, and they were performed at every time step. The simple trapezoidal rule was used for numerical integrations.

While stepping in time the diffusion terms were integrated implicitly by the 'method of splitting' of the variables (r, θ), so that only one-dimensional tridiagonal matrices had to be inverted at each step of the process. The advection terms were calculated explicitly using a conservative scheme that represents each divergence term by the sum of fluxes crossing the four faces (1, 2, 3, 4) of the volume dV_{ij} of the elementary (i, j)-cell of the grid; here the superscripts 1 and 2 will be used instead of $(i - 1/2, j)$ and $(i + 1/2, j)$, and 4, 3 will correspond to $(i, j \pm 1/2)$. Thus

$$-(\nabla \cdot (B_\phi \mathbf{V}_p))_{ij} = (1/dV_{ij})(B_\phi^1 F^1 + B_\phi^2 F^2 + B_\phi^3 F^3 + B_\phi^4 F^4),$$

$$dV_{ij} = (1/3)(r_+^3 - r_-^3)h_\theta \sin\theta_j, \qquad r_\pm = (r_i + r_{i\pm1})/2,$$

$$B_\phi^1 = (B_{\phi i,j} + B_{\phi i-1,j})/2,$$

$$F^1 = \chi_+^1 - \chi_-^1, \qquad \chi_\pm^1 = (\chi_{i,j} + \chi_{i-1,j} + \chi_{i,j\pm1} + \chi_{i-1,j\pm1})/4.$$

Similar expressions can be written for the faces 2,3,4. For example, we have $B_\phi^2 = (B_{\phi i,j} + B_{\phi i+1,j})/2$, etc. The sums of advection terms turn out to be identically zero, $\sum (\nabla \cdot (B_\phi \mathbf{V}_p))_{ij} \, dV_{ij} = \sum B_{ij} (\nabla \cdot (B_\phi \mathbf{V}_p)_{ij} \, dV_{ij} = 0$, where summation is over all grid cells; this makes the conservation properties of the scheme closer to those of the original equations.

Various forms of the α-effect coefficient were tested in computing the evolution of the fields ψ and B_ϕ. Unlike the two-dimensional kinematic problem, it proved to be difficult to find amplitudes, $\mathcal{R}_\alpha$ and $\mathcal{R}_\omega$, such that the solutions were non-attenuating. This was because of displacement (expulsion) of field B_ϕ from the region where $\alpha = 0$. The computations show that during the concurrent evolution of (ψ, B_ϕ) it is B_ϕ that usually adapts first to the existing ψ, and then both fields adjust consistently and more slowly. For different smooth functions α (e.g., $\alpha = 3[4s(1-r)]^2 \cos \theta$ and $\mathcal{R}_\alpha = 50$) it was found that, when the sign of the term $\mathcal{R}_\alpha \alpha B_\phi$ corresponds to intensification of the field $\mathbf{B}_p$, the latter increases initially but then the field B_ϕ weakens rapidly in the region where $\alpha > 0$, and may even change sign in part of this region. After that ψ attenuates and B_ϕ attenuates with it.

An interpretation of B_ϕ-expulsion is given in the previous subsection on the basis of the qualitative picture of the solution predicted by Braginsky (1975) but in fact this effect was discovered 'empirically' during attempts to construct working models. After numerous unsuccessful trials, in which the initial field decayed even though rather large $\mathcal{R}_\alpha$ and $\mathcal{R}_\omega$ were chosen, the following favourable forms for α and ζ_ω were found (they were used in all model-Z computations by Braginsky and by Roberts):

$$G_\alpha = s\alpha = \begin{cases} 20z(1 - z^6/z_1^6)\sin(10\pi(0.9 - s)), & 0.8 < s < 1, \\ 0, & s < 0.8, \end{cases} \qquad (9.3.23a)$$

$$\zeta_\omega = -3s^2(1 - r^2). \qquad (9.3.23b)$$

The driving wind, ζ_ω, corresponds to $\partial_\theta C < 0$, so that the fluid is lighter on the equatorial plane than on the polar axis.

Two variants of the stationary solution were obtained by Braginsky (1978) again using time-stepping methods. The following parameters were used: $\epsilon_\nu = 0.01$, $(\mathcal{R}_\alpha = 50.5, \mathcal{R}_\omega = 500)$ and $(\mathcal{R}_\alpha = 70, \mathcal{R}_\omega = 300)$. Corresponding dynamo numbers are $\mathcal{R}_D = 2.5 \times 10^4$ and $\mathcal{R}_D = 2.1 \times 10^4$. These solutions look very similar, the numbers given below refer to the first one. The parameters $\mathcal{R}_\alpha$ for these solutions were fitted to fulfill the condition $m_1 = 1$, where m_1 is the absolute value of the first Gauss coefficient at the CMB, $g_1^0(1)$; this corresponds to $B_r(1,0) = -2$ for the dipole component of $\mathbf{B}_p$.

In order to monitor the evolution of the system, some characteristic quantities (the mean-square averages of B_z, B_s, B_ϕ, $\partial \psi/\partial t$ and $\partial B_\phi/\partial t$ over the volume of the core, m_1, ζ_0, ζ_{1max}, Q_J, Q_J, etc.) were calculated and printed out after some dimensionless interval of time, usually 10^{-3}. The maximum of $|\partial B_\phi/\partial t|$ and its position were found at each time-step to check the stability of integration. Time steps $h_t = 4 \times 10^{-5}$ and smaller were required for stability. At the beginning and at the end of each run many (about 40) different averages over the grid, and some other characteristic quantities, were calculated and printed out.

The stationary solutions obtained turn out to be of model-Z type. Two quite different regions can be isolated in the pattern of the solution: the main region and an equatorial belt (EB); they are separated approximately by $s = 0.82$. In the main region ($s < 0.82$) we have $B_s \ll B_z$ and a very large positive $\partial_s \zeta_1$, so that a current layer is present there that has a thickness of the order given by (3.12). The geostrophic shear has two extrema: $\zeta_1 \approx -1000$ at $s = 0$ and $\zeta_1 \approx 260$ at $s \approx 0.82$. The second extremum is very sharp. In the equatorial belt the field configuration is complicated, $B_s \sim B_z$, and there is no boundary layer. The field $\mathbf{B}_p$ is generated by the α-effect in the equatorial belt.

The numerical solution gives a root mean square field of $B_{\phi Av} = 0.509$. The Joule and viscous dissipations are $Q_J = 103$ and $Q_f/Q_J = 2.0$. The Joule dissipation coefficient, $\gamma_J = Q_J/B_{\phi Av}^2 \approx 400$, is very large (compare $\gamma_J = 20$ for the free decay of the dipole and $\gamma_J \approx 60$ for the free decay of the quadrupole field B_ϕ); this is associated with extremely large gradients of B_ϕ within the current layer. The viscous dissipation is comparable with the Joule heating; this stands in striking contrast to the Taylor state, for which $Q_f \ll Q_J$. The ratio Q_f/Q_J can be used as a discriminator between model-Z and Taylor-type solutions. A ratio of order unity corresponds to model-Z; a small value indicates a Taylor state.

Many numerical solutions of a model-Z dynamo were obtained by Braginsky & Roberts (1987), in which ϵ_ν and other parameters of the model were varied. Different starting conditions were tried, and they always led to the same stationary solution of model-Z type. The scaling law (2.30) was successfully verified within the range of change of ϵ_ν from 0.05 to 0.004 starting from the solution with $\epsilon_\nu = 0.01$, $\mathcal{R}_\alpha = 50$ and $\mathcal{R}_\omega = 500$. A 'coarse' numerical grid with a total of 1089 grid points and a 'fine' grid with 4225 grid points were employed. The results for these two grids were very close to one another, so confirming the reliability of the numerical solution. Figures 9.1–4, taken from Braginsky & Roberts (1987), show the solution for $\epsilon_\nu = 0.01$, $\mathcal{R}_\alpha = 50$ and $\mathcal{R}_\omega = 500$, obtained with the coarse grid (left half of the pictures) and with the fine grid (right half), while figure 9.5 shows the geostrophic shear for three cases with $\mathcal{R}_\alpha = 50$ and ($\epsilon_\nu = 0.04$, $\mathcal{R}_\omega = 315$), ($\epsilon_\nu = 0.01$, $\mathcal{R}_\omega = 500$) and ($\epsilon_\nu = 0.004$, $\mathcal{R}_\omega = 678.6$), which correspond to the 'asymptotic relations' (3.21). The form of the three curves differ significantly only in scale. The 'asymptotic relations' (3.21) were checked by Braginsky & Roberts

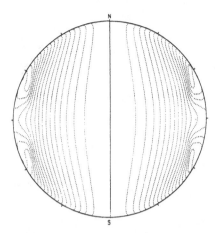

Figure 9.1 The field lines of the axisymmetric meridional field shown inside the conductor on a meridional section through the axis of rotation, NS. The lines on the left half of the circle were obtained using the coarse computational grid (1089 points); those on the right from the fine grid (4225 points). Except near the surface of the sphere, where the field lines must turn in order to match the source-free potential field outisde, the field lines show a strong NS orientation.

(1987). Their results are illustrated by the table 9.1; they are not in serious conflict with (3.21).

9.3.5 Development of Model-Z

Model-Z with magnetic friction. A few variants of the model-Z solution were obtained by Braginsky (1988) with magnetic friction between core and mantle. The solutions were broadly of the same form as they were when the coupling was viscous. A difficulty arises, however, with these models because $\overline{B}_r = 0$ at $\theta = \pi/2$ and so the magnetic coupling (proportional to $\overline{B}_r^2$) is very small in the equatorial region. To overcome this difficulty and to prevent the decay of the dynamo model, it turns out to be necessary to add into the model an additional (rather artificial) magnetic coupling term. It may be interpreted as the ϕ-averaged non-axisymmetric stresses $<B'_\phi B'_r>$ on the boundary. Specifically, the term B_r^2 in (2.25) was replaced by $B_r^2 + \pi_r$ where π_r is an additional parameter of the model. Correspondingly an additional term Π was introduced in the expression (3.4) for χ. This now takes the form

$$\chi = -s \left(B_\phi B_z + s^{-2}\partial_s \left(s^2 \int_0^z B_\phi B_z \, dz \right) + \Pi \right). \tag{9.3.24}$$

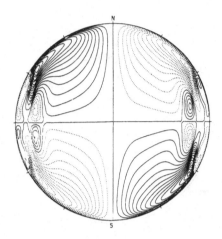

Figure 9.2 The contours of equal field strength for the axisymmetric zonal field, B_ϕ, shown on a meridional section. (See the caption of figure 9.1 for the relationship between left-hand and right-hand sides of the figure.)

The z-dependence of Π was specified in the form of a simple expression

$$\Pi = \Pi_1(s)(z/z_1)^2.$$

This term may be interpreted as resulting from the ϕ-averaged Maxwellian stresses, though the simple z-dependence adopted above is somewhat arbitrary. The shell equilibrium condition (2.22) now takes the form

$$\mathcal{E}\zeta_1 = T_s - T_r, \tag{9.3.25}$$
$$\mathcal{E} = \epsilon_\nu s 2^{-1} z_1^{-1/2} + \epsilon_M s z_1^{-1}(B_{r1}^2 + \pi_r),$$
$$T_s = s^{-2}\partial_s(s^2 T), \qquad T_r = \epsilon_M z_1^{-1} B_{r1}\partial_r B_{\phi 1}.$$

The function $\Pi_1(s)$ can be found using the relation (3.7e) between ζ_1 and χ_1. The latter is determined by the Ekman layer suction which does not depend on magnetic friction. From (3.7e, 24, 25) we obtain

$$\Pi_1(s) = \epsilon_M \pi_r z_1^{-1} s \zeta_1.$$

Model-Z with magnetic friction was computed in the same way as the model with viscous friction. The same spherical and cylindrical numerical grids and the same method of interpolation between them were used as in Braginsky &

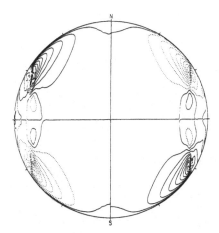

Figure 9.3 The streamlines of the axisymmetric meridional flow shown on a meridional section. (See the caption to figure 9.1 for the relationship between left-hand and right-hand sides of the figure.)

Roberts (1987). Some difficulties arise, however, if the core-mantle coupling is supposed to be rather weak.

The model-Z calculations just described are prone to a specific numerical instability. It develops especially easily (even for extremely small time steps) when core-mantle friction is very weak, that is, when ϵ_ν and ϵ_M are very small. This instability makes model-Z calculations rather difficult. The mechanism of this instability is not clear, but it was observed that it starts from a region close to CMB and near the boundary with the equatorial belt. One may suspect a connection between the instability and sharp gradient of $\partial_s \zeta_1$ in this region. It has proved possible to suppress this instability by applying some smoothing when calculating ζ_1.

Let $\zeta_1 = \zeta_1^0 + u$ where ζ_1^0 is the value of ζ_1 at the preceding time step. Equation (3.25) then takes the form $\mathcal{E}u = T_s - T_r - \mathcal{E}\zeta_1^0$, where $\mathcal{E}(s)$ and the right-hand side are calculated when the fields at the new step have been found. Instead of this expression we shall determine $u(s)$ from

$$\mathcal{E}u - h_t \mathcal{V}(u) = T_s - T_r - \mathcal{E}\zeta_1^0 \qquad (9.3.26)$$

where

$$\mathcal{V}(u) = \nu_N(s)\partial^2 u/\partial s^2,$$

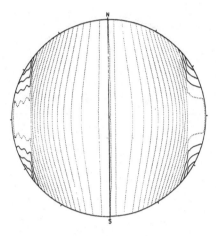

Figure 9.4 The contours of equal zonal shear, $\zeta(s,z)$, shown on a meridional section. (See the caption to figure 9.1 for the relationship between left-hand and right-hand sides of the figure.)

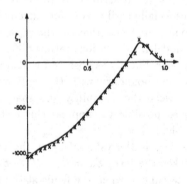

Figure 9.5 The geostrophic shear, $\zeta_1(s)$, as a function of distance, s, from the polar axis for the fine grid 65 × 65 calculation (full curve). The crosses mark values computed from the coarse grid 33 × 33 calculation. The irregularity of the latter values near $s = 0.8$ is paralleled by differences between the left- and right-hand sides of figure 9.4.

ϵ_ν	m_1	B_ϕ	B_z	$\epsilon_\nu^{-1/3}B_s$	$\epsilon_\nu^{1/3}Q_J$	$\epsilon_\nu^{-1/3}\zeta_0$	$\epsilon_\nu^{-1/3}\delta_J$
0.05	1.1	0.69	2.5	2.7	58	73	0.016
0.04	1.1	0.65	2.5	2.7	56	67	0.015
0.03	1.1	0.61	2.5	2.7	55	61	0.014
0.02	1.0	0.56	2.4	2.7	51	56	0.013
0.01	0.93	0.53	2.1	2.7	44	48	0.013
0.01^f	0.91	0.52	2.1	2.8	40	49	0.012
0.008	0.84	0.54	2.0	2.7	42	46	0.012
0.004	0.66	0.64	1.6	2.6	36	40	0.016
0.004^f	0.60	0.68	1.4	2.4	32	45	0.018

Table 9.1 Comparison with the asymptotic scaling law (3.21) according to Braginsky & Roberts (1987). All results are for $\mathcal{R}_\alpha = 50$; the parameter $\mathcal{R}_\omega$ changes as $\mathcal{R}_\omega = 500(0.01/\epsilon_\nu)^{1/3}$. All calculations of the models are performed on the coarse grid except those marked by superscript f, which are made on the fine grid. All the values presented characterizing the solutions are dimensionless. Notation: m_1 is the absolute value of the first Gauss coefficient (axial dipole) at the CMB; B_s, B_z and B_ϕ are the mean squares of the corresponding field components over the core volume; Q_f and Q_J are the viscous dissipation in the Ekman layer and the Joule dissipation; $\delta_J = B_\phi^2/Q_J$, and $\zeta_0 = |\zeta_1(0)|$.

and $\nu_N(s)$ is an artificial (numerical) viscosity introduced in order to smooth $u(s)$. The right-hand side of (3.26) vanishes in the steady state, so that (3.25) is satisfied. The expression $\nu_N(s) = \nu_{N1}s^2(2 - s^2)$ with the parameter $\nu_{N1} = 0.05$ was used by Braginsky (1988). Equation (3.26) was solved with the conditions $\partial_s u(0) = 0$ and $u(1) = 0$. The smoothing algorithm just described allowed stable integration with a step size h_t as large as 10^{-4}.

The 'viscous' smoothing of $\partial_s \zeta_1$ can be achieved by a more natural method suggested to the author by his former student V.V. Vikhrev, in a private discussion at the end of 1978. This method is based on substituting in $T = \int B_\phi B_s\, dz$ the field B_ϕ calculated 'pseudo-implicitly' by using $\zeta_1 = \zeta_1^0 + u$ in equation (3.25). This replaces B_ϕ by $B_\phi + h_t s B_s \partial_s u$ in T, so that (3.25) turns into

$$\mathcal{E}u - h_t s^{-2}\partial_s(s^3 D_s \partial_s u) = T_s - T_r - \mathcal{E}\zeta_1^0, \qquad (9.3.27)$$

$$D_s = \int B_s^2\, dz.$$

This equation is similar to (3.26); both work successfully, in approximately the same way.

Stationary solutions were obtained by Braginsky (1988) for different combinations of (ϵ_M, π_r) and ϵ_ν. Some characteristic parameters of these solutions for $\epsilon_\nu = 0$ ($N = 1, \ldots, 6$) are given in table 9.2. The following variants are also given there: $N = 0$: $\epsilon_\nu = 0.01$, $\epsilon_M = \pi_r = 0$ and $N = 7$: $\epsilon_\nu = 2 \times 10^{-3}$, $\epsilon_M = 0.5 \times 10^{-3}$, $\pi_r = 2$. All solutions presented in table 9.2 correspond to

N	0	1	2	3	4	5	6	7	8	9	10
$10^3\epsilon_\nu$	10	0	0	0	0	0	0	2	1	1	1
$10^3\epsilon_M$	0	2	2	1	1	1	0.5	0.5	1	1	1
π_r	0	1	0.5	4	2	1	2	2	1	1	1
m_1	0.96	0.94	0.75	1.14	0.90	0.64	0.61	0.81	0.76	0.76	0.46
B_z	2.18	2.26	1.81	2.66	2.11	1.55	1.46	1.89	1.80	1.78	1.00
B_s/B_z	0.29	0.32	0.30	0.31	0.28	0.26	0.24	0.26	0.27	0.27	0.21
B_ϕ	0.55	0.78	0.77	0.63	0.62	0.67	0.60	0.57	0.65	0.64	0.84
$B_{\phi 1mm}$	0	0.64	0.50	0.41	0.34	0.26	0.14	0.18	0.29	0.31	0.17
ζ_0	1022	719	773	749	332	990	1170	1003	909	874	872
ζ_{1m}	265	314	314	277	307	297	290	309	312	311	194
Q_J	105	123	123	115	110	116	108	103	112	112	90
Q_J/Q_J	2.09	2.10	1.67	2.42	1.94	1.36	1.27	1.74	1.64	1.69	0.92
$\mathcal{R}$	109	106	66	156	97	49	44	78	69	69	25
B_{pI}	1.92	1.94	2.46	1.60	2.04	2.86	3.00	2.27	2.41	2.41	3.98
$B_{\phi I}$	209	206	163	250	196	140	133	176	166	166	101
Q	3.28	3.76	2.02	5.71	2.90	1.25	1.01	2.04	1.89	1.92	0.41

Table 9.2 Characteristics of the model-Z solution (for $\mathcal{R}_\alpha = 50$, $\mathcal{R}_\omega = 500$), both without an inner core ($N = 0, \ldots, 7$) and with an inner core of radius $r_2 = 0.35$ ($N = 8, 9, 10$). The variant $N = 10$ is for the oscillatory model-Z (see the text) at the moment when the decreasing axial dipole is $m_1 = 0.46$. The notation is the same as in table 9.1, and $B_{\phi 1mm} = \max B_{\phi 1}/\max B_\phi$. The dimensional quantities B_{pI} and $B_{\phi I}$ are given in Gauss, and the total rate of energy dissipation, Q, in 10^{12} W.

$\mathcal{R}_\alpha = 50$ and $\mathcal{R}_\omega = 500$.

When the mantle is conducting, the toroidal field is not zero on the core-mantle boundary. The ratio $B_{\phi 1m}/B_{\phi Av}$ is given in table 9.2; here $B_{\phi 1m}$ is the maximum of $|B_{\phi 1}|$. The magnetic frictional heating, Q_J, represents Joule heat deposited in the lower mantle. Its magnitude is rather large, and its influence on the lower mantle dynamics deserves investigation (see, e.g., Braginsky & Meytlis 1990).

It is interesting to estimate the dimensional quantities in the model. To do this we adopt the value $\eta = 2$ m^2/s. This gives $t_I = 1.93 \times 10^5$ yr, $V_{pI} = 5.74 \times 10^{-5}$ cm/s and $B_* = 20$ G (for $\rho_0 = 10.9$ g/cm^3). For the value of the dipole Gaussian coefficient extrapolated on the CMB we adopt $g_1^0(R_1) = -1.83$ G. Comparing this with the dimensionless coefficient m_1, we find the unit of the poloidal field to be $B_{pI} = 1.83/m_1$ G. The expression $\mathcal{R} = 119.4m_1^2$ follows from $B_{pI} = B_*\mathcal{R}^{-1/2}$; we then find $B_{\phi I} = B_*\mathcal{R}^{1/2} = 218.6m_1$ G, $\zeta_I = t_I^{-1}\mathcal{R} = 3.56 \times 10^{-2}m_1^2$ deg/yr, and the unit of heat production rate is $Q_I = 1.11 \times 10^{10}m_1^2$ W. These values (except for B_{pI}) depend strongly on the chosen diffusivity: $\mathcal{R} \propto \eta^2$, $B_{\phi I} \propto \eta^2$, $\zeta_1 \propto \eta^3$ and $Q_I \propto \eta^3$. The values $\mathcal{R}$, B_{pI}, $B_{\phi I}$ and the total dissipation rate Q (in units of 10^{12} W) are given in table 9.2 for each variant. The sharp dependence of the dissipation on the diffusivity, $Q_I \propto \eta^3 \propto \sigma^{-3}$, places a strict limitation on the conductivity of the core.

A *model-Z with an inner core* involves significant complications; conditions on the inner core boundary are rather complicated (see, e.g., Loper & Roberts 1981). The cylindrical shell equilibrium equation (2.22) and its generalization (3.25) are valid only for $s > r_2$, while for $s < r_2$ the magnetic $(B_{\phi 2} B_{r2})$ and other stresses on the ICB should be taken into account. It is not easy to write them down explicitly. Magnetic stresses are determined by the field distribution, while viscous and topographic coupling may be assumed to be proportional to $\zeta_2 - \zeta_N$ where $\zeta_2 = \zeta(r_2, \theta)$, and ζ_N is the angular velocity of the inner core. With this assumption the condition (3.25), valid for $s > r_2$, is modified for $s < r_2$ into

$$\mathcal{E}\zeta_1 + \mathcal{E}_2(\zeta_2 - \zeta_N) = T_s - T_r - T_2, \qquad (9.3.28)$$
$$T_2 = B_{\phi 2} B_{r2} z_2^{-1}, \qquad z_2 = (r_2^2 - s^2)^{1/2},$$

$$\int (\mathcal{E}_2(\zeta_2 - \zeta_N) + T_2) s^2 \, ds = 0, \qquad 0 \le s < r_2, \qquad (9.3.29)$$

$$\zeta_2 = \zeta_1 + \zeta_{C2} + s^{-2}(B_{\phi 2}^2 - B_{\phi 1}^2), \qquad (9.3.30)$$

where $\mathcal{E}_2 = \mathcal{E}_2(\theta)$ is the friction coefficient, which is very poorly known. The condition (3.28) of shell equilibrium is here complemented by the condition of the inner core equilibrium (3.29). Eliminating ζ_2 by (3.30) we obtain the system of equations for $\zeta_1(\theta)$ and ζ_N. This system is inconvenient; and perhaps even more serious is that $\mathcal{E}_2$ is unknown! Two simplifying limiting assumptions can be imagined, $\mathcal{E}_2 = 0$ and $\mathcal{E}_2 = \infty$. The former provides a condition of inner core equilibrium due to the magnetic forces that determines ζ_N implicitly; then ζ_1 can be found from (3.28). To use the latter we integrate (3.28) from $s = 0$ to $s = r_2$, thus eliminating the difference $\zeta_2 - \zeta_N$ by means of (3.29). Substituting here (3.30) and replacing ζ_2 by ζ_N we obtain expressions for ζ_N and $\zeta_1 = \zeta_N - \zeta_{21}$

$$\zeta_N = \left(T(r_2) + \int (\mathcal{E}_1 \zeta_{21} + T_r) s^2 ds \right) \left(\int \mathcal{E}_1 s^2 \, ds \right)^{-1}, \qquad (9.3.31)$$

where $\zeta_{21} = \zeta_2 - \zeta_1 = \zeta_{C2} + s^{-2}(B_{\phi 2}^2 - B_{\phi 1}^2)$ according to (3.30). Equation (3.31) expresses angular momentum balance for the whole region $s < r_2$ including the inner core. It was used by Braginsky (1989) in calculations of model-Z with the inner core. The function χ at $s < r_2$ was determined from

$$\chi = \chi_1 - s \left(B_\phi B_z - B_{\phi 1} B_{z1} - s^{-2} \int_z^{z_1} \partial_s (s^2 B_\phi B_s) \, dz \right). \qquad (9.3.32)$$

A non-zero value for χ_2 at $r = r_2 + 0$ follows from (3.32) but χ should be zero on the surface of the inner core. A discontinuity of χ arises here, i.e., a thin jet of fluid flowing around the inner core. The functions $\zeta_1(s)$ and $\chi(z, s)$ are determined by different means for $s < r_2$ and $s > r_2$; therefore they are discontinuous at $s = r_2$. The jump of $\zeta_1(s)$ was smoothed artificially over two grid points; the jump of χ was somewhat smoothed in the process of interpolation from the cylindrical to the spherical grid. These discontinuities are not very large and do not influence the calculations significantly.

The variant $N = 9$ in table 9.2 represents a stationary solution with an inner core $(r_2 = 0.35)$ and for the set of parameters $\mathcal{R}_\alpha = 50$, $\mathcal{R}_\omega = 500$, $\epsilon_\nu = 10^{-3}$, $\epsilon_M = 10^{-3}$ and $\pi_r = 1$. It is seen that the main characteristics of model-Z are not changed strongly when the inner core is included. The graphics for ψ, B_ϕ, χ and ζ also are qualitatively similar to those found for $r_2 = 0$.

An oscillatory model-Z that imitates the real (oscillating) geodynamo better than the stationary models was obtained by means of an artificial trick (Braginsky 1989). To turn a stationary solution into an oscillatory one, an 'on-off switch' was added to the generation mechanism. The coefficient $\mathcal{R}_\alpha$ in (3.1) was set to zero when the dipole moment, m_1, exceeded some upper bound, m_{max}, and was restored when the dipole moment fell below some lower bound, m_{min}. These bounds were chosen to agree with the observational (archaeomagnetic) data by McElhinny & Senanayake (1982) to be $m_{max} = 0.58$ and $m_{min} = 0.4$. By comparing the period, $T_0 = 7.7 \times 10^3$ yr, of the observed 'fundamental oscillation of the dynamo' with the nondimensional period $T_{nd} = 4.8 \times 10^{-2}$ defined by the model, it was possible to calculate the magnetic diffusivity of the core by applying the relation $T_0 = T_{nd} R_1^2 / \eta$. The result obtained, $\eta = 2.4$ m^2/s, is not far from the value $\eta = 2$ m^2/s adopted in this paper; the latter could be recovered by small changes in the limits m_{max} and m_{min}. It should be stressed that the reasonable magnitude of the conductivity obtained is a consequence of the rather short non-dimensional period, T_{nd}, which is a characteristic property of an oscillating model-Z.

Typical quantities for the oscillatory model-Z are given in table 9.2 (variant $N = 10$) for a moment of time when m_1 is intermediate between m_{max} and m_{min}. The effective magnetic Reynolds number $\mathcal{R}$ and the dimensional quantities are calculated using $\eta = 2$ m^2/s. For a phase of the oscillation when $m_1 = 0.46$ and $\partial_t m_1 < 0$ (compare the diminishing present dipole of the Earth), it is estimated that $\mathcal{R} = 25$, $<B_\phi^2>^{1/2} = 85$ G, $B_{\phi max} = 192$ G, $s\zeta_{1max} = 0.2$ cm/s and the total rate of dissipation of the dynamo is $Q = 0.41 \times 10^{12}$ W.

From the oscillatory model-Z a relatively moderate value of $\mathcal{R}$ is obtained. It is worth noting that, although the parameter $\mathcal{R}$ is not very large in this case, the dynamo model is definitely of model-Z type: $\zeta_1(s)$ is very large, with $\partial_s \zeta_1 > 0$ over most of the s-interval; the current layer at the CMB is pronounced, and the main feature of model-Z is evident: $Q_J \sim Q_J$.

It can be seen from table 9.1 that the weaker the core-mantle coupling the smaller is the geodynamo energy expenditure. This places some constraint on the intensity of the core-mantle friction. It seems plausible that the results from variant $N = 10$ correspond better to the real dynamo of the Earth than other variants.

The instantaneous switching of the α-effect in the oscillatory model-Z dynamo creates a large perturbation. The calculation by Braginsky (1989) shows that the decay of this perturbation is not monotonic, but is accompanied by oscillations having periods an order of magnitude less than T_{nd}. This shows that the axially symmetric field structure of a model-Z dynamo is stable and is associated with a large effective elasticity. In the real dynamo, such oscillations can interact with MAC waves, which have periods of the same order of magnitude. These oscillations were investigated carefully by Anufriev *et al.* (1993), who confirmed that the oscillations are not an artifact of the method of solution but represent a real property of the AB-model.

Gradual change from Taylor state to model-Z was demonstrated by Roberts (1989*a*). In this paper two families of stationary model-Z solutions were obtained with viscous friction ($\epsilon_\nu = 0.01$ and $\epsilon_\nu = 0.04$) for fixed $\mathcal{R}_w = 500$ and variable $\mathcal{R}_\alpha$. For the maximum value $\mathcal{R}_\alpha = 150$ used, the solution is definitely of model-Z type with a ratio Q_J/Q_J of about 2. With decreasing $\mathcal{R}_\alpha$ the ratio Q_J/Q_J remains nearly the same until $\mathcal{R}_\alpha = 30$ where it starts decreasing and becomes about 1 at $\mathcal{R}_\alpha = 20$. It then drops sharply being only approximately $Q_J/Q_J \sim 10^{-1}$ at $\mathcal{R}_\alpha = 10$. The value of ζ_1 also drops here by about an order of magnitude. Roberts (1989*a*) traced two families of solutions until $\mathcal{R}_\alpha \approx 5\text{--}7$ where their existence terminates, and indicated that subcritical unstable solutions exist with smaller amplitudes. The behaviour of the solution near the generation threshold was investigated in detail by Hofflin & James (1992). They showed that infinitesimal field generation commenced at $\mathcal{R}_\alpha = 10$, and that unstable subcritical solutions with finite amplitude and $\mathcal{R}_\alpha < 10$ exist which transform into stable solutions near $\mathcal{R}_\alpha \approx 5\text{--}7$, their amplitudes increasing with increasing $\mathcal{R}_\alpha$.

Is it possible that both Taylor-type and model-Z type solutions could exist for the same set of input parameters, and that only initial conditions would determine which kind of solution is adopted? This question is so far unanswered in general. The paper by Roberts (1989*a*) gives an example of a situation in which the answer is negative. Here the Taylor-type solution exists near the threshold of generation and at relatively small $\mathcal{R}_\alpha$, but is gradually replaced by a model-Z solution when $\mathcal{R}_\alpha$ becomes large. It is shown in section 9.3.3 that for the existence of the model-Z solution the Archimedean stirring must be large: $\zeta_C \sim \mathcal{R}_w \sim \delta_J^{-1}$. One might expect that gradual changes of solutions would also be observed more generally as $\mathcal{R}_\omega$ is increased from the threshold to large magnitudes. It may be assumed that the transition to model-Z solutions with increasing $\mathcal{R}_D$ is a universal feature, and the character of the geodynamo is determined by the intensity of the energy source driving convection in the Earth's core.

9.4 CONCLUSION

Hydromagnetic dynamo theory is rather far developed at the present time but the creation of a fully-fledged theory of the Earth's hydromagnetic dynamo is still a challenge for geophysicists and mathematicians. Many significant questions are as yet unanswered. In particular it is unclear (1) whether Taylor states or model-Z states typify geodynamo model solutions more faithfully, and (2) which, if either, corresponds better to the actual state of the Earth's core. The 'scouting' numerical investigation of the axisymmetric dynamo model emphasizes the role of weak core-mantle friction in core dynamics. A rigorous and general understanding of the influence of the small parameter on the form of the dynamo solution is still lacking. The following tentative interpretation of the situation in general terms can be presented nevertheless.

The model-Z state of a strong field dynamo is a result of the adaptive reaction of the system to a very powerful source of convection. The ways that linear and nonlinear systems adapt to an increase of excitation are quite different. A linear system reacts simply by adopting a larger amplitude. A nonlinear system can change its structure in an essential way, to increase dissipation. A nonlinear dynamo, working far beyond the threshold of excitation, develops large B_ϕ and also large ζ_G, thus producing greater Q_J and Q_J. If the geostrophic velocity $\zeta_G(s)$ grows until Q_J becomes of the order of Q_J (this is a crucial assumption) then $\zeta_G(s)$ becomes very large. An approximately 'isorotation pattern' develops because of the large $\zeta_G(s)$; therefore a state in which $B_s \ll B_z$ is brought about. A model-Z regime has then become established.

We conclude with some self-criticism and some statements of hope.

It is obvious that intermediate models of every type (not just model-Z) can be accused of being, by their very nature, incomplete. They invite a more general approach to the geodynamo and the construction of more complete models. The model-Z type dynamo may be defined amongst more general classes of models by listing its main features:

(1) The inertia term, $d_t \mathbf{V} = \partial_t \mathbf{V} + (\mathbf{V} \cdot \nabla)\mathbf{V}$, can be ignored in the equation of motion.

(2) The viscosity term, $\nu \nabla^2 \mathbf{V}$, can be ignored in the equation of motion for the bulk of the core

(3) The coupling between the fluid and the mantle is weak, being characterized by a small friction parameter. Nevertheless it is essential and cannot be

ignored. A convenient measure of this coupling is the energy dissipation, Q_f. A model-Z state is characterized by $Q_f \sim Q_J$ (in contrast to $Q_f \ll Q_J$ for the Taylor state).

(4) Taylor's integral is small, $T = \int B_\phi B_s \, dz \ll |B_\phi||\mathbf{B}_p|L$, because $B_s \ll |B_p|$ and/or due to the cancellation of terms in the integrand, but it is much greater than the small parameter of core-mantle friction.

The examples of model-Z considered above belong to the class of 'AB-models' based on equations (2.11) and (2.12). They contain two arbitrary functions, Γ_α and ζ_ω. The arbitrariness of the 'generating wind', ζ_ω, can be eliminated by transition to the more general 'ABC-models'. This necessary step adds equation (2.13) and introduces serious difficulties connected with the estimation of the turbulent diffusivity tensor D_{ij}. To eliminate the arbitrariness of the function Γ_α, it is necessary to turn to a three-dimensional model, and to develop the theory of non-linear MAC-waves jointly with the theory of axisymmetric motion and magnetic field generation. This would be a self-consistent geodynamo theory.

The author hopes that the self-consistent geodynamo will be of model-Z type. This hope is based mostly on the above tentative 'general interpretation' of model-Z, and on the author's belief that the Earth's dynamo is of strong-field type, working far beyond the threshold of excitation. There is some observational evidence indicating that the geodynamo is of strong-field type: the geomagnetic secular variations (SV) with periods of order 10^3 yr and a few times shorter are observed by archaeomagnetic methods. Such variations could be understood as being generated by MAC-waves, and the same waves can produce the α-effect due to the nonaxisymmetry of their fluid velocities. The magnitude of the toroidal magnetic field on which these MAC-waves ride is roughly estimated as about 10^2 G, which indicates that the geodynamo is of strong-field type. These indications, unfortunately, are not quite definite; to make them more certain we need further development of the theory of MAC-waves and its detailed comparison with observation. It should be added also that, if the jump of density at the ICB is an order of magnitude smaller than the commonly used value which was assumed above, then the source of convection would be too weak to produce a geodynamo of model-Z type.

A severe criticism of the model Z dynamo is possible, based on the very large toroidal velocity it requires, $V_\phi \sim 0.2$ cm/s. This large velocity is an essential element of the model. It is about one order of magnitude greater than the observed velocity of the 'westward drift'. The core fluid velocity at the CMB has been estimated by many authors and such a large velocity has never been obtained. How can this contradiction be resolved? There is no definite answer to this question at the present time. The author hopes that a satisfactory answer will emerge from a better understanding of the events near the CMB, and a better theory of the geomagnetic SV will result. It should be noted that the fluid velocity in the core has up to now been deduced from geomagnetic SV data by means of the frozen flux approximation of Roberts & Scott (1965). This approximation is valid only for the short period (decade) SV. The other parts of SV with longer periods of change (of 10^3 yr and more) should be separated and treated in a different way. The events near the CMB are rather complicated and they are still incompletely understood; so that in the problem of determining the fluid velocity on the top of the core there is still much room for discussion (see, e.g., Braginsky 1984).

Acknowledgements

I am grateful to Professor Paul Roberts for improving the presentation of this paper. The major part of the paper was written during my stay in the Isaac Newton Institute for Mathematical Studies of Cambridge University, in the framework of the programme on 'Dynamo Theory'. I am grateful to the programme organizers and INI staff for their hospitality. Partial support from the NASA grant NAGW/2546 is acknowledged.

REFERENCES

Anufriev, A.P., Cupal, I. & Hejda, P. 1993 On the oscillation in model-Z. In *Solar and Planetary dynamos* (ed. M.R.E. Proctor, P.C. Matthews & A.M. Rucklidge), pp. 9–17. Cambridge University Press.

Batchelor, G.K. 1967 *An Introduction to Fluid Dynamics.* Cambridge University Press.

Braginsky, S.I. 1963 Structure of the F layer and reasons for convection in the Earth's core. *Dokl. Akad. Nauk. SSSR* **149**, 1311–1314. (English transl.: *Sov. Phys. Dokl.* **149**, 8–10.)

Braginsky, S.I. 1964a Self-excitation of a magnetic field during the motion of a highly conducting fluid. *Zh. Exp. Teor. Fiz. SSSR* **47**, 1084–1098. (English transl.: *Sov. Phys. JETP* **20**, 726–735 (1965).)

Braginsky, S.I. 1964b Theory of the hydromagnetic dynamo. *Zh. Exp. Teor. Fiz. SSSR* **47**, 2178–2193. (English transl.: *Sov. Phys. JETP* **20**, 1462–1471 (1965).)

Braginsky, S.I. 1964c Kinematic models of the Earth's hydromagnetic dynamo. *Geomag. Aeron.* **4**, 732–747. (English transl.: 572–583.)

Braginsky, S.I. 1964d Magnetohydrodynamics of the Earth's core. *Geomag. Aeron.* **4**, 898–916. (English transl.: 698–712.)

Braginsky, S.I. 1972 Origin of the magnetic field of the Earth and of its secular variations. *Izv., Akad. Nauk SSSR, Fiz. Zemli* no. 1, 3–14. (English transl.: *Izv., Acad. Sci. USSR, Phys. Solid Earth* 649–655.) See also IAGA Bulletin 31: Trans. 15th General Assembly, Moscow, USSR 1971, pp. 41–54.

Braginsky, S.I. 1975 Nearly axially symmetric model of the hydromagnetic dynamo of the Earth, I. *Geomag. Aeron.* **15**, 149–156. (English transl.: 122–128.)

Braginsky, S.I. 1976 On the nearly axially-symmetrical model of the hydromagnetic dynamo of the Earth. *Phys. Earth Planet. Inter.* **11**, 191–199.

Braginsky, S.I. 1978 Nearly axially symmetric model of the hydromagnetic dynamo of the Earth. *Geomag. Aeron.* **18**, 340–351. (English transl.: 225–231.)

Braginsky, S.I. 1984 Short-period geomagnetic secular variation. *Geophys. Astrophys. Fluid Dynam.* **30**, 1–78.

Braginsky, S.I. 1988 The model-Z geodynamo with magnetic friction. *Geomag. Aeron.* **28**, 481–487. (English transl.: 407–412.)

Braginsky, S.I. 1989 A model-Z geodynamo with an internal core, and the oscillations of the geomagnetic dipole. *Geomag. Aeron.* **29**, 121–126 (English transl.: 98–103.)

Braginsky, S.I. 1991 Towards a realistic theory of the geodynamo. *Geophys. Astrophys. Fluid Dynam.* **60**, 89–134.

Braginsky, S.I. & Meytlis, V.P. 1990 Local turbulence in the Earth's core. *Geophys. Astrophys. Fluid Dynam.* **55**, 327–349.

Braginsky, S.I. & Roberts, P.H. 1987 A model-Z geodynamo. *Geophys. Astrophys. Fluid Dynam.* **38**, 327–349.

Childress, S. 1969 A class of solutions of the magnetohydrodynamic dynamo problem. In *The Application of Modern Physics to the Earth and Planetary Interiors* (ed. S.K. Runcorn), pp. 629–648. Wiley, London.

Cowling, T.G. 1934 The magnetic field of sunspots. *Mon. Not. R. Astron. Soc.* **94**, 39–48.

Cupal, I. & Hejda, P. 1989 On the computation of model-Z with electromagnetic core-mantle coupling. *Geophys. Astrophys. Fluid Dynam.* **49**, 161–172.

Dziewonski, A.M. & Anderson, D.L. 1981 Preliminary reference Earth model. *Phys. Earth Planet. Inter.* **25**, 297–356.

Frank, S. 1981 Ascending droplets in the Earth's core. *Phys. Earth Planet. Inter.* **27**, 249–254.

Hofflin, P.W. & James, R.W. 1992 Excitation and stability of the model-Z geodynamo. Lecture presented at the Third SEDI Symposium, Mizusawa, Japan.

Krause, F. & Rädler, K.-H. 1980 *Mean-field magnetohydrodynamics and dynamo theory.* Academy Press, Berlin.

Loper, D.E. & Roberts, P.H. 1981 A study of conditions at the inner core boundary of the Earth. *Phys. Earth Planet. Inter.* **24**, 302–307.

Malkus, W.V.R. & Proctor, M.R.E. 1975 The macrodynamics of α-effect dynamos in rotating fluids. *J. Fluid Mech.* **67**, 417–443.

McElhinny, M.W. & Senanayake, W.E. 1982 Variations in the geomagnetic dipole 1: The past 50000 years. *J. Geomagn. & Geoelectr.* **34**, 39–51.

Moffatt, H.K. 1978 *Magnetic field generation in electrically conducting fluids.* Cambridge University Press.

Moffatt, H.K. 1988 Liquid metal MHD and the geodynamo. In *Liquid Metal Magnetohydrodynamics* (ed. J. Lielpeteris & R. Moreau), pp. 403–412. Kluwer Academic Publishers.

Peyronneau, J. & Poirier, J.P. 1989 Electrical conductivity of the Earth's lower mantle. *Nature* **342**, 537–539.

Proctor, M.R.E. 1977 Numerical solutions of the nonlinear α-effect dynamo equations. *J. Fluid Mech.* **80**, 769–784.

Roberts, P.H. 1972 Kinematic dynamo models. *Phil. Trans. R. Soc. Lond. A* **272**, 663–698.

Roberts. P.H. 1988 Future of geodynamo theory. *Geophys. Astrophys. Fluid Dynam.* **44**, 3–31.

Roberts, P.H. 1989*a* From Taylor state to model-Z? *Geophys. Astrophys. Fluid Dynam.* **49**, 143–160.

Roberts, P.H. 1989*b* Core-mantle coupling. In *Encyclopedia of Geophysics* (ed. D.E. James). Publ. Am. Geophys. Un.

Roberts, P.H. & Scott, S. 1965 On analysis of the secular variation. *J. Geomagn. & Geoelectr.* **17**, 137–151.

Soward, A.M. 1972 A kinematic theory of large magnetic Reynolds number dynamos. *Phil. Trans. R. Soc. Lond. A* **272**, 431–462.

Steenbeck, M., Krause, F. & Rädler, K.-H. 1966 Berechnung der mittleren Lorentz-Feldstärke $\overline{\mathbf{v} \times \mathbf{B}}$ für ein elektrisch leitendendes Medium in turbulenter, durch Coriolis-Kräfte beeinflußter Bewegung. *Z. Naturforsch.* **21a**, 369–376.

Taylor, J.B. 1963 The magneto-hydrodynamics of a rotating fluid and the Earth's dynamo problem. *Proc R. Soc. Lond. A* **274**, 274–283.

Tough, J.G. & Roberts, P.H. 1968 Nearly symmetrical hydromagnetic dynamos. *Phys. Earth Planet. Inter.* **1**, 288–296.

CHAPTER 10

Maps and Dynamos

B. J. BAYLY

Mathematics Department
University of Arizona
Tucson, AZ 85721, USA

10.1 BRIEF REVIEW OF DYNAMO PROBLEMS

The interiors of the Earth, the Sun, and many other large astrophysical objects contain large regions of electrically conducting fluid. If such a fluid is in motion and simultaneously permeated by a magnetic field, the moving material experiences an electric field. The electric field generates currents in the material, which in turn act as a source for new magnetic field. The electrodynamics is a kind of 'triangular' feedback loop (figure 10.1), catalyzed by the mechanical fluid motion. The feedback is leaky; electromagnetic energy is always dissipated when currents flow in a non-superconducting material.

Given the fluid motion, a distribution of electromagnetic fields might arise in which the newly-generated fields reinforce the old. If the reinforcement is strong enough to overcome dissipation, the fields will increase. If this happens, the given flow is called a kinematic dynamo. Electromagnetic fields cannot increase indefinitely, of course — what happens when the field is strong enough is that the electromagnetic forces begin to affect the fluid motion, and the kinematic dynamo becomes a dynamic dynamo. Kinematic dynamo theory, relevant to the growth of weak fields before they become dynamically important, will be the subject of these lectures.

10.1.1 Kinematic Dynamos

If the fluid has a classical Ohmic resistivity, and if the flow is incompressible (see Moffatt 1978 for a more detailed discussion of hypotheses), the feedback described above can be neatly encapsulated in one equation — the *induction equation* — for the magnetic field $\mathbf{B}(\mathbf{x}, t)$:

$$(\partial_t + \mathbf{u} \cdot \nabla)\mathbf{B} = \mathbf{B} \cdot \nabla \mathbf{u} + \eta \nabla^2 \mathbf{B}. \qquad (10.1.1)$$

We assume the fluid occupies a geometrically nice, bounded region V of three-dimensional Euclidean space. The velocity field $\mathbf{u}(\mathbf{x}, t)$, specified as part of the

305

M.R.E. Proctor & A.D. Gilbert (eds.)
Lectures on Solar and Planetary Dynamos, 305–329
©1994 Cambridge University Press.

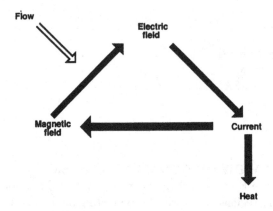

Figure 10.1 Electromagnetic feedback loop in the presence of electrically conducting fluid flow.

problem, is smooth, incompressible and bounded in V. The flow is typically steady, periodic in time, quasiperiodic, random with stationary statistics, or otherwise homogeneous in time. The magnetic diffusivity is $\eta = 1/\mu\sigma$, where μ is the permeability and σ the electrical conductivity of the fluid.

For a steady flow, the kinematic dynamo problem reduces to determining whether the operator

$$\mathcal{L}_\eta(\ ) \equiv (\ ) \cdot \nabla\mathbf{u} - \mathbf{u} \cdot \nabla(\ ) + \eta\nabla^2(\ ) \qquad (10.1.2)$$

has any eigenvalues with positive real part. Because the highest-order derivative is the Laplacian, $\mathcal{L}_\eta$ is elliptic and its spectrum consists of a countable set of eigenvalues, all of finite multiplicity, that accumulate at negative real infinity only. The eigenfunctions are nice, smooth functions. A typical solution to (1.1) will tend to grow at a rate

$$\gamma(\eta) = \max_{\lambda \in \text{Spectrum}(\mathcal{L})} \text{Real}(\lambda) \qquad (10.1.3)$$

as $t \to \infty$, and the spatial form will approach a linear combination of the corresponding eigenfunctions.

For a time-periodic flow, the same remarks hold, except that the λ's are the Floquet exponents rather than eigenvalues of $\mathcal{L}_\eta$. Similar concepts may be defined formally for quasiperiodic or random flows with stationary statistics, but they are not as readily usable as eigenvalues or Floquet exponents. We shall only discuss steady and time-periodic flows in these lectures.

The kinematic dynamo problem is computationally quite simple, provided the flow is not too weird and the diffusivity not too small. The induction operator (1.2) is discretized, perhaps by projecting onto a set of linearly independent smooth functions, and the eigenvalues of the resulting finite matrix found numerically. Such computations were performed on the earliest computers (Bullard & Gellman 1954). It is not trivial, of course, to find velocity fields for which the dominant eigenvalue is positive. In fact, this is a nonlinear problem, since the velocity field appears as a coefficient in $\mathcal{L}_\eta$. A further difficulty is that, by theorems such as those of Cowling (1934), Zel'dovich (1957) and Finn & Ott (1988), dynamo action is impossible in flows with certain geometric simplicities. Therefore flow fields must have a certain level of complexity to give field growth.

10.1.2 Fast Kinematic Dynamos: General Remarks

The fast dynamo problem is concerned with how effectively a flow amplifies magnetic fields when the diffusivity is very small. In this limit the diffusion term is then a singular perturbation, and nontrivial magnetic field structures may be anticipated on the dissipation length scale $l_D = \sqrt{\eta/\|\nabla u\|}$. Naively, let us begin by considering the induction equation at exactly zero diffusivity. The magnetic field vector in this case then behaves exactly like an infinitesimal material line element, being advected by the flow and stretched and rotated by the local shear. The magnetic field evolves in a purely geometric fashion; the field lines are 'frozen into the fluid'.

Formally, we say

$$\mathbf{u}(\mathbf{x}, t) \text{ is a fast dynamo} \iff \limsup_{\eta \to 0} \gamma(\eta) > 0. \qquad (10.1.4)$$

Consider what is necessary to have an exponentially growing distribution of magnetic field in a perfectly conducting fluid. In order to get pointwise growth of magnetic field, there must be a non-empty set of fluid particle trajectories that diverge exponentially rapidly (Vishik 1989). But pure exponential stretching cannot increase quantities like the flux of field through a fixed element of surface, or the integral over a specific subregion. In order to get exponential growth of such integrated measures, there must be fluid particle trajectories that converge exponentially, bringing with them flux from previously separate locations. We call this exponential convergence 'folding'.

Stretching and folding together may still not give net exponential growth of the global field if the folding process brings together regions containing oppositely directed field. In order to get continued growth, the stretched field must be rearranged — by twisting or shearing or something — so as to avoid complete cancellation. Thus, from the diffusionless point of view, there are three properties of a flow apparently essential for fast dynamo action: stretching, folding, and flux rearrangement. These concepts have been enunciated before; everyone is familiar with the stretch–twist–fold (STF) mechanism (Vainshtein & Zel'dovich 1972) and the stretch–fold–shear (SFS) mechanism (Roberts 1972).

The geometric requirements of exponential stretching and folding of material lines suggest that flows whose particle trajectories are chaotic are the best places to seek fast dynamos. Vishik (1992) has shown that

$$\limsup_{\eta \to 0} \gamma(\eta) \leq \text{topological entropy of flow} \qquad (10.1.5)$$

for steady flows and probably also for periodic flows. Since the topological entropy is positive only for flows with chaotic particle trajectories, Lagrangian chaos is necessary for fast dynamo action. It should be noted that this theorem refers only to spatially smooth flows, so there is no contradiction with Soward's (1987) steady fast dynamo.

Another important theorem concerning fast dynamos, given by Moffatt & Proctor (1985), is that the diffusionless operator $\mathcal{L}_0$ has no smooth eigenfunctions belonging to eigenvalues with positive real part. Moffatt and Proctor also showed that if $\limsup_{\eta \to 0} \gamma(\eta) > 0$, then the eigenfunctions of $\mathcal{L}_\eta$ always have substantial structure at the dissipation length l_D. These results are consistent with the essential ingredients of fast dynamos discussed above. The fine structure is presumably created by repeated stretching and folding. We can also think of the fine structure as being a consequence of chaotic mixing.

These theorems show that the fast dynamo problem cannot be formulated as an eigenvalue problem, at least in the ordinary sense. In fact, the best way to formulate the fast dynamo problem is still being debated.

10.1.3 Fast Kinematic Dynamos: Numerical Approaches
In the absence of a theoretical strategy for attacking the fast dynamo problem, many of us have resorted to performing high-resolution numerical simulations of diffusive dynamos at a variety of small diffusivities, from which we try to guess the nature of the zero diffusivity limit by extrapolation. Numerical simulations of dynamos in various flows (Arnol'd & Korkina 1983, Galloway & Frisch 1984, 1986, Otani 1988, Galloway & Proctor 1992) confirm that a reasonable amount of chaos, plus a certain unquantifiable lack of reflectional symmetry, results in apparent fast dynamo action. As the diffusivity in the computations goes to zero, the growth rate levels off at a positive value. This suggests that not only is the supremum limit of the growth rate positive, but that the dominant eigenvalue actually converges as $\eta \to 0$.

Consistent with Moffatt and Proctor's result, the numerically computed eigenfunction develops complicated structure on smaller and smaller scales as the diffusivity goes to zero. In fact, the bulk of the magnetic energy is in the small scales. However, the large scale structures, e.g., the low-wavenumber Fourier modes, do not vary much as the diffusivity diminishes, even as the small scale structures grow up within them.

These observations suggest that as $\eta \to 0$, the dominant eigenfunction approaches a distribution, or generalized function, $\mathbf{B}^*(\mathbf{x})$. This would be a weak eigenfunction, in the sense that

$$\lambda^* \int_V \mathbf{f}(\mathbf{x}) \cdot \mathbf{B}^*(\mathbf{x}) \, dx = \int_V \mathbf{B}^*(\mathbf{x}) \mathcal{L}_0^\dagger \mathbf{f}(\mathbf{x}) \, dx \qquad (10.1.6)$$

for any smooth vector function $\mathbf{f}(\mathbf{x})$. Here $\lambda^* = \lim_{\eta \to 0} \lambda^{\text{dominant}}(\eta)$, $\mathcal{L}_0^\dagger$ is the adjoint of the zero-diffusivity induction operator. The generalized eigenvalue λ^* cannot, of course, be an eigenvalue of $\mathcal{L}_0$, but might lie in some other part of the spectrum (de la Llave 1993, Nuñez 1993).[1] Appealing though this picture is, it has so far been impossible to establish it in all but the simplest examples.

The current situation with regard to the fast dynamo problem is that major conceptual obstacles still exist. The only examples in which fast dynamo action has been proved are overwhelmingly idealized. The examples which are considered realistic by mathematicians, let alone geophysicists, are far too complex for available techniques. But there are many ways in which the problem can be approached. As more and more people from different parts of the mathematical and scientific community get interested in the fast dynamo problem (e.g., Thompson 1990, Collet 1992), it is not so crazy to hope that someone will find a delightfully elegant solution using completely new ideas, at the least expected moment.

10.2 MAP-DYNAMO MODELS

The induction equation (1.1) describes simultaneous advection, stretching, and diffusion of the magnetic field. The combination is too complicated to be grasped intuitively if the flow has nontrivial three-dimensional structure. Furthermore, even without diffusion, following the advection and distortion of the field requires solving third-order systems of coupled nonlinear differential equations for the particle trajectories. Constructing a flow field which performs a desired sequence of stretch, fold, and twist operations on material line elements or magnetic field lines is tricky (Moffatt & Proctor 1985, Bajer & Moffatt 1990).

---

[1] Suppose L is a linear operator on an vector space V over the complex numbers C. The *spectrum* is (roughly) the set of $\lambda \in C$ for which $L - \lambda I$ is not invertible. An *eigenvalue* is a complex number μ for which there exists a nonzero vector v satisfying $Lv = \mu v$. If V is finite dimensional, the set of eigenvalues equals the spectrum. If V is infinite dimensional, the set of eigenvalues must be contained in the spectrum, but the spectrum may contain other complex numbers which are not eigenvalues.

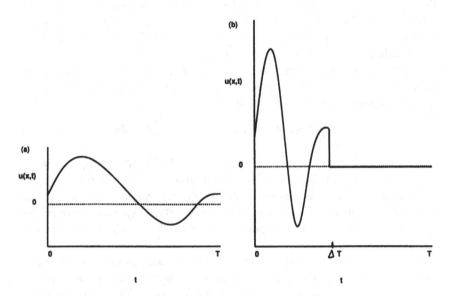

Figure 10.2 Time dependence (schematic) of (a) the original flow $\mathbf{u}$, and (b) the pulsed flow $\tilde{\mathbf{u}}_{\Delta T}$.

From the numerical simulation point of view, equation (1.1) is less than ideal. Because the fields are expected to have nontrivial structure at the length l_D, fine spatial grids are necessary when the diffusivity is very small. In order to avoid numerical instabilities, the time step must be taken correspondingly small.

Map-dynamo models make two substantial modifications to the kinematic dynamo problem that simplify the concepts and streamline the numerics. First, we select ΔT between 0 and T, and define a new velocity field $\tilde{\mathbf{u}}_{\Delta T}$ with the same temporal period T as $\mathbf{u}$ by

$$\tilde{\mathbf{u}}_{\Delta T}(\mathbf{x}, t) = \begin{cases} (T/\Delta T)\mathbf{u}(\mathbf{x}, t(T/\Delta T)) & \text{if } 0 \le t < \Delta T; \\ 0 & \text{if } \Delta T \le t < T, \end{cases} \qquad (10.2.1)$$

and repeated periodically thereafter (figure 10.2). In the flow $\tilde{\mathbf{u}}$, the fluid is in motion for only a part of the original interval. But because the motion is more intense, the flow maps of $\tilde{\mathbf{u}}$ and $\mathbf{u}$ over the interval $0 < t < T$ are exactly the same. If $\Delta T/T$ is short enough, we may neglect diffusion during the motion intervals. Since we can manifestly neglect advection and stretching during the stasis intervals, this trick allows us to consider diffusion separately from the flow.

Neglecting diffusion during the motion intervals implies that the magnetic field at $t = nT + \Delta T$ (n an integer) is related to the field at $t = nT$ by the

Cauchy solution

$$\mathbf{B}(\mathbf{M}(\mathbf{x}), nT + \Delta T) = J(\mathbf{x})\mathbf{B}(\mathbf{x}, nT), \qquad (10.2.2)$$

where $\mathbf{M}$ is the *flow map* for the motion interval. The matrix J, with elements

$$J_{ij} = \frac{\partial M_i(\mathbf{x})}{\partial x_j}, \qquad (10.2.3)$$

is the Jacobian, or derivative, matrix of the flow map. Thus if we know the flow map, there is no necessity to know anything about the velocity field! The flow map of an incompressible flow is always invertible and volume-preserving, and the Jacobian matrix has determinant unity.

Since the duration ΔT of the motion intervals has no effect on the dynamics, we can let $\Delta T \to 0$. The result is a 'flow' in which the fluid is stationary almost all the time, and the only process operating is diffusion. The material jumps rapidly from one configuration to another at the instants $t = T, 2T, 3T, \ldots$, as described by the flow map. The magnetic field evolution is determined solely by the diffusivity η during the stasis intervals, and the flow map and its Jacobian at the mapping instants. Since the map is now the essential quantity, we call such a system a 'map dynamo'. The evolution of the magnetic field in a map dynamo is expressed by

$$\mathbf{B}(\mathbf{x}, (n+1)T) = \int_V H_{\eta T}(\mathbf{x}, \mathbf{M}(\mathbf{y})) J(\mathbf{y}) \mathbf{B}(\mathbf{y}, nT) \, d\mathbf{y}, \qquad (10.2.4)$$

where $H_{\eta T}(\mathbf{x}, \mathbf{y})$ is the Green's function of the diffusion equation.

For conciseness, let $\varepsilon \equiv \eta T$, and denote the linear operator in (2.4) by $\mathcal{G}_\varepsilon$. The kinematic map-dynamo problem is then to find the eigenvalue μ_ε of $\mathcal{G}_\varepsilon$ with largest modulus, and determine whether its modulus is greater than unity. The fast map-dynamo problem is to determine whether

$$\limsup_{\varepsilon \to 0} |\mu_\varepsilon| > 1. \qquad (10.2.5)$$

As described so far, the map that operates on the fluid at $t = nT$ is the flow map of some time-periodic velocity field. But there is no formal necessity for this always to be the case. The science of map dynamos is greatly enriched if we allow maps which are not the flow maps of any flow. In later sections, we will consider some map dynamos using discontinuous maps, which can of course never arise as flow maps of well-behaved flows. Of course, we still require any proposed map to be invertible and volume-preserving.

Map dynamos have a conceptual advantage over flow dynamos in that the map specifies a finite deformation of the fluid while a flow only specifies the infinitesimal deformation. We can therefore build quite explicit combinations of stretching, folding and twisting into maps with great ease. Map dynamos also have computational advantages over continuous-time flow dynamos. Since the diffusion equation is easy to solve numerically in simple geometries like spheres or cubes, the computational effort needed to compute one map-diffusion cycle in the map dynamo is comparable to the effort of computing one time step in a flow dynamo. Thus map dynamos can be computed much faster than flow dynamos. This is especially true because there are no 'time step' restrictions on map dynamo computations; the time step in a map dynamo is always T! In fact, because of the way the map dynamo is formulated, there are no computational errors associated with finite time steps, as there necessarily are with flow dynamos.

10.3 PULSED BELTRAMI WAVE DYNAMOS
So-called ABC flows have been used extensively in kinematic dynamo studies. There are several definitions in the literature; we shall use the form

$$\mathbf{u}(\mathbf{x}) = \begin{pmatrix} C\cos(z - \phi_3) - B\sin(y - \phi_2) \\ A\cos(x - \phi_1) - C\sin(z - \phi_3) \\ B\cos(y - \phi_2) - A\sin(x - \phi_1) \end{pmatrix} \qquad (10.3.1)$$

with A, B, C, ϕ_1, ϕ_2, and ϕ_3 constants.[2] Any ABC flow of this form has $\nabla \times \mathbf{u} = \mathbf{u}$, maximal (positive) relative helicity, and is an exact steady solution of the inviscid incompressible fluid equations. When A, B, and C are all nonzero, the flow has extensive regions of chaotic particle trajectories and thus seems a natural candidate for a fast dynamo. Computations of dynamo action in flows of this type were performed by Arnol'd & Korkina (1983), Galloway & Frisch (1984, 1986) and Gilbert (1991).

Map dynamos based on chaotic ABC flows cannot be directly constructed because such flows are non-integrable and there is no nice analytic formula for the finite-time flow map if A, B, and C are all non-zero. But if one coefficient vanishes, the particle trajectories may be found in terms of elliptic integrals, and if two coefficients vanish the particle trajectories may be expressed in terms of elementary functions. Such flows are nothing but right-handed circularly polarized plane waves. Since they have the Beltrami property, they are called Beltrami waves.

For example, if $B = C = 0$, then

$$\mathbf{u}(\mathbf{x}) = \begin{pmatrix} 0 \\ A\cos(x - \phi_1) \\ -A\sin(x - \phi_1) \end{pmatrix}$$

---

[2] This form is subtly different from the form used by Roberts (chapter 1) and Soward (chapter 6). This form has been kept here because it was used in generating the computer simulations described below, but Roberts' and Soward's forms would generate very similar results.

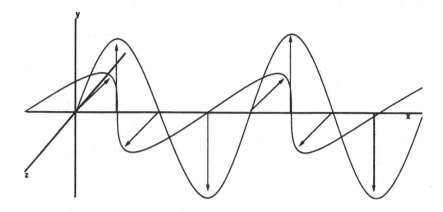

Figure 10.3 Velocity field of ABC flow with $B = C = 0$.

(figure 10.3). The flow map is

$$\mathbf{M}^A(\mathbf{x}) = \begin{pmatrix} x \\ y + AT\cos(x - \phi_1) \\ z - AT\sin(x - \phi_1) \end{pmatrix}, \qquad (10.3.2a)$$

with Jacobian matrix

$$J^A(\mathbf{x}) = \begin{pmatrix} 1 & 0 & 0 \\ -AT\sin(x - \phi_1) & 1 & 0 \\ -AT\cos(x - \phi_1) & 0 & 1 \end{pmatrix}. \qquad (10.3.2b)$$

Flow maps $\mathbf{M}^B$ with $A = C = 0$ and $\mathbf{M}^C$ with $A = B = 0$ are defined analogously.

Each flow map $\mathbf{M}^A$, $\mathbf{M}^B$, $\mathbf{M}^C$ is too simple to display kinematic dynamo action by itself. But the composition

$$\mathbf{M}^{ABC}(\mathbf{x}) \equiv \mathbf{M}^C(\mathbf{M}^B(\mathbf{M}^A(\mathbf{x}))), \qquad (10.3.3a)$$

whose Jacobian matrix is

$$J^{ABC}(\mathbf{x}) = J^C(\mathbf{M}^B(\mathbf{M}^A(\mathbf{x})))J^B(\mathbf{M}^A(\mathbf{x}))J^A(\mathbf{x}), \qquad (10.3.3b)$$

can. We can think of the map $\mathbf{M}^{ABC}$ as representing an ABC flow in which the coefficients are 'turned on' in sequence, or 'pulsed', each for time T, instead of being turned on simultaneously all the time. The flow maps obtained this way are suitable for fast, high resolution map-dynamo simulations of the kind described in section 10.2.

General ABC map dynamos were introduced by Finn & Ott (1988) in a completely diffusionless model. Bayly & Childress (1988) made extensive investigations of a quasi-two-dimensional pulsed Beltrami wave model that used only the A and B waves. This device allowed numerical investigations resolving only the structures in the x-y plane, with the z dependence being proportional to e^{ikz} for real k. Quasi-two-dimensionality of the flow means that although there is nontrivial motion in all three directions of space, the structure only depends on two coordinates. Quasi-two-dimensional dynamos are therefore exempt from Zel'dovich's antidynamo theorem, and the results of Bayly and Childress showed that strong, apparently fast, dynamo action occurs in these flows. Soward (chapter 6) has also used these flows, with large AT and BT, in an asymptotic investigation of dynamo action.

An interesting variant on the pulsed helical flow model is the random pulsed helical flow dynamo (Gilbert & Bayly 1992). Based on the 'random renewing flow' ideas of Dittrich *et al.* (1984), the flow is assumed to consist of a sequence of instantaneous maps of the form (3.3). But instead of fixing the phases ϕ_1, ϕ_2, ϕ_3, we choose them randomly from a uniform distribution on $[0, 2\pi]$ at each mapping time. This yields a spatially homogeneous ensemble of random flows.

The spatial homogeneity of the ensemble of random pulsed helical flows implies that we may seek random magnetic fields whose ensemble-average is proportional to a *single* Fourier mode

$$\left\langle \mathbf{B}^{(n)}(\mathbf{x}) \right\rangle = \hat{\mathbf{B}}^{(n)} \exp(i\mathbf{k} \cdot \mathbf{x}). \tag{10.3.4}$$

The statistical independence of each mapping from the magnetic field before the mapping allows us to evaluate analytically $\hat{\mathbf{B}}^{(n+1)}$ directly from $\hat{\mathbf{B}}^{(n)}$. The growth rates of higher-order statistics, indicating the intermittency of the magnetic field, can also be calculated by essentially analytical means. For details, we refer the reader to Gilbert and Bayly.

10.4 ESSENTIALLY ONE-DIMENSIONAL MAP DYNAMOS

The pulsed-flow dynamos of the last section are not much different from the dynamos of Otani (1988) and Galloway & Proctor (1992) that have smooth time-dependence. Although pulsed-flow map dynamos are easier to compute, they display no less complex structure. The flows are not chaotic everywhere, and the KAM regions interpenetrate wildly. The magnetic field points in all different directions. These features seem to be unavoidable in dynamos constructed with nontrivial two- or three-dimensional flow structure.

Map dynamos with simpler geometry can be obtained at the cost of using maps which are not smooth — perhaps even discontinuous. With this freedom, map dynamos can be devised in which the magnetic field everywhere points in the same direction, with strength depending on only one spatial

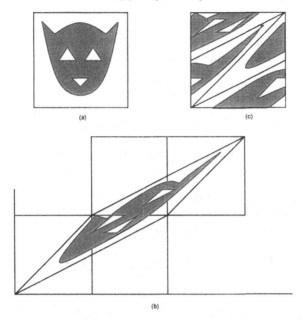

Figure 10.4 The cat map: (*a*) a cat on the two-torus, (*b*) stretched out, and (*c*) placed back on the torus.

coordinate. Such models must be cautiously interpreted. But if judiciously used, they can illuminate specific aspects of generic dynamos.

10.4.1 Cat Dynamo

The cat dynamo of Arnol'd *et al.* (1981) is a beautiful example that illustrates the power of simplicity. It was presented as a steady-flow dynamo on a manifold of peculiar geometry, but it is equivalent to an appropriately constructed map dynamo on the square $[0,1] \times [0,1]$. In this context, the map is the cat map (Arnol'd & Avez 1968),

$$\mathbf{M}^{\text{cat}} \begin{pmatrix} x \\ y \end{pmatrix} = \begin{pmatrix} 2x + y \\ x + y \end{pmatrix} \quad (\text{mod } 1). \qquad (10.4.1)$$

The cat map is a simple example of an Anosov system, a dynamical system with idealized chaotic properties. The action of the cat map on a cat face is illustrated in figure 10.4.

As defined here, the cat map is discontinuous — pieces of the transformed square must be cut and rearranged in the original square. But if the opposite edges of the square are identified, yielding a 2-torus, the cat map turns out to be perfectly smooth everywhere. Even on the torus, however, the cat map cannot be realized as the flow map of a flow; this would require defining a special three-dimensional manifold (Arnol'd *et al.* 1981). For the purposes of these lectures, we will use the cat map as an example of a dynamo model in which the map itself is given a priori.

The Jacobian matrix of the cat map is constant:

$$J^{cat} = \begin{pmatrix} 2 & 1 \\ 1 & 1 \end{pmatrix}, \tag{10.4.2}$$

hence the magnetic field evolution under the cat map is

$$\mathbf{B}^{after} \begin{pmatrix} 2x + y \\ x + y \end{pmatrix} = \begin{pmatrix} 2 & 1 \\ 1 & 1 \end{pmatrix} \mathbf{B}^{before} \begin{pmatrix} x \\ y \end{pmatrix}. \tag{10.4.3}$$

After the mapping the field is allowed to diffuse for time T. Then the map is applied again, the field diffused again, and the cycle repeated.

Arnol'd et al. showed that for generic initial data,

$$\mathbf{B}^{(n)} \begin{pmatrix} x \\ y \end{pmatrix} \sim (\text{const.}) \times \mu^n \xi \qquad \text{as } n \to \infty, \tag{10.4.4}$$

where

$$\mu = \frac{3 + \sqrt{5}}{2}, \qquad \xi = \begin{pmatrix} (1 + \sqrt{5})/2 \\ 1 \end{pmatrix} \tag{10.4.5}$$

is the dominant eigenvalue-eigenvector pair of the Jacobian matrix J, and the constant depends on the initial magnetic field. ξ is the stretching direction of the cat map, and μ the factor by which material line elements grow in length under the map. Since $|\mu| > 1$ we have exponential stretching, which is consistent with the strong chaos of the cat map as an Anosov dynamical system.

Amazingly, there is no effect of diffusion at all on the dominant eigensolution in the cat dynamo. Hence it is a fast dynamo. Diffusion does appear, as Arnol'd et al. describe, in the way the field approaches its asymptotic behavior.

There is no twisting or shearing per se in the cat map. The roles of these processes in flow dynamos are played here by cutting and rearranging the pieces of the square after it has been stretched. The cutting and rearranging also plays the role of folding flux back into the stretching region, so there is no necessity for an explicit folding operation. The cat dynamo does not display any small-scale field variation in the eigenfunction. The apparent contradiction with the Moffatt–Proctor result is resolved by noting that since there is no flow on the torus that realizes the map, the hypotheses of the Moffatt–Proctor theorem are not satisfied and it makes no predictions.

The cat dynamo may appear trivial, but it illustrates how exponential stretching of material lines can yield fast dynamo action. The main omission of the cat dynamo is cancellation.

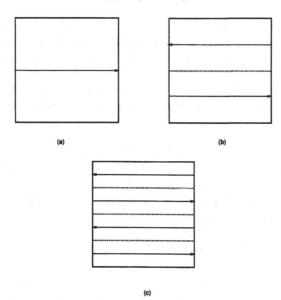

Figure 10.5 The Baker's map: (a) before, (b) after one application and (c) after two applications.

10.4.2 Baker's Non-dynamo

An even simpler map on the square $[0,1] \times [0,1]$ is the Baker's transformation

$$\mathbf{M}^{\text{Baker}} \begin{pmatrix} x \\ y \end{pmatrix} = \begin{cases} \begin{pmatrix} 2x \\ y/2 \end{pmatrix} & \text{if } 0 \le x < \tfrac{1}{2} \\ \begin{pmatrix} 2 - 2x \\ 1 - y/2 \end{pmatrix} & \text{if } \tfrac{1}{2} \le x \le 1. \end{cases} \tag{10.4.6}$$

(Actually, how the map is defined at $x = 0$, $1/2$ and 1 is not important.) The Jacobian matrix of the Baker's map is

$$J = \text{sign}(1/2 - x) \begin{pmatrix} 2 & 0 \\ 0 & \tfrac{1}{2} \end{pmatrix}. \tag{10.4.7}$$

This map has the effect shown in figure 10.5.

We can use the Baker's map to define a map dynamo just as we did with the cat map. The discontinuity of the mapped field at $y = 1/2$ after each mapping is unphysical but, as defined here, there is no actual singular behavior. In any case, the field discontinuity gets smoothed out during the diffusive stage of the cycle.

Despite the stretching and folding, the map dynamo defined by the Baker's map is not actually a dynamo — that is, the overall magnetic field at any positive diffusivity does not increase. The reason is that after one application of the map, there is perfect cancellation of flux over the entire cube, after the second mapping there is perfect cancellation over horizontal strips of width 1/2, after the third there is cancellation over strips of width 1/4, and so on. Eventually there is perfect cancellation right down at the diffusion scale, and everything vanishes. The reason why the Baker's map fails to yield dynamo action is that there is no rearrangement of flux to prevent catastrophic cancellation.

10.4.3 Finn–Ott Generalized Baker Dynamos

These dynamos are based on maps similar to the Baker's map and the cat map in that they involve stretching the material in the unit square, cutting it in pieces and rearranging it in the original square (Finn & Ott 1988). Using more sophisticated dissection rules, the effects of cancellation and non-uniform stretching can be modeled. These models also provide a framework in which magnetic field intermittency can be studied, and related concepts such as quantitative measurement of the amount of cancellation in the small scale fields (Du & Ott 1993). Although two-dimensional, these models escape Zel'dovich's antidynamo theorem because they are based on maps rather than flows. A similar model of intermittency has been developed by Poezd, Galeeva & Sokoloff (1992).

The Finn–Ott dynamo model begins by dividing the square into a number N of vertical slices with boundaries given by

$$x = a_j \quad \text{for} \quad 0 = a_0 < a_1 < \ldots < a_N = 1. \qquad (10.4.8)$$

We call the rectangle

$$\{a_{j-1} < x < a_j, \quad 0 < y < 1\}$$

the j'th vertical strip. Then each vertical strip is squashed downward, while preserving its area, until its width is 1 and its height is its previous width. Finally, the strips — now horizontal — are piled on top of each other, filling the original square perfectly. The orientation of the j'th strip may be reversed or not. To determine the pattern of strip reversals we specify $s_j = -1$ if the j'th strip is to be reversed and $s_j = 1$ otherwise. The Finn–Ott map is illustrated in figure 10.6.

Some care is necessary in defining the Finn–Ott maps for points on the boundaries of strips. But since these subtleties do not affect the conclusions for dynamos, we shall intentionally omit them.

The effect of a Finn–Ott map on a general magnetic field is moderately complicated. But if the field in the material before the map is applied is everywhere parallel to the x-axis, with strength depending only on y, then the field after the mapping will have the same form. Explicitly, if

$$\mathbf{B}^{\text{before}}(x, y) = \begin{pmatrix} b^{\text{before}}(y) \\ 0 \end{pmatrix},$$

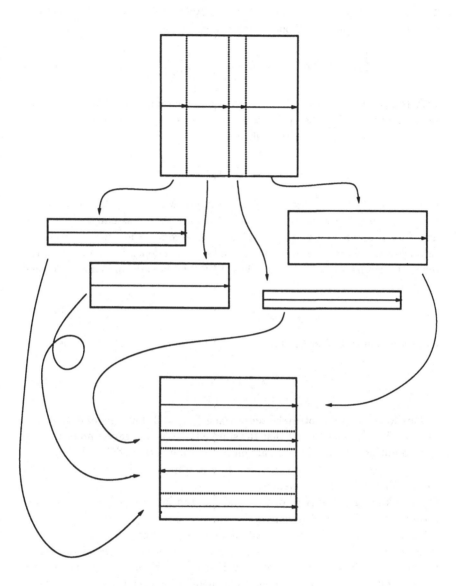

Figure 10.6 Finn–Ott generalized Baker map with $N = 4$.

then

$$\mathbf{B}^{\text{after}}(x, y) = \begin{pmatrix} b^{\text{after}}(y) \\ 0 \end{pmatrix}.$$

The vector field operator taking $\mathbf{B}^{\text{before}}$ to $\mathbf{B}^{\text{after}}$ now reduces to an operator G^{FO} on scalar functions of one variable, defined by

$$b^{\text{after}}(y) = [G^{\text{FO}} b^{\text{before}}](y) \equiv \frac{s_j}{a_j - a_{j-1}} \, b\left(\frac{y - a_{j-1}}{a_j - a_{j-1}}\right), \qquad (10.4.9)$$

with (a_{j-1}, a_j) understood to be the interval containing y. Again, we don't worry about defining the operator precisely on strip boundaries.

Before the mapping, the total flux is

$$\Phi = \int_0^1 b(y)\, dy$$

through any vertical section. The squashing-down process does not change the flux through any strip. When the strips are then reassembled, the total new flux through any vertical cross section becomes the sum of the fluxes through the now-horizontal strips, multiplied by ± 1 depending on whether the relevant strip has been reversed. Thus the total flux ends up being multiplied by the factor

$$\mu = \sum_{j=1}^{N} s_j. \qquad (10.4.10)$$

Diffusion does not alter the total flux, therefore the growth rate is just

$$\lambda = \frac{1}{T} \log \left| \sum_{j=1}^{N} s_j \right|. \qquad (10.4.11)$$

This formula does not make sense when $\sum s_j = 0$, but this case is a non-dynamo for the same reason that the simple Baker's map gave a non-dynamo. The condition for fast dynamo action in the Finn–Ott model is that $\mu > 1$.

The Finn–Ott dynamos exemplify several important features which seem to be typical of all fast dynamos:

(1) Thinking of the growth of the total flux (an integral measure of field) is a very profitable line of reasoning.

(2) The maximum value of the field increases by a factor of $1/\min_j\{a_j - a_{j-1}\}$ at every iteration of the map. This is always at least as big as the factor by which the flux grows — equality holds only when the a_j's are evenly spaced and the s_j's are all equal. We therefore expect plenty of magnetic field intermittency at very low diffusivity.

(3) When there are reversals of direction of magnetic field vector, the growth rate of the total flux may be substantially smaller than the growth of the actual values of the magnetic field at all points. There is an overwhelming amount of cancellation between neighboring regions of strong but oppositely directed field.

The last observation indicates that even in a fast dynamo, the cancellation is almost perfect. There is only a subtle difference between three- and two-dimensional systems — recall that in two dimensions geometrical considerations guarantee exact cancellation. We like to think that by going to three dimensions we get a totally new range of phenomena, but we still have almost perfect cancellation.

10.4.4 Stretch–Fold–Shear Model

This map was conceived by Bayly & Childress (1988) as a model of the processes operating in the Glyn Roberts (1972) dynamo. In the Glyn Roberts flow, magnetic field is stretched and folded into tongue-like structures at the periphery of the flow cells, and sheared in the direction of the cell axes. Diffusion acts at the same time.

The SFS map reproduces these processes sequentially, as shown in figure 10.7. The 'flow domain' is the cube $0 < x, y, z < 1$. We suppose that the initial magnetic field is aligned purely in the x-direction, pointing in the positive direction in one half of the cube (divided vertically) and in the negative direction in the other half. The map begins by stretching the fluid in the x-direction and compressing it in the y-direction, and folding it over, just as in the two-dimensional Baker's map. Although the field intensity is everywhere doubled, the average of the field on any vertical section is exactly zero because of the folding.

But now comes the shear. The top of the cube is shifted by an amount $\alpha/2$ in the positive z-direction, and the bottom shifted by an equal amount in the negative z-direction, and all interior points shifted proportionately. The result is that vertical field averages are now no longer zero everywhere. Some cancellation is unavoidable, but — we hope — the losses due to cancellation may be smaller than the factor-of-two intensification achieved during the stretching phase. The amount of cancellation obviously depends on the shear parameter α, with perfect cancellation when $\alpha = 0$.

The SFS system has the property that if the field before the mapping has the form $(b(y)e^{2\pi iz}, 0)^{\mathrm{T}}$, then so does the field after the mapping. As in the Finn–Ott model, we can therefore represent the evolution of the field under the map by the scalar operator G^{SFS}, defined by

$$
b^{\text{after}}(y) = [G^{\mathrm{SFS}} b^{\text{before}}](y)
$$

$$
\equiv 2e^{-2\pi i\alpha(y-1/2)} \begin{cases} b^{\text{before}}(2y) & \text{if } 0 < y < 1/2; \\ -b^{\text{before}}(2-2y) & \text{if } 1/2 < y < 1. \end{cases} \qquad (10.4.12)
$$

The power of the SFS model is the ability to vary α (i.e., the amount of cancellation) continuously — which is impossible in Finn–Ott models — and see how the growth rate and other aspects of the dynamo change.

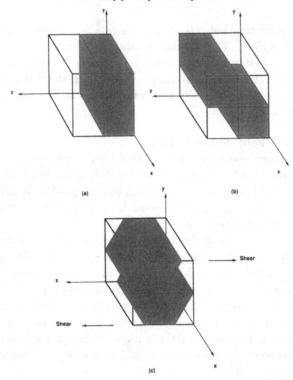

Figure 10.7 The SFS map: (a) before, (b) after stretch and fold, (c) after shear.

Unlike the Finn–Ott model, we cannot solve analytically for the dynamo growth rate. It is easy, however, to discretize G^{SFS} and find the eigenvalues and eigenvectors of the resulting matrix. In figure 10.8 we show the computed growth rate (i.e., the logarithm of the modulus of the dominant eigenvalue) as a function of the shear rate α. The good behavior of the growth rate is in stark contrast to the crazy structure of the discrete eigenfunctions, shown (for $\alpha = 1$) in figure 10.9.

10.5 ANALYSIS OF FINN–OTT AND SFS MODELS

The growth rate of the Finn–Ott dynamos could be found analytically because the total flux, an *integrated* quantity, satisfied a natural recursion relation. The idea that integrals of the magnetic field may be robust in the limit of small diffusivity even if the field itself oscillates wildly was enunciated forcefully by Finn & Ott (1988) and by Bayly & Childress (1988). Of course the idea was not new even then, but the Finn–Ott and SFS models dramatically illustrated its importance. Finn and Ott envisioned measuring dynamo growth rates in terms of the growth of flux through surfaces fixed in space, and Bayly and

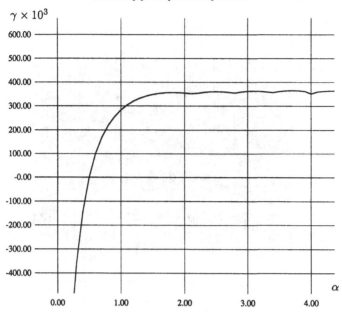

Figure 10.8 Growth rate of the SFS dynamo as a function of shear α. The computation is based on discretization of the interval $0 < y < 1$ into $N = 1024$ equal cells.

Childress contemplated integrating the scalar product of the field with a fixed smooth vector function, but the basic idea was the same. In fact, the result is exactly the same for the scalar models.

Following Bayly & Childress (1988), we choose a specific smooth vector function $\mathbf{f}(\mathbf{x})$ and consider the behavior as $t \to \infty$ of

$$\int_V \mathbf{f}(\mathbf{x}) \cdot \mathbf{B}(\mathbf{x}, t)\, d\mathbf{x}.$$

If the diffusivity is finite, we expect that for generic functions $\mathbf{f}$ the integral will grow at the same rate as the dominant eigenfunction(s), hence

$$\gamma(\eta) = \lim_{t \to \infty} \frac{1}{t} \log \int_V \mathbf{f}(\mathbf{x}) \cdot \mathbf{B}(\mathbf{x}, t)\, d\mathbf{x}. \qquad (10.5.1)$$

This formula can be extended to the case of exactly zero diffusivity, at which we *define*

$$\gamma^* = \limsup_{t \to \infty} \frac{1}{t} \log \int_V \mathbf{f}(\mathbf{x}) \cdot \mathbf{B}(\mathbf{x}, t)\, d\mathbf{x}. \qquad (10.5.2)$$

The author knows no definite reason why the supremum limit should be a true limit for general flows, nor why γ^* should equal $\limsup_{\eta \to 0} \gamma(\eta)$.

In the case of map dynamos, we can equivalently say that

$$\gamma(\eta) = \lim_{n \to \infty} \frac{1}{nT} \log \int_V \mathbf{f}(\mathbf{x}) \cdot \mathbf{B}^{(n)}(\mathbf{x})\, d\mathbf{x}. \qquad (10.5.3)$$

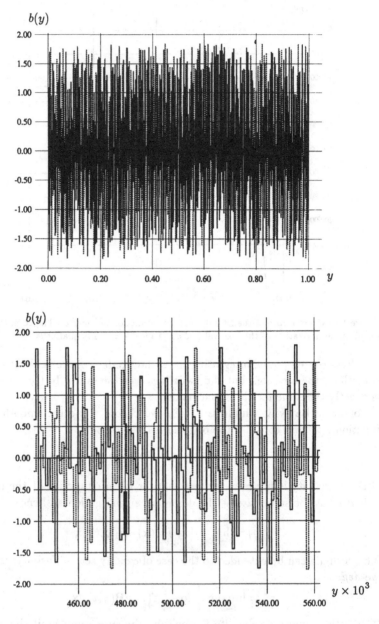

Figure 10.9 Dominant eigenfunction of discretized SFS operator (real part solid, imaginary part dotted) at $\alpha = 1$ and $N = 1024$ cell resolution. (b) shows the detailed structure of the central part of (a).

Since $\mathbf{B}^{(n)} = \mathcal{G}_\eta^n \mathbf{B}^{(0)}$, we can equally well write

$$\gamma(\eta) = \lim_{n \to \infty} \frac{1}{nT} \log \int_V \left[\mathcal{G}_\eta^{\dagger n} \mathbf{f} \right](\mathbf{x}) \cdot \mathbf{B}^{(0)}(\mathbf{x}) \, d\mathbf{x}. \tag{10.5.4}$$

If we are solely interested in the dynamo growth rate, it is sufficient to have information about the eigenvalues and eigenfunctions of $\mathcal{G}_\eta^\dagger$. For general maps and flows the adjoint operator $\mathcal{G}_\eta^\dagger$ is no nicer an operator than the original operator $\mathcal{G}_\eta$ (Roberts 1967). But for the Finn–Ott and SFS models, the adjoint formulation results in much more tractable problems. The trick is to work with the one-dimensional 'reduced' operators that describe the evolution of the y-structure of a purely x-directed field. We shall leave out diffusion.

Let $f(y)$ be a smooth function and consider the integral

$$I = \int_0^1 f(y) \left[\mathcal{G}_0^{\text{FO}} b \right](y) \, dy \tag{10.5.5}$$

where $\mathcal{G}_0^{\text{FO}}$ is the diffusionless Finn–Ott operator (4.9). Using the definition, we have explicitly

$$I = \sum_{j=1}^N \int_{a_{j-1}}^{a_j} f(y) \left[\frac{s_j}{a_j - a_{j-1}} b \left(\frac{y - a_{j-1}}{a_j - a_{j-1}} \right) \right] dy. \tag{10.5.6}$$

Setting $z = (y - a_{j-1})/(a_j - a_{j-1})$ gives

$$I = \sum_{j=1}^N s_j \int_0^1 f \left(a_{j-1} + z(a_j - a_{j-1}) \right) b(z) \, dz. \tag{10.5.7}$$

Note that the field stretching factor $1/(a_j - a_{j-1})$ gets exactly cancelled by dy/dz when changing variables in the integral. We can now define the adjoint Finn–Ott operator $\mathcal{G}_0^{\text{FO}\dagger}$ by

$$\left[\mathcal{G}_0^{\text{FO}\dagger} f \right](z) \equiv \sum_{j=1}^N s_j f \left(a_{j-1} + z(a_j - a_{j-1}) \right). \tag{10.5.8}$$

The adjoint of the diffusive operator is also easily found; where the original operator involves convolution with a heat kernel after the mapping, the adjoint involves convolution with the heat kernel before the mapping.

Inspection of (5.8) reveals that $f(z) \equiv 1$ is an eigenfunction of $\mathcal{G}_0^{\text{FO}\dagger}$ with eigenvalue $\lambda = \sum_{j=1}^N s_j$. This is the quantity we identified as the total flux growth rate in the original Finn–Ott model, but in that case there was no eigenfunction if $\eta = 0$, and only a very complicated one with $\eta > 0$. Working with the adjoint operator has magically picked out what appears to be exactly the right physical solution.

In this case it is easy to see what the adjoint operator is doing. Its eigenfunction $f(y) \equiv 1$ is the answer to the question: what function f, if any, has the property that

$$\int_0^1 f(y)\, [\mathcal{G}_0^{\text{FO}} b](y)\, dy = \mu \int_0^1 f(y) b(y)\, dy$$

for all smooth $b(y)$? Our initial discussion of the Finn–Ott operator pointed out that whatever the initial field $b(y)$, the total flux after the map is greater than the initial flux by a factor $\mu = \sum_{j=1}^N s_j$ — thus μ is the adjoint eigenvalue. The adjoint eigenfunction is simply the function which, when multiplied by $b(y)$ and integrated over the interval, yields the flux. I.e., the dominant adjoint eigenfunction is just the constant function.

Next consider the SFS operator. The adjoint operator (in the diffusionless case) is defined by considering the integral

$$
\begin{aligned}
I &= \int_0^1 f(y)\, [\mathcal{G}_0^{\text{SFS}} b](y)\, dy \\
&= \int_0^{1/2} f(y)\, 2 e^{2i\pi\alpha(y-1/2)}\, b(2y)\, dy \\
&\quad + \int_{1/2}^1 f(y)\, (-2) e^{2i\pi\alpha(y-1/2)}\, b(2-2y)\, dy.
\end{aligned}
\tag{10.5.9}
$$

Changing variables to $z = 2y$ in the first integral and $z = 2 - 2y$ in the second yields

$$I = \int_0^1 b(z) \left[f(z/2)\, e^{i\pi\alpha(z-1)} - f(1-z/2)\, e^{-i\pi\alpha(z-1)} \right] dz. \tag{10.5.10}$$

Hence the adjoint zero-diffusivity SFS operator may be identified as

$$\left[\mathcal{G}_0^{\text{SFS}\dagger} f \right](z) \equiv e^{i\pi\alpha(z-1)} f(z/2) - e^{-i\pi\alpha(z-1)} f(1-z/2). \tag{10.5.11}$$

Just as we could not analytically find the eigenvalues of the original SFS operator, we cannot solve the eigenproblem for the adjoint SFS operator. But numerical computations reveal that the dominant eigenfunctions are beautifully smooth functions (figure 10.10).

The interpretation of the adjoint eigenfunctions is the same as in the Finn–Ott example. The eigenfunctions of the adjoint operator answer the question: what function $f(y)$ has the property that

$$\int_0^1 f(y)\, [\mathcal{G}_0 b](y)\, dy = \mu \int_0^1 f(y) b(y)\, dy$$

for any function $b(y)$? The adjoint eigenfunction we obtain is not constant, so we cannot make an identification with a simple physical concept like total

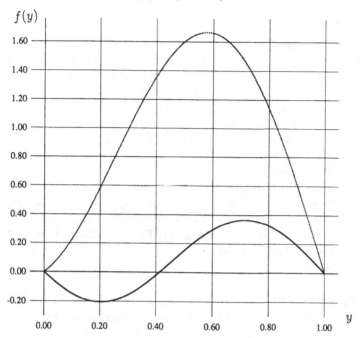

Figure 10.10 Dominant eigenfunction (real part solid, imaginary part dotted) of the adjoint SFS map at $\alpha = 1$. The computation uses $N = 1024$ cells, as in previous figures.

flux. But we can interpret the result as saying that the *weighted* flux (weighted with the adjoint eigenfunction) increases by a factor μ with every iteration of the operator.

The success of the adjoint formulation here is surprising, as the adjoint formulation results in hardly any simplification for more complicated maps and flows which are continuous in time. Indeed, Roberts (1967) shows that the adjoint of the magnetic field induction equation is just the evolution equation of the vector potential, with time going in the opposite direction. In general, the adjoint eigenfunctions are just as badly behaved as the original eigenfunctions, except that the orientation of the small-scale structure is in the dilating direction of the original flow rather than the contracting direction.

In the models of this section, however, the dilating direction of the map is along the x-axis, which *coincides* with the direction along which we were able to assume fields constant. Therefore there is no structure in the x-direction to get squeezed as the adjoint in effect advects the vector potential backwards in time. It is this fortuitous coincidence of directions that results in adjoint eigenfunctions with smooth dependence on y (which is the dilating direction for the time-reversed flow) and no dependence on x.

10.6 SUMMARY

The fast dynamo problem is still far from being solved, whether for flows or maps. Map dynamos have played an important role in reaching the current understanding of fast dynamos. They have facilitated advances both on the theoretical front and numerical computation. In the future, when the fast dynamo problem has been solved, we expect that the classic textbook examples will be also given in terms of maps, and remarks made to the effect that there exist continuous flows which do the same things!

Acknowledgments

I am most grateful to the Director and staff of the Isaac Newton Institute for their warm support and hospitality during my residence in the summer of 1992. The work on which these lectures are based is supported by the USA National Science Foundation under grants EAR-8902759 and DMS-9057124, and by the USA Air Force Office of Scientific Research under Contract No. FQ8671-900589.

REFERENCES

Arnol'd, V.I. & Avez, A. 1968 *Ergodic Problems of Classical Mechanics*. Benjamin.

Arnol'd, V.I. & Korkina, E.I. 1983 The growth of a magnetic field in a steady incompressible flow. *Vest. Mosk. Un. Ta. Ser. 1, Math. Mec.* **3**, 43–46.

Arnol'd, V.I., Zel'dovich, Ya.B., Ruzmaikin, A.A. & Sokoloff, D.D. 1981 A magnetic field in a stationary flow with stretching in Riemannian space. *Zh. Eksp. Teor. Fiz.* **81**, 2052–2058. (English transl.: *Sov. Phys. JETP* **54**, 1083–1086 (1981).)

Bajer, K. & Moffatt, H.K. 1990 On a class of steady confined Stokes flows with chaotic streamlines. *J. Fluid Mech.* **212**, 337–363.

Bayly, B.J. & Childress, S.C. 1988 Construction of fast dynamos using unsteady flows and maps in three dimensions. *Geophys. Astrophys. Fluid Dynam.* **44**, 211–240.

Bullard, E.C. & Gellman, H. 1954 Homogeneous dynamos and terrestrial magnetism. *Proc R. Soc. Lond. A* **247**, 213–278.

Collet, P. 1992. Personal communication.

Cowling, T.G. 1934 The magnetic field of sunspots. *Mon. Not. R. Astron. Soc.* **94**, 39–48.

Dittrich, P., Molchanov, S.A., Sokoloff, D.D. & Ruzmaikin, A.A. 1984 Mean magnetic field in renovating random flow. *Astron. Nachr.* **305**, 119–125.

Du Y. & Ott, E. 1993 Fractal dimensions of fast dynamo magnetic fields. *Physica D* **67**, 387–417.

Finn, J.M. & Ott, E. 1988 Chaotic flows and fast magnetic dynamos. *Phys. Fluids* **31**, 2992–3011.

Galloway, D. & Frisch, U. 1984 A numerical investigation of magnetic field generation in a flow with chaotic streamlines. *Geophys. Astrophys. Fluid Dynam.* **29**, 13–18.

Galloway, D. & Frisch, U. 1986 Dynamo action in a family of flows with chaotic streamlines. *Geophys. Astrophys. Fluid Dynam.* **36**, 53–83.

Galloway, D. & Proctor, M.R.E. 1992 Numerical calculations of fast dynamos for smooth velocity fields with realistic diffusion. *Nature* **356**, 691–693.

Gilbert, A.D. 1991 Fast dynamo action in a steady chaotic flow. *Nature* **350**, 483–485.

Gilbert, A.D. & Bayly, B.J. 1992 Intermittency and fast dynamo action in random helical waves. *J. Fluid Mech.* **241**, 199–214.

de la Llave, R. 1993 Hyperbolic dynamical systems and generation of magnetic fields by perfectly conducting fluids. *Geophys. Astrophys. Fluid Dynam.* **73**, 123–131.

Moffatt, H.K. 1978 *Magnetic Field Generation in Electrically Conducting Fluids.* Cambridge University Press.

Moffatt, H.K. & Proctor, M.R.E. 1985 Topological restrictions associated with fast dynamo action. *J. Fluid Mech.* **154**, 493–507.

Nuñez, M. 1993 Localised magnetic fields in a perfectly conducting fluid. In *Solar and Planetary dynamos* (ed. M.R.E. Proctor, P.C. Matthews & A.M. Rucklidge), pp. 225–228. Cambridge University Press.

Otani, N.J. 1988 Computer simulation of fast kinematic dynamos. *EOS, Trans. Amer. Geophys. Union* **69**(44), abstract SH51-15, p. 1366.

Poezd, A.D., Galeeva, R. & Sokoloff, D.D. 1992 Discrete model of a dynamo with diffusion. *Chaos* **2**, 253–256.

Roberts, G.O. 1972 Dynamo action of fluid motions with two-dimensional periodicity. *Proc R. Soc. Lond. A* **271**, 411–454.

Roberts, P.H. 1967 *An Introduction to Magnetohydrodynamics.* Elsevier.

Soward, A.M. 1987 Fast dynamo action in a steady flow. *J. Fluid Mech.* **180**, 267–295.

Thompson, M. 1990 Kinematic dynamo in random flow. *Mat. Aplic. Comp.* **9**(3), 213–245.

Vainshtein, S.I. & Zel'dovich, Ya.B. 1972 Origin of magnetic fields in astrophysics. *Sov. Phys. Usp.* **15**, 151–172.

Vishik, M.M. 1989 Magnetic field generation by the motion of a highly conducting fluid. *Geophys. Astrophys. Fluid Dynam.* **48**, 151–167.

Vishik, M.M. 1992 Personal communication.

Zel'dovich, Ya.B. 1957 The magnetic field in the two-dimensional motion of a conducting turbulent fluid. *Sov. Phys. JETP* **4**, 460–462.

CHAPTER 11

Bifurcations in Rotating Systems

E. KNOBLOCH

Department of Physics
University of California
Berkeley, CA 94720, USA

11.1 INTRODUCTION

This chapter is devoted to understanding the nature of the transitions that
are possible in rotating systems. Rotation is implicated in most instabilities
of astrophysical and geophysical interest. These include, for example, the
baroclinic instability responsible for the formation of weather fronts in the
earth's atmosphere, the instability that forms the spiral arms of galaxies,
and of course the dynamo instability. The approach we take emphasizes
generic, i.e., model-independent, behaviour. As a result the discussion that
follows focuses on the symmetries of the system which are often responsible
for much of the observed behaviour. As such we do not address specific
physical mechanisms that give rise to the instabilities, or even specific model
equations that might be used to describe them. Nonetheless we find that the
approach used provides a number of new insights into the type of dynamics
that are characteristic of rotating systems. In addition it points out the
shortcomings of local studies of rotating systems that have been used to
simplify the analysis. Moreover, since the results are model-independent,
they apply to any system sharing the same symmetry properties. Thus our
results shed light not only on the possible transitions in dynamo theory, but
also on those occurring in baroclinic and other rotating flows.

We begin by pointing out that a rotating cylinder and a rotating sphere
have the *same* symmetry: both are invariant under proper rotations about the
rotation axis. In fact any figure of revolution rotating about its axis has this
symmetry. For a solid body the meaning of this statement is quite intuitive.
It says that the motion of a point in the body looks the same in one frame as
in another one rotated by a fixed angle θ relative to it. In the following we
shall be interested in studying *motions* in rotating bodies and so it pays to
be a little more precise. Let R_θ denote a rotation of the cartesian axes fixed
in the body through an angle θ about the rotation axis z, and suppose that

M.R.E. Proctor & A.D. Gilbert (eds.)
Lectures on Solar and Planetary Dynamos, 331–372
©1994 Cambridge University Press.

a point P in the body has coordinates (x, y, z) in the first coordinate system and (x', y', z') in the rotated system. Then $(x', y', z') = R_\theta(x, y, z)$ with R_θ given by

$$R_\theta = \begin{pmatrix} \cos\theta & \sin\theta & 0 \\ -\sin\theta & \cos\theta & 0 \\ 0 & 0 & 1 \end{pmatrix}. \tag{11.1.1}$$

If, in addition, $\mathbf{u} \equiv (u, v, w)$ and $\mathbf{u}' \equiv (u', v', w')$ denote the velocities in the two coordinate systems,

$$\mathbf{u}' = R_\theta \mathbf{u}(R_\theta^{-1}(x', y', z')). \tag{11.1.2a}$$

The statement that the system is invariant under rotations means that if (u, v, w) is a solution of the governing equations (and boundary conditions!), so is (u', v', w'). In contrast a scalar quantity such as the pressure $p(x, y, z)$ transforms into

$$p' = p(R_\theta^{-1}(x', y', z')). \tag{11.1.2b}$$

It is important to bear this distinction in mind.

A nonrotating cylinder has an additional reflection symmetry which we denote by R_1. This is a reflection in any vertical plane through the rotation axis. In terms of cylindrical polar coordinates, $R_1(r, \phi, z) = (r, -\phi, z)$, and a system is invariant under R_1 if $R_1(u(R_1^{-1}(r, \phi, z)), v(R_1^{-1}(r, \phi, z)), w(R_1^{-1}(r, \phi, z)))$, i.e., $(u(r, -\phi, z), -v(r, -\phi, z), w(r, -\phi, z))$, is a solution of the governing equations if $(u(r, \phi, z), v(r, \phi, z), w(r, \phi, z))$ is one. Here (u, v, w) now denote the velocity components in polar coordinates. Note that we are always talking about the symmetries of the solutions of a given set of equations. Thus the system itself is not subjected to a reflection, only the solutions are. In particular when we are interested in the effects of the reflection R_1 on a solution of the rotating problem we do *not* change the direction of rotation of the system. If the specification of the state of the system involves other fields like the temperature, pressure or magnetic field appropriate transformations must be applied to these quantities as described by equations (1.2).

Symmetries are classified in terms of the groups of symmetries they produce. The symmetry with respect to proper rotations is called SO(2) symmetry, and the corresponding symmetry group consists of all rotations of the form R_θ. The symmetry of a nonrotating cylinder is described by a group that consists of the rotations R_θ, the reflections R_1, and their products. This group is called O(2). The group of symmetries of a nonrotating sphere is even larger, and is called O(3). We shall meet other groups of symmetries as we go along. The fact that the symmetries of a rotating cylinder and sphere are the same implies that the dynamical behaviour in both systems will be qualitatively similar. In particular it is the symmetry SO(2) that is responsible for the ubiquity of precessing patterns in rotating systems. By a precessing pattern we shall mean a pattern that drifts in the rotating frame. An obvious example is the westward drift of the earth's magnetic field or the motion of weather fronts in the atmosphere.

The rotating cylinder and sphere have another symmetry as well: they are invariant under reflection in the equatorial plane. This reflection, which we denote by R_3 generates the group Z_2, i.e., $Z_2 = \{1, R_3\}$. Thus if $(u(r, \phi, z)$, $v(r, \phi, z), w(r, \phi, z))$ is a solution of the equations so is

$$R_3 \left(u(R_3^{-1}(r, \phi, z)), v(R_3^{-1}(r, \phi, z)), w(R_3^{-1}(r, \phi, z)) \right)$$
$$\equiv (u(r, \phi, -z), v(r, \phi, -z), -w(r, \phi, -z)).$$

Consequently the symmetry of a rotating cylinder or sphere is actually the group $SO(2) \times Z_2$. It is important to appreciate that although this group also consists of rotations and reflections, it differs from the group $O(2)$. In the former the rotations and reflections act independently of one another. In the latter this is not the case. It will prove useful to bear the above examples in mind when reading the remainder of this chapter.

The solutions of equations that have a symmetry need not share that symmetry. Instabilities that break the symmetry of a system are called symmetry breaking instabilities, and are associated with the formation of patterns. For example the solar magnetic activity cycle, as summarized by the 'butterfly diagram', breaks the equatorial symmetry of the Sun. On a finer scale the instability that causes the eruption of toroidal magnetic fields that form sunspots breaks the rotational symmetry of the Sun. Of course if one has a system with symmetry and finds a solution with lower symmetry then one can use the elements of the symmetry group to generate distinct but equivalent solutions. This multiplicity of solutions is the most important feature of systems with symmetry. Additional background and information may be found in the review by Crawford & Knobloch (1991).

In the following section we explore the origin of precessing patterns from the above point of view. Section 11.3 is devoted to studying mode interaction in systems with $SO(2)$ and $SO(2) \times Z_2$ symmetry. The conclusions of the abstract analysis are illustrated throughout with examples drawn from various problems of geophysical and laboratory interest, as well as from the dynamo problem.

11.2 ORIGIN OF PRECESSION

In this section we explain why precession is so ubiquitous in rotating systems and why it is absent in many models of such systems. Although the discussion may appear at first sight to be somewhat abstract we will see that it ultimately throws light on the type of approximations that can be made without losing the essence of the problem, and others that should not be made, when modelling such systems. Throughout this section we use a simple system, convection in a rotating cylinder, to illustrate both the theory and the effects of various approximations on both the linear and nonlinear properties of this system.

In order to appreciate the distinction between rotating and nonrotating systems it is important to have first a thorough understanding of the nonrotating system. Suppose that such a system possesses a time-independent equilibrium that is invariant under the full symmetry of the system, i.e., under rotations about the vertical axis of the cylinder and reflections in any vertical plane through the axis. Such a state will be called the trivial solution. It shares the full symmetry of the problem. In convection the conduction state is trivial in this sense. As a parameter is varied the trivial state may lose stability either to another state with full symmetry or to a state of lower symmetry. For example, in a circular cylinder the conduction solution may lose stability to a pattern of concentric rolls. Such a pattern is still invariant under both rotations and reflections, and hence is also a trivial state. However, if the conduction solution loses stability to a nonaxisymmetric pattern, the new pattern is no longer invariant under rotations, and hence has a lower degree of symmetry.

Since the instability takes place in a system invariant under rotations and reflections, we know that if we have a solution of the governing equations and subject it to a reflection or an arbitrary rotation we will generate another solution of the equations. This solution may either be the same as the original one or it may be distinct. In the above example the rotation of a nonaxisymmetric solution generates a new solution. It is not hard to capture this requirement mathematically. For this purpose it is most convenient to select a scalar property of the system, like the temperature, and ask how the solution transforms under the operations that leave the system unchanged. Dealing with a component of the velocity is a little more confusing since it is not a scalar (compare equations (1.2a) and (1.2b)). Suppose the instability takes place when a parameter μ vanishes, and that at $\mu = 0$ a single real eigenvalue passes through zero. For example if the destabilizing parameter is the Rayleigh number, we take $\mu \equiv (R - R_c)/R_c$, where R_c is the critical Rayleigh number for the instability of the mode in question. Because of periodicity in the azimuthal direction the marginally stable mode is characterized by a well defined azimuthal wavenumber m. At $\mu = 0$ the temperature eigenfunction thus takes the form

$$\Theta_m(r, \phi, z, t) = \Re\{a_m(t)e^{im\phi}f_m(r, z)\}, \qquad (11.2.1a)$$

where

$$\dot{a}_m = 0. \qquad (11.2.1b)$$

Here (r, ϕ, z) are cylindrical coordinates, $f_m(r, z)$ is the eigenfunction of the mode m and $a_m(t)$ is its (complex) amplitude. When $m = 0$ the eigenfunction is axisymmetric; when $m \neq 0$ it is nonaxisymmetric. The difference between these two cases is fundamental. In the following we shall be interested in understanding the behaviour of the instability for slightly supercritical μ. In order to understand the saturation of the instability for $\mu > 0$ we must add to equation (2.1b) not only terms that are responsible for the growth of the instability, but also the nonlinear terms that result in its saturation. The structure of both types of terms is restricted by the symmetries of the system. When the instability breaks azimuthal symmetry the requirement that a rotated solution also be a solution translates into the requirement that the equation satisfied by the amplitude a_m be unchanged under the operation

$$\text{rotations} \quad \phi \to \phi + \theta: \quad a_m \to a_m e^{im\theta}, \qquad (11.2.2a)$$

while reflection symmetry leads to the requirement that the equation be invariant under complex conjugation:

$$\text{reflections} \quad \phi \to -\phi: \quad a \to \bar{a}. \qquad (11.2.2b)$$

These requirements follow directly from the structure of the eigenfunction. Consequently the nature of the eigenfunction, i.e., the solution of the linear problem, determines to a large extent its *nonlinear* evolution. The most general equation that is invariant under the operations (2.2a, b) has the form

$$\dot{a}_m = g_m(\mu, |a_m|^2) a_m, \qquad (11.2.3)$$

where the function g_m is forced by (2.2b) to be real. Since μ and hence a_m is small, we may expand g_m in a Taylor series,

$$\dot{a}_m = \mu a_m + \alpha_m |a_m|^2 a_m + \cdots. \qquad (11.2.4)$$

Here α_m is a real coefficient that may be taken to be independent of μ provided $\alpha_m(\mu = 0) \neq 0$. The resulting equation governs the evolution of any nonaxisymmetric instability in a nonrotating circular cylinder. It is helpful to write it in terms of a real amplitude A_m and a phase Φ_m, defined by $a_m = A_m e^{i\Phi_m}$:

$$\dot{A}_m = \mu A_m + \alpha_m A_m^3 + \cdots, \quad \dot{\Phi}_m = 0. \qquad (11.2.5)$$

The equation $\dot{\Phi}_m = 0$ is forced by the reflection symmetry of the system and embodies the requirement that the resulting pattern be neutrally stable with respect to rotations. The resulting bifurcation is called a pitchfork of revolution. Note that at $\mu = 0$ the linear stability problem has in fact two zero eigenvalues. This is because the symmetry (2.2a) requires the use of a complex amplitude. In contrast, if the initial instability is to an axisymmetric mode, no symmetry is broken and the circular nature of the system exerts no influence on the dynamics. In this case the amplitude a_0 is real and obeys the equation

$$\dot{a}_0 = g_0(\mu, a_0). \tag{11.2.6}$$

In many applications the eigenfunction $f_0(r, z)$ of the mode that is first unstable satisfies the condition $f_0(r, -z) = -f_0(r, z)$. As a result the equation for a_0 inherits the additional requirement that it be invariant under the operation $a_0 \to -a_0$. In this case a_0 satisfies

$$\dot{a}_0 = \mu a_0 + \alpha_0 a_0^3 + \cdots \tag{11.2.7}$$

and the resulting bifurcation is a pitchfork. If this is not the case the bifurcation to an axisymmetric mode will be a transcritical one.

When the cylinder is rotated about the vertical with angular velocity Ω the reflection symmetry (2.2b) is broken. Consequently the function g_m in equation (2.3) acquires an imaginary part, and (2.4) becomes

$$\dot{a}_m = (\mu + i\Omega\delta_m)a_m + (\alpha_m + i\Omega\beta_m)|a_m|^2 a_m + \cdots, \tag{11.2.8}$$

where δ_m, α_m and β_m are now real-valued functions of Ω^2. In terms of the real variables A_m, Φ_m we now have

$$\dot{A}_m = \mu A_m + \alpha_m A_m^3 + \cdots, \qquad \dot{\Phi}_m = \Omega(\delta_m + \beta_m A_m^2 + \cdots). \tag{11.2.9}$$

Consequently the broken reflection symmetry turns the steady-state bifurcation in the nonrotating system into a Hopf bifurcation in the rotating system (cf. Ruelle 1973). Moreover, since $\dot{\Phi}_m$ is the rate of change of the azimuthal phase, it is to be identified with the precession frequency ω_m in the rotating frame. In particular, Θ_m takes the form

$$\Theta_m = \Re\left\{A_m e^{i(m\phi + \omega_m t)} f_m(r, z)\right\} + \cdots. \tag{11.2.10}$$

The bifurcation is thus to a *rotating wave* (Chossat 1982, Rand 1982), i.e., the resulting pattern is invariant under rotations followed by a suitable evolution either forwards or backwards in time. Another way to state this result is that the pattern rotates rigidly. There is therefore a comoving reference frame in which the pattern appears steady. A more abstract derivation of this result is given by Golubitsky, Stewart & Schaeffer (1988). The resulting pattern is also neutrally stable with respect to rotations. A simple consequence of equations (2.9) is that the precession frequency of a steady-state pattern is given by

$$\omega_m = \Omega\left(\delta_m - \frac{\beta_m}{\alpha_m}\mu\right) + O(\mu^2). \tag{11.2.11}$$

For convection in a rotating cylinder the dependence of ω_m on Ω and μ suggested by the above theory has been verified experimentally (Ecke, Zhong

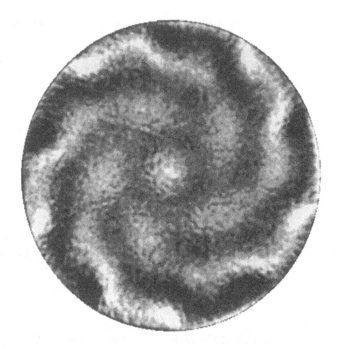

Figure 11.1 Shadowgraph image of an $m = 5$ state at $\mu = 2.64$ in a $\Gamma = 1$ cylinder rotating with dimensionless angular velocity $\Omega = 2145$. The fluid is water ($\sigma = 6.4$). The cylinder rotates in the anticlockwise direction; the pattern precesses in the clockwise direction. (Courtesy R. Ecke.)

& Knobloch 1992), and the experimental fits can be used to determine the values of δ_m and β_m/α_m as a function of Ω. In the following we denote the precession frequency at onset ($\mu = 0$) by $\omega_c^{(m)}(\equiv \Omega\delta_m)$.

The discussion presented above shows that whenever a pattern breaks the rotational invariance of a rotating circularly symmetric system, it will necessarily precess in the rotating frame. Only at exceptional parameter values can such a pattern remain steady in the rotating frame (cf. Goldstein *et al.* 1993b). Note that the patterns necessarily lack reflection symmetry in vertical planes, i.e., they take the form of generalized spirals. We illustrate this point in figure 11.1 with a shadowgraph image of convection in a rotating cylinder from Ecke *et al.* (1992). Spiral patterns have also been found in convection in rotating spheres (Zhang 1992).

11.2.1 Effect of Rotation on Other Bifurcations

We now consider the case of a Hopf bifurcation in a system with circular symmetry. We consider two cases, first the nonrotating case with O(2) symmetry followed by the rotating case with SO(2) symmetry. We do not discuss the group O(3). In a nonrotating system, an axisymmetric mode at onset can be written in the form

$$\Theta_0(r, z, t) = \Re\{a_0(t) f_0(r, z)\}, \qquad (11.2.12)$$

where

$$\dot{a}_0 = i\omega_c^{(0)} a_0 \qquad (11.2.13)$$

and $\omega_c^{(0)}$ is the Hopf frequency. Since the mode does not break the O(2) symmetry of the nonrotating system the dynamics of the amplitude a_0 are not affected by the symmetry of the container. The bifurcation at $\mu = 0$ is therefore the standard Hopf bifurcation, and is described by the normal form equation

$$\dot{a}_0 = (\mu + i\omega_0)a_0 + \alpha_0|a_0|^2 a_0 + \cdots, \qquad (11.2.14)$$

where $\omega_0 - \omega_c^{(0)} = O(\mu)$, and α_0 is a complex constant, independent of μ (provided $\Re\alpha_0(\mu = 0) \neq 0$). Equation (2.14) is identical to equation (2.8) but the reason for this is entirely different. The symmetry responsible for the structure of (2.14) is the so-called 'normal form' symmetry. This is a phase shift symmetry

$$t \to t + \tau: \quad a_0 \to a_0 e^{i\omega\tau}, \qquad (11.2.15)$$

where $2\pi/\omega$ is the period of the periodic solution created in the Hopf bifurcation. This symmetry is manifest for any periodic solution. In normal form the same symmetry is shared by the *vector field*, i.e., by the right-hand side of equation (2.14). The normal form is constructed in a series of near-identity nonlinear transformations of the dynamical equation for a_0 chosen so as to lead to the simplest possible equation for a_0 (see, for example, Guckenheimer & Holmes 1986). For the present problem this is equation (2.14).

When the instability breaks the rotational symmetry of the system equation (2.12) is replaced by

$$\Theta_m(r, \phi, z, t) = \Re\{(v_m(t) + w_m(t))e^{im\phi} f_m(r, z)\}, \qquad (11.2.16)$$

where

$$\begin{pmatrix} \dot{v}_m \\ \dot{w}_m \end{pmatrix} = \begin{pmatrix} i\omega_c^{(m)} & 0 \\ 0 & -i\omega_c^{(m)} \end{pmatrix} \begin{pmatrix} v_m \\ w_m \end{pmatrix}. \qquad (11.2.17)$$

Equation (2.16) is an arbitrary superposition of anticlockwise ($v_m = 0$) and clockwise ($w_m = 0$) waves. Since the instability now breaks the symmetry of the container, the symmetry constrains the dynamics of the amplitudes (v_m, w_m) near the bifurcation. Rotations and reflection act on (v_m, w_m) as follows:

$$\text{rotations} \quad \phi \to \phi + \theta: \quad (v_m, w_m) \to e^{im\theta}(v_m, w_m), \qquad (11.2.18a)$$

$$\text{reflection} \quad \phi \to -\phi: \quad (v_m, w_m) \to (\bar{w}_m, \bar{v}_m). \qquad (11.2.18b)$$

The most general amplitude equation commuting with these operations takes the form

$$\begin{pmatrix} \dot{v}_m \\ \dot{w}_m \end{pmatrix} = \begin{pmatrix} g_1 & g_2 \\ \bar{g}_2 & \bar{g}_1 \end{pmatrix} \begin{pmatrix} v_m \\ w_m \end{pmatrix}, \tag{11.2.19}$$

where $g_j = g_j(\sigma_1, \sigma_2, \sigma_3)$, $j = 1, 2$, are complex-valued functions of $\sigma_1 \equiv |v_m|^2 + |w_m|^2$, $\sigma_2 \equiv v_m \bar{w}_m$ and $\sigma_3 \equiv \bar{v}_m w_m$, as well as of the bifurcation parameter μ. As before these equations can be simplified by an appropriate near-identity nonlinear change of variables. Applied at $\mu = 0$ this procedure identifies the essential nonlinear terms as the ones that cannot be removed by any such transformation. These terms commute with the normal form symmetry, $(v_m, w_m) \rightarrow (e^{i\omega_m \tau} v_m, e^{-i\omega_m \tau} w_m)$, the generalization of (2.15) appropriate to this problem. Restoring the small linear terms now yields

$$\begin{aligned} \dot{v}_m &= (\mu + i\omega_m + \beta_m |w_m|^2 + \alpha_m A_m^2) v_m + \cdots \\ \dot{w}_m &= (\mu - i\omega_m + \bar{\beta}_m |v_m|^2 + \bar{\alpha}_m A_m^2) w_m + \cdots, \end{aligned} \tag{11.2.20}$$

where α_m and β_m are complex constants and A_m is the total amplitude defined by $A_m^2 \equiv |v_m|^2 + |w_m|^2$. In terms of the real variables defined by $v_m \equiv r_1 e^{i\theta_1}$ and $w_m \equiv r_2 e^{i\theta_2}$ one now finds

$$\dot{r}_1 = (\mu + \Re \beta_m r_2^2 + \Re \alpha_m A_m^2) r_1 + \cdots \tag{11.2.21a}$$

$$\dot{r}_2 = (\mu + \Re \beta_m r_1^2 + \Re \alpha_m A_m^2) r_2 + \cdots \tag{11.2.21b}$$

$$\dot{\theta}_1 = \omega_m + \Im \beta_m r_2^2 + \Im \alpha_m A_m^2 + \cdots \tag{11.2.21c}$$

$$\dot{\theta}_2 = -\omega_m - \Im \beta_m r_1^2 - \Im \alpha_m A_m^2 + \cdots. \tag{11.2.21d}$$

Note that the equations for the phases θ_1 and θ_2 decouple from the equations for the amplitudes r_1 and r_2. This is a consequence of the invariance of equations (2.20) under the symmetry transformation (2.18a) *together* with the normal form symmetry. As a result the dynamics of (2.20) are completely specified by the two-dimensional system (2.21a, b). When $\Re \alpha_m \neq 0$, $\Re \beta_m \neq 0$ and $\Re(2\alpha_m + \beta_m) \neq 0$ this system can be truncated at third order. There are then three nontransient solutions, the trivial solution $(|v_m|, |w_m|) = (0, 0)$, the rotating wave $(|v_m|, |w_m|) = (A_m, 0)$ or $(0, A_m)$, and the standing wave $(|v_m|, |w_m|) = (A_m, A_m)/\sqrt{2}$. Note that here two solution branches, one of rotating waves and the other of standing waves, bifurcate simultaneously from the trivial solution. This is a consequence of the reflection R_1 in the symmetry group $O(2)$. This symmetry also identifies the clockwise and anticlockwise travelling waves, so that neither one or other direction is preferred. It is now easy to show (Knobloch 1986a) that at most one of these solutions is stable, and that a stable solution exists if and only if both the rotating and standing waves bifurcate supercritically. The stable solution is the one with the largest amplitude A_m.

Note that at the bifurcation at $\mu = 0$ the multiplicity of the purely imaginary eigenvalues is two. This, and the resulting multiple bifurcation, is a consequence of $O(2)$ symmetry, and arises because the reflection symmetry in $O(2)$ implies that if $e^{im\phi}$ is an eigenfunction, so is $e^{-im\phi}$. Rotation, however, breaks the reflection symmetry, and consequently the two pairs of eigenvalues split into two separate Hopf bifurcations. One can show that for $|\Omega| \ll 1$ the system is now described by equations that, without loss of generality, can be written in the form

$$\dot{v}_m = (\mu + \Omega\delta_{1m} + i\omega_m + \beta_m|w_m|^2 + \alpha_m A_m^2)v_m + \cdots \qquad (11.2.22a)$$

$$\dot{w}_m = (\mu + \Omega\bar{\delta}_{2m} - i\omega_m + \bar{\beta}_m|v_m|^2 + \bar{\alpha}_m A_m^2)w_m + \cdots, \qquad (11.2.22b)$$

where δ_{1m} and δ_{2m} are complex constants. Note that the resulting equations break the reflection invariance (2.18b) (if $\delta_{1m} \neq \delta_{2m}$), but respect the rotational invariance (2.18a). The linear terms we have added change the linear stability problem for the trivial solution $(0,0)$: the clockwise rotating wave now sets in at $\mu = -\Omega\Re\delta_{1m}$, while the anticlockwise one requires $\mu = -\Omega\Re\delta_{2m}$. Consequently, in the presence of rotation one wave or the other is preferred. The following conclusions are not difficult to establish for the nonlinear solutions (van Gils & Mallet-Paret 1986, Crawford & Knobloch 1988a). Since the reflection symmetry is now absent standing waves cannot be a solution of the problem; instead their role is taken by a branch of *quasiperiodic* (i.e., two frequency) waves that bifurcates in a secondary bifurcation from one or other of the rotating wave branches. Owing to the rotational invariance of the system, there can be no frequency locking in such waves. This is because in such a system one can go into a comoving reference frame in which one of the frequencies vanishes. In that frame there is then nothing for the remaining frequency to lock to. In figure 11.2 we show two typical bifurcation diagrams that arise in a nonrotating system and the way they break up when rotation is introduced.

In the above case it was sufficient to include the effects of rotation at linear order in (v_m, w_m). Near the degeneracies $\Re\alpha_m = 0$, $\Re\beta_m = 0$ or $\Re(2\alpha_m + \beta_m) = 0$ it is necessary, however, to include symmetry breaking effects in the coefficients of the nonlinear terms as well (Crawford & Knobloch 1988a).

Finally, we consider the Takens–Bogdanov (TB) bifurcation. This bifurcation arises when the Hopf frequency in the above example vanishes. When this is the case the linear problem then has two zero eigenvalues, each of which is doubled if the mode breaks the $O(2)$ symmetry of the system. To obtain such a bifurcation it is necessary to have a second parameter that can be tuned independently of μ. Consequently the TB bifurcation is an example of a codimension-two bifurcation. The general form of the linear problem is now

$$\Theta_m(r, \phi, z, t) = \Re\{v_m(t)e^{im\phi}f_m(r, z)\}, \qquad (11.2.23)$$

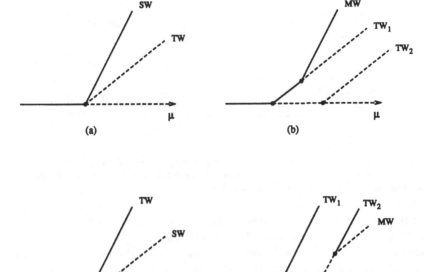

Figure 11.2 The effect of rotation on the bifurcation diagrams for the Hopf bifurcation with O(2) symmetry. (a) The transition to stable standing waves in the nonrotating case, and (b) the corresponding bifurcation diagram in the rotating case. (c) and (d) show the same but for stable travelling waves. The solutions labelled MW are quasiperiodic. The diagrams show the square of the total amplitude as a function of μ. The solid (broken) lines indicate stable (unstable) branches.

where

$$\begin{pmatrix} \dot{v}_m \\ \dot{w}_m \end{pmatrix} = \begin{pmatrix} 0 & 1 \\ 0 & 0 \end{pmatrix} \begin{pmatrix} v_m \\ w_m \end{pmatrix}. \qquad (11.2.24)$$

When $m \neq 0$ the O(2) symmetry of the system acts on (v_m, w_m) as follows:

$$\text{rotations} \quad \phi \to \phi + \theta : \quad (v_m, w_m) \to e^{im\theta}(v_m, w_m), \qquad (11.2.25)$$

$$\text{reflection} \quad \phi \to -\phi : \quad (v_m, w_m) \to (\bar{v}_m, \bar{w}_m). \qquad (11.2.26)$$

The use of the same kind of techniques already alluded to now yields the truncated normal form equations

$$\dot{v}_m = w_m \qquad (11.2.27a)$$

$$\dot{w}_m = \mu v_m + \nu w_m + A|v_m|^2 v_m + B|w_m|^2 v_m + C(v_m \bar{w}_m + \bar{v}_m w_m)v_m + D|v_m|^2 w_m, \qquad (11.2.27b)$$

where μ, ν are two real unfolding parameters specifying the distance from the TB point in the two-dimensional parameter space, and A, B, C and D

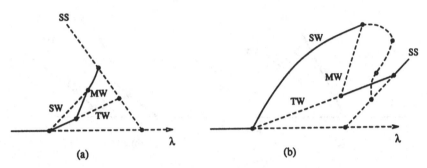

Figure 11.3 Two typical bifurcations diagrams arising in the unfolding of the Takens-Bogdanov bifurcation with O(2) symmetry. The solid dots indicate local bifurcations, while the open circles indicate global bifurcations.

are real coefficients. The dynamics of the resulting equations are understood in complete detail (Dangelmayr & Knobloch 1987a). Depending on the parameters μ, ν the trivial solution undergoes either a Hopf or a steady state bifurcation. Near these points the system behaves as already described. However, the proximity of the Hopf and steady state bifurcations implies that in addition to this behaviour new secondary bifurcations occur at small amplitudes and are captured in the unfolding (2.27). These bifurcations enable us to describe the interaction among the rotating waves, the standing waves and steady motion. Two of the possible scenarios are shown in figure 11.3. Note in particular that the rotating waves terminate on the branch of steady states. As this happens the precession frequency of the wave decreases to zero as the square root of the distance from its termination point. In addition secondary bifurcations to quasiperiodic waves are possible. In the present case such waves can propagate in either direction, in contrast to similar waves arising in the Hopf bifurcation with broken O(2) symmetry discussed above.

In the presence of rotation equations (2.27) become, without loss of generality,

$$\dot{v}_m = w_m \qquad (11.2.28a)$$

$$\dot{w}_m = (\mu + i\Omega\gamma_m)v_m + (\nu + i\Omega\delta_m)w_m + A|v_m|^2 v_m + B|w_m|^2 v_m + \\ + C(v_m\bar{w}_m + \bar{v}_m w_m)v_m + D|v_m|^2 w_m, \qquad (11.2.28b)$$

where now μ, ν, γ, δ, A, B, C, D are all functions of Ω^2. Locally, near the Hopf and steady state bifurcations, the bifurcation diagrams deform as already described. The standing waves are no longer present and their presence is taken by quasiperiodic waves produced in a secondary bifurcation from one of the rotating wave branches. The steady states are also no longer present, and the corresponding solutions precess. These connect smoothly to one or other of the rotating wave branches created from the break up of the Hopf bifurcation (see figure 11.4). For a full discussion of this problem, see

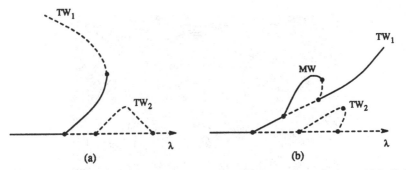

Figure 11.4 The effect of rotation on the TW and SS branches shown in figure 11.3. Only some of the quasiperiodic states are shown.

Hirschberg & Knobloch (1993a).
We saw above that precessing patterns can exist in nonrotating systems provided there is no preferred direction of precession. In these examples such solutions came about through a Hopf bifurcation. However, precessing patterns can arise spontaneously even in the absence of such a bifurcation. The general mechanism was elucidated by Crawford & Knobloch (1988b): an axisymmetric pattern loses stability to a reflection symmetric pattern as a parameter is increased. Owing to the circular symmetry of the system this pattern is neutrally stable with respect to rotations. If, with further increase in the parameter, this pattern loses stability to a perturbation that breaks its reflection symmetry the pattern will start to drift, either clockwise or anti-clockwise, depending on the perturbation. This behaviour is beautifully illustrated in the experiments of Steinberg, Ahlers & Cannell (1985) on Rayleigh–Bénard convection in a container. Under certain conditions Rayleigh–Bénard convection in a nonrotating circular container can take the form of beautiful spirals (Bodenschatz et al. 1991). An example is shown in figure 11.5. Since these spirals break reflection symmetry these too must precess. In the experiments the spirals have precessed rigidly, without change of shape, for a number of days. Only in nongeneric circumstances is it possible for such spirals to be stationary (cf. Bestehorn et al. 1992). The above mechanism also explains the nature of the secondary bifurcation with which the branch of precessing patterns terminates on the steady state branch in figure 11.3.

11.2.2 Convection in a Rotating Cylinder
In this section we discuss how to calculate the precession rate at onset of a nonaxisymmetric pattern, i.e., the coefficient δ_m in equation (2.8). We do not discuss here the nonlinear problem except to emphasize that the precession frequency (2.11) is a function of the amplitude of the pattern. We consider Boussinesq convection in a vertical right circular cylinder of radius d and height h, filled with a pure fluid and rotating with constant and uniform

344 E. *Knobloch: Bifurcations in Rotating Systems*

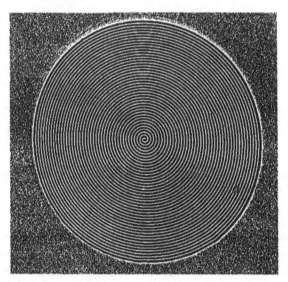

Figure 11.5 A precessing $m = 1$ spiral pattern in non-Boussinesq convection in a nonrotating circular container. The spiral terminates in a defect that rotates with the same angular velocity as the tip. (Courtesy G. Ahlers.)

angular velocity Ω_{phys} about its axis. We denote by Γ its aspect ratio d/h. For simplicity we suppose that the rotation rate is sufficiently small that the Froude number $d\Omega_{phys}^2/g$ is small and the buoyancy force continues to act in the vertical direction. The linearized, nondimensionalized equations of motion take the form (Chandrasekhar 1961)

$$\left(\frac{1}{\sigma}\partial_t - \nabla^2\right)\mathbf{u} = -\nabla p + R\Theta \mathbf{1}_z + 2\Omega \mathbf{u} \times \mathbf{1}_z, \quad (11.2.29a)$$

$$(\partial_t - \nabla^2)\Theta = w, \quad (11.2.29b)$$

$$\nabla \cdot \mathbf{u} = 0, \quad (11.2.29c)$$

where $\mathbf{u} = u\mathbf{1}_r + v\mathbf{1}_\phi + w\mathbf{1}_z$ is the velocity field, Θ and p are the departures of the temperature and pressure from their conduction profiles, and $\mathbf{1}_z$ is the unit vector in the vertical direction. The quantities $\Omega \equiv \Omega_{phys}h^2/\nu$, $R \equiv g\alpha\Delta T h^3/\kappa\nu$, and $\sigma \equiv \nu/\kappa$ denote, respectively, the dimensionless angular velocity, the Rayleigh number, and the Prandtl number. In these equations, length is in units of the layer thickness, h, and time is in units of the vertical thermal diffusion time, h^2/κ.

For simplicity we use free-slip, impenetrable, perfectly conducting boundary conditions at the top and bottom, and rigid, impenetrable, insulating boundary conditions along the curved sidewall:

$$\partial_z u = \partial_z v = w = \Theta = 0 \quad \text{on } z = 0, 1, \quad (11.2.30a)$$

$$u = v = w = \partial_r\Theta = 0 \quad \text{on } r = \Gamma. \quad (11.2.30b)$$

These boundary conditions are mathematically convenient in that they allow separation of variables in the vertical and azimuthal directions. We use them here in order to compare the essentially exact results obtained by Goldstein et al. (1993a, b) for the above problem with existing solutions to various approximations to the above system.

To solve the problem (2.29, 30) recall that the conduction solution, $u = v = w = \Theta = 0$, is stable to small perturbations with azimuthal wavenumber m below some critical value of the Rayleigh number, $R_c^{(m)}$, which depends, in general, on the aspect ratio, the rotation rate, the Prandtl number, as well as the boundary conditions and the wavenumber m. In other words, if we write the time dependence of a solution to the linear problem as $e^{s_m t}$, then $\Re(s_m) < 0$ for all solutions when $R < R_c^{(m)}$. At $R = R_c^{(m)}$ there is for the first time a neutrally stable solution to the linear problem with wavenumber m, i.e., $\Re(s_m) = 0$. If $\Im(s_m) = 0$, the bifurcation is steady-state, and if $\Im(s_m) = \omega_c^{(m)} \neq 0$, we have a Hopf bifurcation with Hopf frequency $\omega_c^{(m)}$.

11.2.3 Method of Solution
In this section we provide a brief description of the method used by Goldstein et al. (1993a) to obtain an essentially exact solution of the problem (2.29, 30). Equations (2.29) are written as a single equation for Θ:

$$\left[\left(\frac{1}{\sigma}\partial_t - \nabla^2\right)^2 (\partial_t - \nabla^2)\nabla^2 - R\nabla_h^2\left(\frac{1}{\sigma}\partial_t - \nabla^2\right) + 4\Omega^2\partial_z^2(\partial_t - \nabla^2)\right]\Theta = 0,$$
(11.2.31)

where ∇_h^2 is the two-dimensional Laplacian in the horizontal coordinates. In view of the boundary conditions (2.30a) the solution of (2.31) takes the form (cf. equation (2.1a))

$$\Theta_{km}(r, \phi, z; k) = \Re\{J_m(kr)e^{i(m\phi+\omega t)}\}\sin(\pi z),$$
(11.2.32)

where m is a non-zero integer and $k \equiv k(R, \omega, \Omega, \sigma)$ is a solution to the dispersion relation,

$$(\frac{i\omega}{\sigma}+k^2+\pi^2)^2(i\omega+k^2+\pi^2)(k^2+\pi^2)-Rk^2(\frac{i\omega}{\sigma}+k^2+\pi^2)+4\Omega^2\pi^2(i\omega+k^2+\pi^2) = 0,$$
(11.2.33)

obtained by substituting (2.32) into (2.31). The solution (2.32) satisfies the boundary conditions at the top and bottom of the layer but not on the sidewall. In order to satisfy the latter, we note that equation (2.33) is quartic in k^2, and so has eight solutions $\pm k_j$, $j = 1, 2, 3, 4$. We can therefore write Θ_m as the linear combination

$$\Theta_m = \sum_{j=1}^{4} A_j J_m(k_j r)e^{i(m\phi+\omega t)}\sin(\pi z),$$
(11.2.34a)

where the A_j are complex amplitudes to be determined by the boundary conditions. It is understood here and in what follows that the physical quantities are the real parts of the given solutions. Note that we need not explicitly use the solutions corresponding to the $-k_j$, since $J_m(-z) = (-1)^m J_m(z)$, and those solutions are therefore not linearly independent.

The velocity field corresponding to (2.34a) is

$$u_m = \pi \sum_{j=1}^{4} A_j \gamma_j \left(\beta_j \partial_r + 2\Omega \frac{im}{r} \right) J_m(k_j r) e^{i(m\phi + \omega t)} \cos(\pi z), \quad (11.2.34b)$$

$$v_m = \pi \sum_{j=1}^{4} A_j \gamma_j \left(-2\Omega \partial_r + \beta_j \frac{im}{r} \right) J_m(k_j r) e^{i(m\phi + \omega t)} \cos(\pi z), \quad (11.2.34c)$$

$$w_m = \sum_{j=1}^{4} A_j \alpha_j J_m(k_j r) e^{i(m\phi + \omega t)} \sin(\pi z), \quad (11.2.34d)$$

where

$$\alpha_j \equiv i\omega + k_j^2 + \pi^2, \quad \beta_j \equiv \frac{i\omega}{\sigma} + k_j^2 + \pi^2, \quad \text{and} \quad \gamma_j = \frac{\alpha_j}{k_j^2 \beta_j}. \quad (11.2.35)$$

We now impose the boundary conditions (2.30b) at the sidewall, $r = \Gamma$, to obtain a set of linear, homogeneous equations for the A_j of the form

$$\mathcal{M}(R, \omega, \Omega, \sigma, m, \Gamma) \begin{pmatrix} A_1 \\ A_2 \\ A_3 \\ A_4 \end{pmatrix} = 0, \quad (11.2.36)$$

where $\mathcal{M}$ is a 4×4 complex-valued matrix which depends on the parameters both explicitly and implicitly through the k_j. The system (2.36) has a nontrivial solution for the A_j only if $\det(\mathcal{M}) = 0$. Since $\mathcal{M}$ is complex-valued, we must make the real and imaginary parts of the determinant vanish simultaneously. To find the nontrivial solutions to the linear problem with the required boundary conditions, we therefore hold the four parameters Ω, σ, m and Γ fixed and solve the system

$$\Re[\det(\mathcal{M}(R, \omega; \Omega, \sigma, m, \Gamma))] = 0, \quad (11.2.37a)$$

$$\Im[\det(\mathcal{M}(R, \omega; \Omega, \sigma, m, \Gamma))] = 0 \quad (11.2.37b)$$

numerically for $R = R_c^{(m)}$ and $\omega = \omega_c^{(m)}$. We emphasize that while these equations cannot be solved analytically, the implicit solutions solve the linear boundary-value problem *exactly*.

The solutions reveal that there are in fact two types of unstable nonaxisymmetric modes. The first type, called a wall mode, peaks near the sidewall and has a low amplitude in the interior of the container. This mode is essentially a convectively destabilized inertial oscillation; as a result the mode precesses relatively rapidly. The second type we call a body mode. This mode is essentially a convective mode; it fills the whole cylinder, has a low amplitude near the sidewall, and it precesses slowly. The wall modes are fundamentally nonaxisymmetric and are preferred in small aspect-ratio containers or for sufficiently fast rotation rates. Figure 11.6(a) shows the critical Rayleigh numbers $R_c^{(m)}$ for the wall modes as a function of the dimensionless rotation rate for several values of m and $\sigma = 7.0$, $\Gamma = 1$. Figure 11.6(b) shows the corresponding precession frequencies. Note that in general the precession frequency is larger for smaller azimuthal wavenumbers. This is because the low m eigenfunctions are less symmetrical under reflection in vertical planes than the high m eigenfunctions. This is also the reason why the wall modes precess faster than the body modes. The mean asymmetry in the latter, integrated over the container, is much smaller than for the wall modes. This fact is a consequence of the structure of the modes (see figure 11.7).

11.2.4 The Shortcomings of Local Approximations
Let us now compare these results with those obtained by Chandrasekhar (1961) and Davies-Jones & Gilman (1971) for ostensibly the same problem. Both sets of authors study rotating convection by making different types of local approximations, the former ignoring the presence of sidewalls entirely while the latter consider the narrow gap limit of an annulus. We shall refer to these two approximations as the unbounded problem and the 'channel' problem, respectively. Both formulations suffer from serious shortcomings and yield qualitatively misleading results.

Chandrasekhar (1961) considers an unbounded horizontal layer rotating uniformly about the vertical. With the boundary conditions (2.30a) the resulting stability problem is easily solved. One finds that for Prandtl numbers $\sigma > 0.68$ the first instability is a *steady state* one. Only for low Prandtl numbers and sufficiently high rotation rates is the first instability a Hopf bifurcation. These results appear to be inconsistent with the theory presented above, which shows that the first instability is generically a Hopf bifurcation. The resolution of this apparent inconsistency was provided by Goldstein *et al.* (1993a). These authors pointed out that the Chandrasekhar procedure in effect computes only *axisymmetric* solutions, the steady parallel rolls found by Chandrasekhar for $\sigma > 0.68$ being a local approximation to a pattern of concentric rolls in the cylinder. Since these solutions are axisymmetric they do not precess. The oscillations that are possible for $\sigma < 0.68$ are oscillations in the radial direction. Since they are also axisymmetric no precession is associated with them. The exact solution for the rotating cylinder demonstrates that both types of solutions do indeed exist (Goldstein *et al.* 1993a, b); in some cases the axisymmetric state is in fact the first to set in. How-

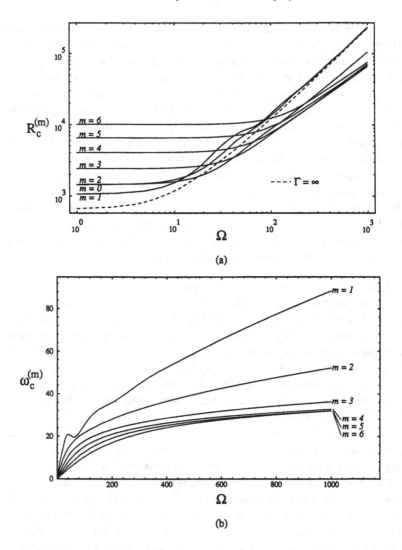

(a)

(b)

Figure 11.6 Onset of convection in a rotating cylinder with $\Gamma = 1$ and $\sigma = 7.0$. (a) The critical Rayleigh numbers for different m values as a function of the dimensionless rotation rate Ω. (b) The corresponding precession frequencies.

ever, the exact solution also reveals that typically it is a nonaxisymmetric and hence a precessing mode that is the one that first becomes unstable. For small aspect ratios these modes are wall modes. It follows therefore that the sidewall is *destabilizing*, since it is responsible for a mode of instability that is absent in the unbounded problem. As the aspect ratio Γ of the system is

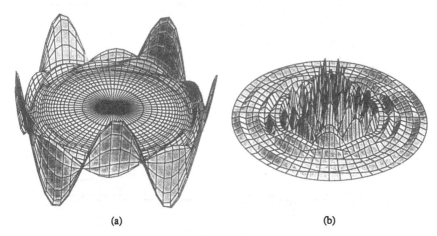

(a)   (b)

Figure 11.7 The eigenfunctions $\Theta_m(r, \phi, z = \frac{1}{2})$ for (a) an $m = 5$ wall and (b) an $m = 5$ body mode when $\Gamma = 4.39$, $\Omega = 500$ and $\sigma = 7.0$. At this value of the aspect ratio both modes have the same critical Rayleigh number, $R_c^{(5)} = 92486$. The wall mode precesses with frequency $\omega_c^{(5)} = 60.1$, the body mode with $\omega_c^{(5)} = 0.00344$.

increased, the critical Rayleigh number for each body mode falls below that for the corresponding wall mode so that for large aspect ratios it is the body modes that are preferred. This is illustrated in figure 11.7 which shows the eigenfunctions for the $m = 5$ wall and body modes at the cross-over aspect ratio. With increasing aspect ratio the critical Rayleigh number for each body mode approaches the Chandrasekhar result, as shown in figure 11.8 for $m = 1$ and $m = 2$. At the same time the precession frequency decreases to zero, albeit in an oscillatory manner. In this sense the Chandrasekhar solution does indeed describe the onset of convection in a large aspect ratio rotating container. This appealing picture holds, however, only for moderate or high Prandtl numbers. For small Prandtl numbers the distinction between the wall and body modes becomes less clear and the picture becomes much more complicated (Goldstein *et al.* 1993*b*). Note that without the circular symmetry there is no distinction between axisymmetric and nonaxisymmetric states.

A traditional method for studying problems in rotating systems is to recognize the importance of sidewalls in breaking translation invariance in the radial direction, and employ in effect the 'narrow gap' limit. In this limit one considers a narrow annular region between two walls, and describes the system using local cartesian coordinates. This approximation results in a considerable simplification of the problem but at a serious cost. Consider once again convection in a rotating layer but now confined to such an annular region. This system retains the SO(2) symmetry and hence one expects any pattern

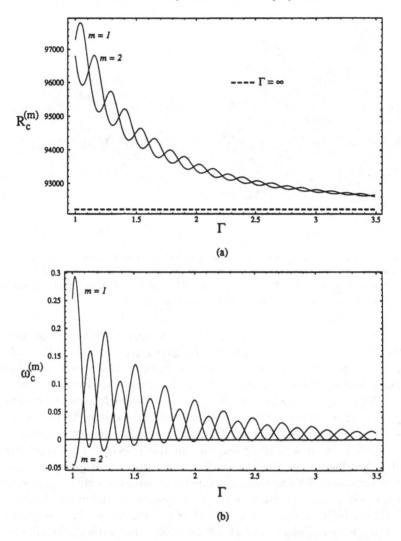

(a)

(b)

Figure 11.8 As for figure 11.6 but for the body modes with $m = 1$ and $m = 2$. The value of R_c for an unbounded layer is indicated by a dashed line.

that breaks this symmetry to precess in the rotating frame. However, in the locally cartesian coordinates one finds that this is not the case. The results of Davies-Jones & Gilman (1971) for a channel with stress-free and conducting boundary conditions at top and bottom and no-slip insulating boundary conditions at the sidewalls (i.e., the boundary conditions (2.30)) show that for fast enough rotation rates the initial instability is indeed a Hopf bifurcation even for Prandtl numbers in excess of 0.68. From the point of view of this article this comes as no surprise. What does come as a surprise is that for slow rotation rates the instability is to a 'nonaxisymmetric' time-*independent* state. The reason this is surprising is that the channel problem ought to have the same symmetry as the annulus, periodic boundary conditions in the 'downstream' direction mimicking mathematically the periodicity in the azimuthal direction.

To understand what is going on we compare the analysis of the channel problem by Davies-Jones & Gilman (1971) with that summarized in section 11.2.3. This is quite easy since the boundary conditions in both cases are identical. Observe first that mathematically the reason that the patterns precess is a direct consequence of the fact that the 'dispersion relation' (2.37) is complex. The imaginary terms arise from those in expressions (2.34) for the velocity components u_m and v_m. Both of these terms are proportional to the azimuthal wavenumber m; this is the reason why axisymmetric patterns do not precess. To see exactly what happens when one passes from the annulus problem to the channel problem one can solve the annulus problem exactly. The solution procedure is very similar to that described in section 11.2.3. The only difference is that in order to satisfy the boundary conditions on the inner sidewall one must now employ the Bessel functions $Y_m(kr)$ as well as $J_m(kr)$. The resulting dispersion relation (2.37) continues to be complex. In contrast, for the corresponding channel problem the matrix $\mathcal{M}$ is complex, but the resulting dispersion relation is now real. The reason for this remarkable fact can be traced to the symmetry properties of the channel equations. It turns out that these equations have an additional symmetry that forces the dispersion relation to be real thereby preventing the selection of a preferred direction. Moreover, the presence of the symmetry guarantees the existence of both steady solutions and standing waves in the *nonlinear* regime as well. Such nonlinear solutions have in fact been obtained by Gilman (1973). To see this observe that with the boundary conditions adopted by Davies-Jones & Gilman (1971) or those employed by Gilman (1973) the system is invariant under a reflection R_1 in a cross-stream plane (i.e., under $x \to -x$, where x is the stream-wise or azimuthal coordinate) *followed* by a second reflection R_2 in the stream-wise midplane (i.e., under $y \to -y$, where y is the cross-stream coordinate). More explicitly, if $(u(x,y,z), v(x,y,z), w(x,y,z), \Theta(x,y,z))$ is a solution to the problem, then so is $(-u(-x,-y,z), -v(-x,-y,z), w(-x,-y,z), \Theta(-x,-y,z))$. Note that we are *not* changing the direction of rotation. We shall denote this symmetry by R, i.e., $R \equiv R_2 \circ R_1$. The operation R is a reflection symmetry, since R is not the identity, but $R \circ R$ is. It follows therefore that the symmetry of the channel problem is in fact the group O(2) and not SO(2)! At the linear level it is this fact that forces the presence of a steady state bifurcation and

352   E. Knobloch: Bifurcations in Rotating Systems

the coalescence of the Hopf bifurcations to waves in the two possible directions. At the nonlinear level it explains the presence of tilted but steady cells arising from the steady state bifurcation, and the presence of standing waves arising from the Hopf bifurcation (Gilman 1973). However, it also shows that the latter bifurcation produces a competing branch of waves that precess in one or other direction. Gilman (1973) did not consider this possibility, and as a result did not test the stability of his solutions with respect to rotating wave perturbations. It is possible, however, that the standing waves are in fact unstable to rotating waves (cf. figure 11.2c), much as in thermohaline convection (Knobloch et al. 1986). It is now clear why the channel solutions do not precess, while those in the annulus do. It is the presence of *curvature* effects that destroys the symmetry R and hence leads to a preference of one direction over the other. Thus the channel approximation has inadvertently changed the symmetry of the problem, and thereby fundamentally changed its dynamics.

The symmetry R described above is present only because one can combine a reflection in x with a reflection in y to generate a symmetry of the system. This in turn requires that the boundary conditions at the two sidewalls be *identical*. Any variation of the problem that destroys this symmetry in the boundary conditions will lead to precession in addition to that caused by the curvature terms. For example if one models the thermal boundary conditions more realistically by including finite values of the Biot number, with different Biot numbers at the two sidewalls, precession will result. Similar considerations apply to nonaxisymmetric perturbations of Taylor vortices in the Taylor-Couette system. In circular geometry these result in several different precessing states (wavy vortices, twisted vortices and vortices with wavy inflow and outflow boundaries) while in the narrow gap limit they arise through a steady state instability (Nagata 1986, 1988).

A different example of the above ideas is provided by convection in a rotating sphere. Busse (1970) showed that for large rotation rates the convection takes the form of a 'cartridge belt' of Taylor columns parallel to the rotation axis (figure 11.9). In this limit, the problem can be modelled by a rotating cylinder with *sideways* heating and gravity. This situation thus differs from that discussed at the beginning of this section. Busse showed that with a horizontal and stress-free top and bottom the pattern he found did not precess, even though it was nonaxisymmetric. Instead the solution started to precess only after including the sloping top and bottom that are characteristic of a sphere. In this example the absence of precession is due to a *degeneracy* in the problem rather than symmetry constraints. The equations of motion do break the symmetry R_1 and may break R_2 as well if curvature effects are retained. In spite of this precession does not take place because the stress-free boundary conditions allow the flow to be strictly two-dimensional. Any

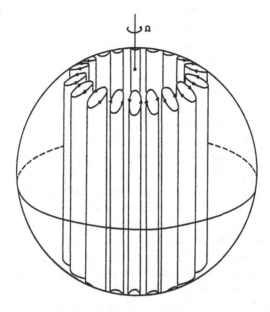

Figure 11.9 Columnar convection in a rotating sphere. (After Busse 1970.)

effect that perturbs this two-dimensional flow, be it sloping walls at the top and bottom or no-slip boundary conditions at the top and bottom, will cause precession, as will the presence of a thermal wind.

It should be mentioned that many models of geophysical processes suffer from similar defects (see, e.g., the Eady model of the baroclinic instability, Pedlosky 1987). In some cases curvature effects are critical even for axisymmetric instabilities. An example of this curious behaviour was recently discovered by Knobloch (1992) in the context of magnetic field instabilities in differentially rotating accretion disks. In this example a local analysis reveals a steady-state axisymmetric instability while a global study that takes fully into account curvature shows that such an instability cannot in fact be present. Instead the instability *must* be a Hopf bifurcation.

The above arguments also apply in situations where there is no *a priori* SO(2) symmetry. In a recent paper Matthews *et al.* (1992) study Boussinesq and compressible convection in an *oblique* magnetic field with periodic boundary conditions in the horizontal. Because of the boundary conditions in the horizontal the system is invariant under SO(2), the tilt of the magnetic field breaking the reflection symmetry in vertical planes. For this case the generic theory discussed above shows that the onset of convection should be through a Hopf bifurcation leading to a pattern drifting preferentially in one or other direction, depending on the tilt of the field. In fact in the Boussinesq case the initial instability is a steady state one and no drifts are found; in the nonlinear regime steady but tilted cells are present. The reason this nongeneric behaviour takes place is related, as above, to the presence of an additional reflection symmetry. This is the symmetry under the reflection R_3 in the midplane of the layer. The argument is identical to that already presented. Indeed, Matthews *et al.* find that if the Boussinesq approximation is relaxed, so that this reflection symmetry is lost, the tilted cells begin to drift, as predicted by the analysis of the generic case. An identical explanation accounts for the onset of overstability in rotating compressible convection with or without an imposed axisymmetric magnetic field (Jones, Roberts & Galloway 1990). In this example the system is treated in the channel approximation with identical boundary conditions at the sidewalls; however, it is the radial stratification that breaks the mid-channel reflection symmetry, leading to precession. The final example we mention arises in the stability of two-dimensional shear flows. With periodic boundary conditions in the downstream direction such flows have SO(2) symmetry. For generic profiles one therefore expects instability to set in at a Hopf bifurcation, giving rise to waves with a nonzero phase velocity relative to the shear flow. Djordjevic & Redekopp (1990) find that this is indeed so, and point out that if the vorticity profile is symmetric under mid-channel reflection the bifurcation becomes stationary.

11.3 MODE INTERACTIONS IN ROTATING SYSTEMS

We now turn attention to the possible interactions among precessing patterns. In many experiments one finds transitions from one pattern to another with a different azimuthal wavenumber, as the Rayleigh number increases. Such transitions have been studied both in the rotating annulus experiments with sideways heating (Hide & Mason 1975, Hignett 1985, Read *et al.* 1992), and in the rotating cylinder with uniform heating from below (Zhong, Ecke & Steinberg 1991). A beautiful illustration is provided by Rabaud & Couder (1983) and Chomaz *et al.* (1988) in their experiments on vortex mergers in circular geometry, and in the experiments of Meyers, Sommeria & Swinney (1989) on the coalescence of different potential vorticity patches. All these systems have an obvious SO(2) symmetry. In other situations, such as in simulations of the von Kármán vortex street, the use of periodic boundary conditions in the downstream direction also introduces the group SO(2) into the problem.

We begin by considering the interaction between two modes with azimuthal wavenumbers m and n in a rotating system. The wavenumbers are assumed to be positive and relatively prime. To understand the possible types of interaction, we seek parameter values where such interactions will take place at small amplitude. Thus we consider varying a second parameter, typically the rotation rate or the aspect ratio of the system, such that the trivial state loses stability simultaneously to both modes as the Rayleigh number is increased. The corresponding point in the two-dimensional parameter space is an example of another codimension-two point. At this point the solution to the linear stability problem takes the form

$$\Theta_{mn}(r, \phi, z, t) = \Re\{a_m(t)e^{im\phi}f_m(r, z) + a_n(t)e^{in\phi}f_n(r, z)\}. \tag{11.3.1}$$

The structure of the equations describing the competition between the complex amplitudes (a_m, a_n) is constrained by the $SO(2)$ symmetry of the system:

$$\text{rotations} \quad \phi \to \phi + \theta : \quad (a_m, a_n) \to (a_m e^{im\theta}, a_n e^{in\theta}). \tag{11.3.2}$$

Thus

$$\dot{a}_m = g_1 a_m + g_2 a_n^m \bar{a}_m^{n-1} \tag{11.3.3a}$$

$$\dot{a}_n = g_3 a_n + g_4 a_m^n \bar{a}_n^{m-1}, \tag{11.3.3b}$$

where the functions g_j, $j = 1, 2, 3, 4$, are complex-valued functions of the four invariants $\sigma_1 \equiv |a_m|^2$, $\sigma_2 \equiv |a_n|^2$, $\sigma_3 \equiv a_m^n \bar{a}_n^m$ and $\sigma_4 \equiv \bar{a}_m^n a_n^m$, as well as of the two unfolding parameters μ_m, μ_n. If the two Hopf frequencies are nonresonant, one can now proceed to put these equations into normal form, by requiring that the equations commute in addition with the phase shift symmetry

$$t \to t + \tau : \quad (a_m, a_n) \to (a_m e^{i\omega_m \tau}, a_n e^{i\omega_n \tau}), \tag{11.3.4}$$

where ω_m and ω_n are the precession frequencies of the two modes. When ω_m and ω_n are not rationally related this has the effect of shifting the spatially resonant terms appearing in (3.3) beyond all orders. Since we wish to truncate the equations, and elucidate the role played by different spatial resonances we do not use the normal form of the amplitude equations.

Equations (3.3) are identical to those for the resonant double Hopf bifurcation with no symmetry. This is because in normal form the equations for such a resonant bifurcation commute with the phase shift symmetry (3.4). If $\omega_c^{(m)}/\omega_c^{(n)} = m/n$ the symmetry acts in the same way as spatial rotations. When $m + n > 5$ the resonance is called weak. In that case it is known that the bifurcation produces an invariant two-torus. The condition for this torus to be attracting is not hard to find and is the same as that for the nonresonant case (Knobloch 1986b). When this is the case all trajectories of the system approach the invariant torus as $t \to \infty$; the dynamics on the torus is complicated, however, by both weak linear resonances (if present) and by nonlinear resonances (Guckenheimer & Holmes 1986). In spite of this we can draw a number of conclusions about the interaction between two modes with weak spatial resonance $(m + n > 5)$. In this case equations (3.3) reduce to

$$\dot{a}_m = (\mu_m + i\omega_m + a|a_m|^2 + b|a_n|^2)a_m + \cdots \tag{11.3.5a}$$

$$\dot{a}_n = (\mu_n + i\omega_n + c|a_m|^2 + d|a_n|^2)a_n + \cdots, \tag{11.3.5b}$$

where a, b, c, d are complex coefficients. In terms of the real variables $a_j \equiv r_j e^{i\theta_j}$, $j = m, n$, equations (3.5) take the form

$$\dot{r}_m = (\mu_m + a_r r_m^2 + b_r r_n^2) r_m + \cdots \qquad (11.3.6a)$$

$$\dot{r}_n = (\mu_n + c_r r_m^2 + d_r r_n^2) r_n + \cdots \qquad (11.3.6b)$$

$$\dot{\theta}_m = \omega_m + a_i r_m^2 + b_i r_n^2 + \cdots \qquad (11.3.6c)$$

$$\dot{\theta}_n = \omega_n + c_i r_m^2 + d_i r_n^2 + \cdots, \qquad (11.3.6d)$$

where the subscripts r, i indicate the real and imaginary parts. Note that the two phase equations (3.6c, d) decouple. This decoupling is not exact, however, the rotational invariance of the system being responsible only for the decoupling of the overall phase $m\theta_n - n\theta_m$. The remaining equations, (3.6a, b), are easily solved (see, for example, Knobloch & Guckenheimer 1983, Guckenheimer & Holmes 1986). One finds that there are three non-trivial solutions. Solutions of the form $(r_m, r_n) = (r, 0)$ and $(0, r)$ will be referred to, respectively, as pure m and n modes; solutions of the form (r_m, r_n), $r_m > 0, r_n > 0$, are mixed-mode solutions. The pure modes are both supercritical if $a_r < 0$, $d_r < 0$. The mixed modes bifurcate in a steady-state bifurcation from the pure m modes when $c\mu_{1r} = a\mu_{2r}$ and from the pure n modes when $d\mu_{1r} = b\mu_{2r}$. When $a_r < 0$, $d_r < 0$ the mixed modes are stable if, in addition, $\Delta \equiv a_r d_r - b_r c_r > 0$; if $\Delta < 0$ they are unstable. In this case the mixed modes cannot undergo any tertiary bifurcations. This is not the case when $a_r d_r < 0$, in which case a tertiary Hopf bifurcation from the mixed modes is always present. Note that if one restores equations (3.6c, d) a steady-state bifurcation from the pure modes introduces a new frequency into the dynamics, i.e., in the original phase space such a bifurcation is actually a Hopf bifurcation, and the mixed modes that it gives rise to are *quasiperiodic*. As before, because of the rotational invariance, no frequency locking can take place. Similarly, the tertiary Hopf bifurcation from the mixed mode state produces waves with three frequencies, i.e., a three torus in phase space. The stability of this torus is not determined by the truncation of the amplitude equations at third order, however, and additional terms must be retained (Guckenheimer & Holmes 1986). Frequency locking now becomes possible, followed typically by a period-doubling cascade and a transition to chaotic waves, as the destabilizing parameter is increased further. It remains to consider now the structure of the bifurcation diagrams as a physical parameter is varied. Since in general both μ_{1r} and μ_{2r} are linearly related to the two physical parameters that have to be tuned to arrange for the mode interaction, varying one parameter while keeping the other one fixed corresponds to tracing a line through the (μ_{1r}, μ_{2r}) plane. Typical bifurcation diagrams describing the sequence of transitions along this line (i.e., as the parameter λ, say, varies) are shown in figures 11.10(a, b). For example, if both pure modes

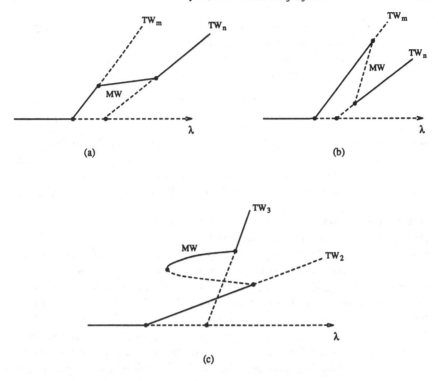

Figure 11.10 Theoretical bifurcation diagrams describing mode interactions in a rotating system. (a) A continuous transfer of stability. (b) Hysteretic transfer of stability. Both diagrams are drawn for the case of a weak spatial resonance ($m+n > 5$). (c) The bifurcation diagram suggested by the results shown in figure 11.11 for the (2,3) interaction.

bifurcate supercritically the branch of pure m-modes may lose stability at a secondary bifurcation as λ increases, forming a secondary branch of mixed modes. This branch may either go off to large amplitudes, or it may terminate in another secondary bifurcation on the branch of pure n-modes. In the latter case there are two possibilities. If the mixed-mode branch is stable, stability is transferred continuously from the pure m-modes to the pure n-modes, as μ increases. If it is unstable the stability transfer is abrupt and hysteretic. For an explicit example, see Knobloch & Guckenheimer (1983).

The situation is more involved in the case of strong spatial resonance ($m + n \leq 5$). Many aspects of such resonances are still not well understood. The best studied is the 2:1 resonance (Knobloch & Proctor 1988, Hughes & Proctor 1992). The strong spatial resonances differ in that no invariant torus is created in the bifurcation. This means that other types of dynamical behaviour are now possible, the most interesting ones involving homoclinic or heteroclinic orbits. But even without going into details one can see one very elementary consequence of equations (3.3). For $m + n > 3$ there is always a branch of pure m-modes and pure n-modes that bifurcate in succession from the trivial solution. This is not the case, however, when $m = 1$ and $n = 2$. In this case equations (3.3) become

$$\dot{a}_1 = g_1 a_1 + g_2 a_2 \bar{a}_1 \tag{11.3.7a}$$

$$\dot{a}_2 = g_3 a_2 + g_4 a_1^2. \tag{11.3.7b}$$

Observe that no solution of the form $(a_1, 0)$ is now possible unless $g_4(0, 0, 0, 0) = 0$, a condition that does not hold generically. Consequently there is a primary bifurcation to a pure 2-mode but *not* to a pure 1-mode, and one expects the $(1,2)$ mode interaction to be quite different from the other ones.

11.3.1 The Rotating Annulus Experiments

The annulus experiments with sideways heating, studied as a model of baroclinic instability in the atmosphere, provide a nice example of these results. The heating of the annulus is axisymmetric, so that the annulus has SO(2) symmetry. The system is specified by two easily controllable parameters, the rate of heating and the rotation rate, in addition to the radius ratio, Prandtl number, etc. Detailed measurements reported by Hignett (1985) show that (m, n) mode interactions are present at certain points in this two-parameter plane (see figure 11.11). Out of these codimension-two points emanate wedges containing quasiperiodic oscillations referred to as amplitude vacillations. Hignett describes in some detail the sequence of transitions that takes place as the rotation rate is varied for two different temperature differences across the annulus. The transition from $m = 2$ to $m = 3$ appears to be typical. When the rotation rate is too low the basic state is axisymmetric; with increasing rotation rate this state loses stability to an $m = 2$ wave that drifts in the azimuthal direction. For larger rotation rates one encounters a region in parameter space in which both pure $m = 2$ waves and an amplitude vacillation involving a mixture of $m = 2$ and $m = 3$ can be found depending on the experimental protocol. The transition between the two states is strongly hysteretic. If the branch of vacillations is followed the wavenumber content simplifies and the state becomes more and more a pure $m = 3$ wave. In fact for a somewhat larger rotation rate a pure $m = 3$ is found. The latter transition is apparently nonhysteretic, although the pure $m = 2$ mode continues to coexist with the $m = 3$ wave.

Figure 11.10(c) indicates a plausible interpretation of Hignett's results. The amplitude vacillations are identified with the $(2,3)$ mixed mode that bifurcates subcritically from the pure $m = 2$ waves and hence is initially unstable. The mixed mode branch is in fact very strongly subcritical so that by the time it gains stability at a tertiary saddle-node bifurcation it looks more like a pure $m = 3$ wave than the pure $m = 2$ wave that it

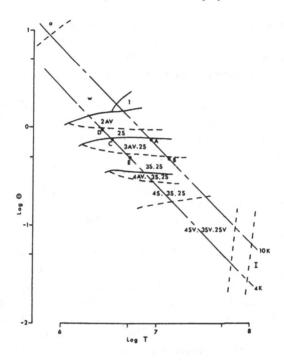

Figure 11.11 Experimental observation of mode interactions in a rotating annulus with sideways heating. The notation S denotes stationary waves, while AV denotes vacillating modes. The integers indicate the dominant azimuthal wavenumber. (After Hignett 1985.)

resembled near the subcritical bifurcation. The theory predicts that the $m = 2$ content of the vacillation should decrease along the mixed mode branch until it vanishes at its termination on the branch of pure $m = 3$ waves, in qualitative agreement with the experiment. The theory also predicts that if the secondary bifurcation with which the mixed mode branch terminates is supercritical (in an appropriate sense) then the transfer of stability from the mixed mode branch to the branch of pure $m = 3$ waves will be nonhysteretic (see figure 11.10a). In fact the bifurcation diagram shown in figure 11.10(c) is a simple deformation of the theoretical bifurcation diagrams near the mode interaction point (figures 11.10a, b). Such a deformation would not be unusual, given that the experiment is not conducted very close to the codimension-two point. As already mentioned, however, the (2,3) mode interaction is in fact a strong spatial resonance. The work of Dangelmayr (1986) for the case with $O(2)$ symmetry shows that for this interaction one typically finds a tertiary saddle-node bifurcation on the mixed mode branch, in contrast to interactions with $m + n > 5$. Consequently for the (2,3) interaction the hysteresis may persist all the way to the codimension-two point. The experimental observation that the amplitude vacillation is apparently quasiperiodic is also in accord with the theory, as is the absence of frequency locking. The observations are thus in qualitative agreement with the mode interaction analysis presented above in which stability is transferred from one pure mode to another one via a stable branch of quasiperiodic mixed modes. Moreover, the modulated amplitude vacillations recently observed by Read *et al.* (1992) find a ready explanation in a tertiary Hopf bifurcation on the mixed mode branch; in the O(2) symmetric case such bifurcations are known to occur if one of the pure modes (either $m = 2$ or $m = 3$) bifurcates subcritically. Finally, the experiments of Hignett also reveal that the (1,2) transition is qualitatively different from the rest (see figure 11.11), again as one would expect from the analysis, but in this case the transition was apparently not studied in sufficient detail to establish its nature. Nonetheless it is clear that the transition sketched in figure 11.11 is unlikely to be correct.

It is interesting to compare the above experiment with an experiment on vortex mergers in shear flow in circular geometry conducted by Rabaud and Couder (1983) and Chomaz *et al.* (1988). Both experiments have SO(2) symmetry. The transitions among different azimuthal wavenumbers are again found to be strongly subcritical, much as in the annulus experiments. There are important differences, however. The nature of the transition depends on whether the azimuthal wavenumber is even or odd and on the experimental protocol. When the stress is varied quasistatically the observed transitions with increasing shear stress are always from m to $m - 1$. When m is odd the transition is via a vacillating state. This state loses stability with further increase in the stress to an $m - 1$ state. When m is even the transition takes place with a secondary bifurcation to a *stationary* mixed mode with wavenumbers m and $m/2$. This mode is stable over a narrow interval of stress before losing stability to an $m - 1$ state. This latter transition is also highly hysteretic. The origin of these differences would make an interesting subject for further study. In a recent attempt (Churilov & Shukhman 1992) the general form of the problem (3.3) was not recognized.

11.3.2 Other Mode Interactions

It is also of interest to consider the case where one has a mode interaction between two modes with the same wavenumbers. Although this does not apparently happen in the annulus experiments, an example of such an interaction was found by Zhang & Busse (1987) in their study of low Prandtl number convection in rotating spheres. This interaction is also a double Hopf bifurcation, but with identical frequencies. Consequently it is a codimension-three bifurcation, since two parameters must be varied to get the two modes to bifurcate simultaneously, while a third parameter is required to match their frequencies. Indeed Zhang and Busse vary both the rotation rate and

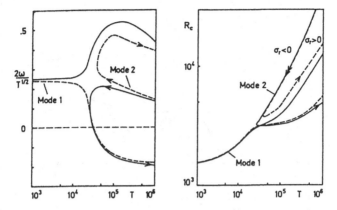

Figure 11.12 The 'switch-over' phenomenon of Zhang & Busse (1987). At the critical Prandtl number σ_c (0.15 < σ < 0.20) the two $m = 6$ modes labelled 1 and 2 have the same precession frequency at onset.

the Prandtl number, as well as the Rayleigh number, in order to locate this mode interaction (see figure 11.12). Since the wavenumbers are also identical the interaction is the non-semisimple 1:1 resonant double Hopf bifurcation. What this means is that at the codimension-two point in parameter space the linear problem takes the form

$$\begin{pmatrix} \dot{v}_m \\ \dot{w}_m \end{pmatrix} = \begin{pmatrix} i\omega_c^{(m)} & 1 \\ 0 & i\omega_c^{(m)} \end{pmatrix} \begin{pmatrix} v_m \\ w_m \end{pmatrix}. \qquad (11.3.8)$$

As before the SO(2) rotational symmetry of the problem is identical to the phase shift symmetry introduced into the normal form by the normal form transformations. Consequently the analysis of van Gils, Krupa & Langford (1990) applies directly to the interaction found by Zhang and Busse. In the unfolding one finds both two- and three-frequency states, as well as period-doubling bifurcations, depending on the coefficients of the nonlinear terms. For smaller Prandtl numbers the neutral stability curves for the two modes cross but their onset frequencies move apart; the resulting mode interaction problem is then of codimension-two and is of the form (3.3) with $n = m$.

Note that if the above problems were described in the 'channel' approximation the resulting equations would have O(2) symmetry. For the (m, n) mode interactions the corresponding analysis was carried out by Dangelmayr (1986). In this case the pure modes do not precess, the secondary branch of mixed modes is a steady branch, and only tertiary bifurcations lead to time dependence. There are two types of such bifurcations, a Hopf bifurcation that gives rise to standing waves, and a steady state bifurcation that gives rise to rotating waves. The latter we have already met. The mixed mode always has a reflection symmetry. The nature of this symmetry depends on the relative parity of m and n (cf. Crawford, Knobloch & Riecke 1990). Because of

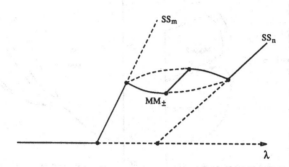

Figure 11.13 A typical bifurcation diagram for a weak mode interaction with O(2) symmetry. The primary branches correspond to time-independent pure modes. There are *two* branches of time-independent mixed modes. Under appropriate conditions these may be connected by a branch of travelling waves. *Such waves appear in a tertiary pitchfork bifurcation that breaks the reflection symmetry of the mixed modes.* The resulting diagram should be compared with the corresponding one for SO(2) symmetry, shown in figure 11.10(a).

the O(2) symmetry this reflection-symmetric mixed mode is neutrally stable with respect to rotations. If there is a tertiary instability that breaks the reflection symmetry the mode will start to precess. The (2,1) mode interaction with O(2) symmetry is special, as might be expected from the discussion of equations (3.7). Here additional phenomena, such as structurally stable heteroclinic orbits, are present (Armbruster, Guckenheimer & Holmes 1988, Proctor & Jones 1988). Strong spatial resonances with SO(2) symmetry have by and large not yet been studied. But the different scenarios pertaining to the weak resonances (see figure 11.13) indicate clearly the importance of using equations with the correct symmetry properties.

11.3.3 Parity Breaking Bifurcations
Thus far we have only considered the effects of SO(2) symmetry on the dynamics of rotating systems. However, in many cases there are additional symmetries that also play a role. The sun, like the planets, has a reflection symmetry in the equatorial plane. This implies that the symmetry group of the Sun is actually SO(2)×Z_2. In order to understand the role played by the additional symmetry we first discuss the interaction between two precessing waves with the same azimuthal wavenumber, one of which is even and the other of which is odd with respect to reflections in the equator. At the codimension-two point the linear problem has an eigenfunction of the form

$$\Theta_{eo}(r, \phi, z, t) = \Re\{a_e(t)e^{im\phi}f_e(r, z) + a_o(t)e^{im\phi}f_o(r, z)\}, \qquad (11.3.9)$$

where (r, ϕ, z) are cylindrical polar coordinates. The group $SO(2) \times Z_2$ acts on the amplitudes a_e, a_o of the even and odd modes as follows:

$$\text{rotations} \quad \phi \to \phi + \theta : \quad (a_e, a_o) \to e^{im\theta}(a_e, a_o), \tag{11.3.10a}$$

$$\text{reflection} \quad z \to -z : \quad (a_e, a_o) \to (a_e, -a_o). \tag{11.3.10b}$$

Note that this action is different from the action (2.25, 26) of the group $O(2)$: as already mentioned, the groups $SO(2) \times Z_2$ and $O(2)$ are not the same. It follows that the amplitudes must satisfy equations of the form

$$\dot{a}_e = g_1 a_e + g_2 a_o^2 \bar{a}_e \tag{11.3.11a}$$

$$\dot{a}_o = g_3 a_o + g_4 a_e^2 \bar{a}_o, \tag{11.3.11b}$$

where the functions g_j, $j = 1, 2, 3, 4$, are complex-valued functions of $\sigma_1 \equiv |a_e|^2$, $\sigma_2 \equiv |a_o|^2$, $\sigma_3 \equiv a_e^2 \bar{a}_o^2$ and $\sigma_4 \equiv \bar{a}_e^2 a_o^2$, as well as of the two unfolding parameters. Assuming appropriate nondegeneracy conditions on the coefficients of the nonlinear terms we may expand the functions g_j, $j = 1, 2, 3, 4$, in a Taylor series about the origin. We obtain

$$\dot{a}_e = (\mu_e + i\omega_e + \alpha_e |a_e|^2 + \beta_e |a_o|^2)a_e + \gamma_e a_o^2 \bar{a}_e \tag{11.3.12a}$$

$$\dot{a}_o = (\mu_o + i\omega_o + \beta_o |a_e|^2 + \alpha_o |a_o|^2)a_o + \gamma_o a_e^2 \bar{a}_o, \tag{11.3.12b}$$

where the coefficients $\alpha_{e,o}$, $\beta_{e,o}$ and $\gamma_{e,o}$ are all complex. A general analysis of the above equations is currently in progress. Hirschberg & Knobloch (1993b) have discussed the case in which all the coefficients are purely real and $\omega_o = \omega_e = 0$. This case affords an important simplification since the dynamics of the phases become trivial. This is the situation that arises in the 'channel' approximation. In terms of the real variables $a_{e,o} \equiv r_{e,o}e^{i\theta_{e,o}}$ one finds

$$\dot{r}_e = (\mu_e + \alpha_e r_e^2 + \beta_e r_o^2 + \gamma_e r_o^2 \cos 2\psi)r_e \tag{11.3.13a}$$

$$\dot{r}_o = (\mu_o + \beta_o r_e^2 + \alpha_o r_o^2 + \gamma_o r_e^2 \cos 2\psi)r_o \tag{11.3.13b}$$

$$\dot{\psi} = -(\gamma_e r_o^2 + \gamma_o r_e^2)\sin 2\psi, \tag{11.3.13c}$$

where ψ is the relative phase $\theta_e - \theta_o$. Consequently the dynamics of ψ is very simple indeed. Two primary branches corresponding to pure even and pure odd modes always bifurcate in succession from the trivial state. Under appropriate circumstances there are secondary bifurcations that produce modes of mixed parity. There are two types of such mixed parity modes distinguished by the relative phase ψ between the even and odd components, either $\psi = 0$ or $\psi = \pi/2$. Tertiary bifurcations can give rise to either rotating waves or to standing waves, much as in the (m, n) mode interaction with $O(2)$ symmetry discussed by Dangelmayr (1986) for $m > n \geq 1$. The above construction is readily extended to similar parity breaking interactions involving different azimuthal wavenumbers.

11.3.4 Towards the Solar Dynamo

The approach discussed in the preceding section suffers from the disadvantage that the patterns propagate only in the azimuthal direction. The solar magnetic activity cycle shows, however, that the patterns of activity propagate towards the equator from higher latitudes. In order to be able to describe such propagation in latitude it is useful to imagine that the Sun is (nearly) translation invariant in latitude. This allows one to introduce periodic boundary conditions in latitude, enlarging the group of symmetries to $O(2) \times SO(2)$. The effects of sphericity are then introduced as a translation symmetry breaking perturbation. This approach has met with a great deal of success in explaining the effects of distant sidewalls on travelling wave convection (Dangelmayr, Knobloch & Wegelin 1991) and of the effects of ends on spiral vortices in the Taylor-Couette system (Knobloch & Pierce 1992). To describe such propagating patterns one needs to study the Hopf bifurcation with $O(2) \times SO(2)$ symmetry. This is in fact a very simple proposition. Suppose that at onset

$$\Theta_{km}(r, \phi, z, t) = \Re\left\{(v_{km}(t)e^{ikz+im\phi} + w_{km}(t)e^{ikz-im\phi})f_{km}(r)\right\}, \quad (11.3.14a)$$

where, as before,

$$\begin{pmatrix} \dot{v}_{km} \\ \dot{w}_{km} \end{pmatrix} = \begin{pmatrix} i\omega_c^{(km)} & 0 \\ 0 & -i\omega_c^{(km)} \end{pmatrix} \begin{pmatrix} v_{km} \\ w_{km} \end{pmatrix}. \quad (11.3.14b)$$

Here k and m are the latitudinal and azimuthal wavenumbers, respectively. The expression $(3.14a)$ represents an arbitrary superposition of left- and right-handed helical waves. The symmetries act on (v_{km}, w_{km}) as follows:

translations $z \to z + d$: $(v_{km}, w_{km}) \to e^{ikd}(v_{km}, w_{km})$, $(11.3.15a)$

reflection $z \to -z$: $(v_{km}, w_{km}) \to (\bar{w}_{km}, \bar{v}_{km})$, $(11.3.15b)$

rotations $\phi \to \phi + \theta$: $(v_{km}, w_{km}) \to (e^{im\theta}v_{km}, e^{-im\theta}w_{km})$. $(11.3.15c)$

Note that the rotation symmetry $(3.15c)$ is the same as the phase shift symmetry present in the normal form of the Hopf bifurcation with $O(2)$ symmetry. Thus the Hopf bifurcation with $O(2) \times SO(2)$ symmetry is described by the same equations as the Hopf bifurcation with $O(2)$ symmetry. The difference is that in the present case equations (2.20) are exact, and no normal form transformations are necessary. In particular the decoupling of the phases from the amplitudes is exact, unlike that in equations (2.21). The possible solutions to equations (2.20) were described in section 11.2.1. All that is necessary is to reinterpret them in the light of $(3.14a)$. The solutions $(|v_{km}|, |w_{km}|) = (A_{km}, 0)$ or $(0, A_{km})$ correspond to pure left- and right-handed helical waves; the solution $(|v_{km}|, |w_{km}|) = (A_{km}, A_{km})/\sqrt{2}$ corresponds to a standing wave pattern that is reflection symmetric with respect to the equator. When the translation invariance in the z-direction is broken by sphericity effects, the governing equations can be reduced to the amplitude equations

$$\dot{v}_{km} = (\mu + i\omega_{km} + \beta_{km}|w_{km}|^2 + \alpha_{km}A_{km}^2)v_{km} + \epsilon\bar{w}_{km} \qquad (11.3.16a)$$

$$\dot{w}_{km} = (\mu - i\omega_{km} + \bar{\beta}_{km}|v_{km}|^2 + \bar{\alpha}_{km}A_{km}^2)w_{km} + \bar{\epsilon}\bar{v}_{km}, \qquad (11.3.16b)$$

where ϵ is a complex coefficient specifying the strength of the symmetry breaking (Dangelmayr et al. 1991). The resulting equations have been analyzed by Dangelmayr & Knobloch (1991). Simple group theory shows that the effects of broken translation invariance in z split the two pairs of purely imaginary eigenvalues present at $\mu = 0$ when $\epsilon = 0$. As a result the double Hopf bifurcation splits into two successive simple Hopf bifurcations, each of which gives rise to a standing and hence *reflection symmetric* wave pattern (see figure 11.14). These two patterns differ in the phase relationship between the incident waves at the equator. The waves can either be in phase at $z = 0$, or exactly out of phase. In the terminology of Dangelmayr & Knobloch (1987b, 1991) these two solutions are referred to as SW_0 and SW_π. These correspond, respectively, to the quadrupole and dipole solutions described recently by Schmitt & Schüssler (1989), Brandenburg et al. (1989a), Brandenburg, Tuominen & Moss (1989b) and Jennings & Weiss (1991).

It remains to consider the *appearance* of these solutions. Once the translation symmetry in z is broken, plane waves no longer solve the problem. Instead the eigenfunctions in z appearing in (3.14a) are modified, subject to the boundary conditions that they vanish at $z = \pm\pi/2$. Typically the eigenfunctions for the left helical wave peaks in $z > 0$ and that for the right helical wave peaks in $z < 0$, or vice versa. In the former case the solution therefore consists of waves travelling towards the *equator*, while in the latter case they travel towards the *poles*. Both solutions are, however, symmetric under reflection in $z = 0$. For an explicit example, see Dangelmayr et al. (1991). This example demonstrates that quite complicated 'butterfly' diagrams can be contructed from simple nonlinear but ordinary differential equations for the amplitudes of the correct linear theory eigenmodes.

The above analysis thus accounts naturally for the two basic types of solutions that have been observed in dynamo models. In addition it includes the azimuthal variation of the waves of activity, that has only recently begun to be taken into account in dynamo models of the solar cycle. The analysis of equations (3.16) also shows that there are interesting *secondary* bifurcations (see figure 11.14). The steady-state bifurcations lead to stationary states in which the reflection symmetry in the equator is lost, with the wave of one or other helicity dominating (figure 11.14d). Examples of such solutions were reported

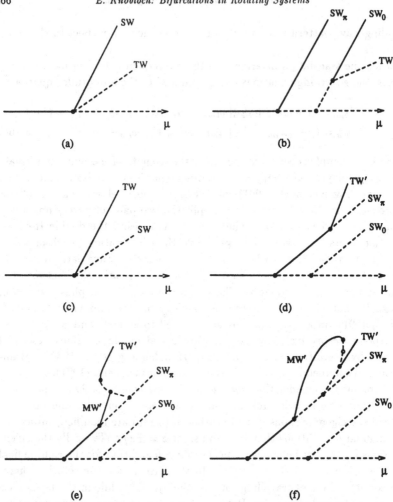

Figure 11.14 The effect of breaking translation invariance on the Hopf bifurcation with $O(2) \times SO(2)$ symmetry. (a, c) show two possible scenarios in the unperturbed case $\epsilon = 0$. (b) shows the effect on the transition to stable standing waves. (d, e, f) show some of the possible bifurcation diagrams when travelling waves are preferred. (d) shows the transition from a dipole solution to a stationary nonsymmetric solution; (e) shows the appearance and disappearance of symmetric oscillations about the dipole state, while (f) shows the global bifurcation by which the symmetric oscillations may become nonsymmetric.

by Schmitt & Schüssler (1989) and by Jennings & Weiss (1991). In addition under certain circumstances it is possible to find secondary Hopf bifurcations leading either to oscillations in the amplitudes of the two competing waves or to oscillations about one of the stationary asymmetric states (Dangelmayr & Knobloch 1991). In the former case in the first half of the cycle the waves travel (mostly) in one direction, while in the second half they travel (mostly) in the reverse direction. Near the Hopf bifurcation the solution retains reflection symmetry in the equator. With increasing bifurcation parameter (the dynamo number) the period of these oscillations increases, becoming infinite at a heteroclinic bifurcation connecting the two stationary but nonsymmetric waves. At this point a hysteretic transition to a steady but nonsymmetric state takes place (figure 11.14e). Alternatively the symmetric oscillatory solution can increase in period and undergo a transition to an oscillation about one or other of the two nonsymmetric stationary states via a so-called 'gluing' bifurcation. With increasing parameter these oscillations terminate on the nonsymmetric stationary solution in the second type of secondary Hopf bifurcation (see figure 11.14f). Such oscillations are therefore nonsymmetric oscillations. The nonsymmetric oscillations reported by Brandenburg et al. (1989a, b) in their dynamo models appear to be of this type, although their connection to the primary branches remains to be elucidated. The calculations of Brandenburg et al. suggest that they arise from a Hopf bifurcation of a stationary mixed polarity state. Note that both types of oscillations produce quasiperiodic (two-frequency) states since the amplitude oscillation is superimposed on a wave pattern propagating with an unrelated frequency. The frequency of the amplitude oscillation is necessarily small compared to the pattern frequency since it is related to the amplitude of the $SW_{0,\pi}$ state at which the secondary Hopf bifurcation takes place. Note also that it is the loss of translation invariance due to sphericity effects that allows such oscillations to exist, and that they are only present in a limited range of the parameter values (cf. Brandenburg et al. 1989a, b). Moreover the oscillations are not always present; whether or not the secondary Hopf bifurcations exist depends on the argument of the symmetry breaking coefficient ϵ (see Dangelmayr & Knobloch 1991). These results summarize the possible behaviour near the onset of an appropriate symmetry-breaking Hopf bifurcation in a system with the symmetry of a rotating star.

We have seen above that an approach based on symmetries is capable of elucidating much of the phenomenology discovered by integrating specific dynamo models. Consequently the behaviour of the models is to a large extent forced by their symmetries, and is therefore independent of the details of the model. Given the uncertainties in current dynamo models this is perhaps a reassuring observation.

11.4 DISCUSSION

In this chapter we have summarized the most important results from bifurcation theory that are relevant to the study of instabilities in rotating systems. We have indicated with specific examples how the abstract results relate to specific systems, and indicated where the theory still remains to be developed. Undoubtedly the most fundamental distinction between nonrotating and rotating systems is the absence of steady-state bifurcations in the latter. As a result patterns in rotating systems invariably precess. We have seen that this fact is related to the absence of reflection symmetry in vertical planes in such systems, and showed explicitly that if reflection symmetry is inadvertently restored in modelling the system then precession is lost. This observation has shed new light on the origin of precessing patterns in rotating spheres and spherical shells, since it shows that sloping walls are not necessary for precession — patterns in rotating cylinders with a horizontal top and bottom precess provided curvature effects are retained. Consequently care must be taken in interpreting observations of, say, the westward drift of the geomagnetic field in terms of the frequency generated in $\alpha\omega$ dynamos. This is because a nonaxisymmetric 'steady' solution will precess in the rotating frame. This precession frequency is unrelated to the α and ω parameters of the model. Instead it is determined by the degree of asymmetry (with respect to reflection in vertical planes) of the dynamo mode, and in particular its azimuthal wavenumber. Only in the case where one has an oscillatory dynamo in the axisymmetric case could the drift of the nonaxisymmetric features be related to the α and ω parameters.

That the absence of reflection symmetry in vertical planes has fundamental consequences for the dynamics should come as no surprise. The same fact is responsible for the Küppers–Lortz instability of steady rolls in rotating convection (Küppers & Lortz 1969), and for similar instabilities of travelling and standing waves (Knobloch & Silber 1992).

The approach we have adopted focuses on the symmetries of the problem and the structure of the amplitude equations that are consistent with it. As such the results apply to any system with the specified symmetry properties undergoing the particular instability in question. Indeed we have illustrated our results from a variety of systems of laboratory and geophysical interest. This approach has much to recommend it since it captures model-independent behaviour. In addition it can be used to ascertain whether a model is suitable, in the sense that it preserves the symmetries of the system. On the other hand the study of specific models becomes essential in order to discriminate among various possible scenarios predicted by the amplitude equations. This typically involves the calculation of the coefficients in the amplitude equations from the model equations. Although such computations can be laborious they are straightforward. Consequently we have not dwelt on this aspect of the problem.

Acknowledgements

The preparation of this article was supported by a Rosenbaum Fellowship at the Isaac Newton Institute for Mathematical Sciences, and an SERC travel grant. I am grateful to J.D. Crawford and K. Zhang for valuable discussions.

REFERENCES

Armbruster, D., Guckenheimer, J. & Holmes, P.J. 1988 Heteroclinic cycles and modulated travelling waves in systems with O(2) symmetry. *Physica D* **29**, 257–282.

Bestehorn, M., Fantz, M., Friedrich, R., Haken, H. & Pérez-García, C. 1992 Spiral patterns in thermal convection. *Z. Phys. B* **88**, 93–94.

Bodenschatz, E., de Bruyn, J.R., Ahlers, G. & Cannell, D.S. 1991 Transition between patterns in thermal convection. *Phys. Rev. Lett.* **67**, 3078–3081.

Brandenburg, A., Krause, F., Meinel, R., Moss, D. & Tuominen, I. 1989a The stability of nonlinear dynamos and the limited role of kinematic growth rates. *Astron. Astrophys.* **213**, 411–422.

Brandenburg, A., Tuominen, I. & Moss, D. 1989b On the nonlinear stability of dynamo models. *Geophys. Astrophys. Fluid Dynam.* **49**, 129–141.

Busse, F.H. 1970 Thermal convection in rapidly rotating systems. *J. Fluid Mech.* **44**, 441–460.

Chandrasekhar, S. 1961 *Hydrodynamic and Hydromagnetic Stability*. Dover.

Chomaz, J.M., Rabaud, M., Basdevant, C. & Couder, Y. 1988 Experimental and numerical investigation of a forced circular shear layer. *J. Fluid Mech.* **187**, 115–140.

Chossat, P. 1982 Interactions entre bifurcations par brisure partielle de symétrie sphérique. *Ann. Scient. Éc. Norm. Sup.* **4**, 117–145.

Churilov, S.M. & Shukhman, I.G. 1992 Weakly nonlinear theory of the alternation of modes in a circular shear flow. *J. Fluid Mech.* **243**, 155–170.

Crawford, J.D. & Knobloch, E. 1988a On degenerate Hopf bifurcation with broken O(2) symmetry. *Nonlinearity* **1**, 617–652.

Crawford, J.D. & Knobloch, E. 1988b Symmetry-breaking bifurcations in O(2) maps. *Phys. Lett. A* **128**, 327–331.

Crawford, J.D. & Knobloch, E. 1991 Symmetry and symmetry-breaking bifurcations in fluid dynamics. *Annu. Rev. Fluid Mech.* **23**, 341–387.

Crawford, J.D., Knobloch, E. & Riecke, H. 1990 Period-doubling mode interactions with O(2) symmetry. *Physica D* **44**, 340–396.

Dangelmayr, G. 1986 Steady-state mode interactions in the presence of O(2) symmetry. *Dynamics & Stability of Systems* **1**, 159–185.

Dangelmayr, G. & Knobloch, E. 1987a The Takens–Bogdanov bifurcation with O(2) symmetry. *Proc R. Soc. Lond. A* **322**, 243–279.

Dangelmayr, G. & Knobloch, E. 1987b On the Hopf bifurcation with broken O(2) symmetry. In *The Physics of Structure Formation: Theory and Simulation* (ed. W. Güttinger and G. Dangelmayr), pp. 387–393. Springer.

Dangelmayr, G. & Knobloch, E. 1991 Hopf bifurcation with broken circular symmetry. *Nonlinearity* 4, 399–427.

Dangelmayr, G., Knobloch, E. & Wegelin, M. 1991 Dynamics of travelling waves in finite containers. *Europhys. Lett.* 16, 723–729.

Davies-Jones, R.P. & Gilman, P.A. 1971 Convection in a rotating annulus uniformly heated from below. *J. Fluid Mech.* 46, 65–81.

Djordjevic, V.D. & Redekopp, L.G. 1990 The effect of profile symmetry on the nonlinear stability of mixing layers. *Studies in Appl. Math.* 83, 287–317.

Ecke, R.E., Zhong, F. & Knobloch, E. 1992 Hopf bifurcation with broken reflection symmetry in rotating Rayleigh–Bénard convection. *Europhys. Lett.* 19, 177–182.

Gilman, P.A. 1973 Convection in a rotating annulus uniformly heated from below. Part 2. Nonlinear results. *J. Fluid Mech.* 57, 381–400.

van Gils, S. & Mallet-Paret, J. 1986 Hopf bifurcation and symmetry: travelling and standing waves on the circle. *Proc. R. Soc. Edinburgh A* 104, 279–307.

van Gils, S., Krupa, M. & Langford, W.F. 1990 Hopf bifurcation with non-semisimple 1:1 resonance. *Nonlinearity* 3, 825–850.

Goldstein, H.F., Knobloch, E., Mercader, I. & Net, M. 1993a Convection in a rotating cylinder I: Linear theory for moderate Prandtl numbers. *J. Fluid Mech.* 248, 583–604.

Goldstein, H.F., Knobloch, E., Mercader, I. & Net, M. 1993b Convection in a rotating cylinder II: Linear theory for low Prandtl numbers. *J. Fluid Mech.* (in press).

Golubitsky, M., Stewart, I. & Schaeffer, D.G. 1988 *Singularities and Groups in Bifurcation Theory*. Springer.

Guckenheimer, J. & Holmes, P. 1986 *Nonlinear Oscillations, Dynamical Systems and Bifurcations of Vector Fields*. Springer.

Hide, R. & Mason, P.J. 1975 Sloping convection in a rotating fluid. *Adv. Phys.* 24, 47–100.

Hignett, P. 1985 Characteristics of amplitude vacillation in a differentially heated rotating fluid annulus. *Geophys. Astrophys. Fluid Dynam.* 31, 247–281.

Hirschberg, P.C. & Knobloch, E. 1993a, b (in preparation).

Hughes, D.W. & Proctor, M.R.E. 1992 Nonlinear three-wave interaction with non-conservative coupling. *J. Fluid Mech.* 244, 583–604.

Jennings, R.L. & Weiss, N.O. 1991 Symmetry breaking in stellar dynamos *Mon. Not. R. Astron. Soc.* 252, 249–260.

Jones, C.A., Roberts, P.H. & Galloway, D.J. 1990 Compressible convection in the presence of rotation and a magnetic field. *Geophys. Astrophys. Fluid*

Dynam. **53**, 145–182.

Knobloch, E. 1986*a* Oscillatory convection in binary mixtures. *Phys. Rev. A* **34**, 1538–1549.

Knobloch, E. 1986*b* Normal form coefficients for the nonresonant double Hopf bifurcation. *Phys. Lett. A* **116**, 365–369.

Knobloch, E. 1992 On the stability of magnetized accretion discs. *Mon. Not. R. Astron. Soc.* **225**, 25P–28P.

Knobloch, E. & Guckenheimer, J. 1983 Convective transitions induced by varying aspect ratio. *Phys. Rev. A* **27**, 408–417.

Knobloch, E. & Pierce, R. 1992 Spiral vortices in finite cylinders. In *Ordered and Turbulent Patterns in Taylor–Couette Flow* (ed. C.D. Andereck and F. Hayot), pp. 83–90. Plenum.

Knobloch, E. & Proctor, M.R.E. 1988 The double Hopf bifurcation with 2:1 resonance. *Proc R. Soc. Lond. A* **415**, 61–90.

Knobloch, E. & Silber, M. 1992 Oscillatory convection in a rotating layer. *Physica D* **63**, 213–232.

Knobloch, E., Deane, A., Toomre, J. & Moore, D.R. 1986 Doubly diffusive waves. *Contemp. Math.* **56**, 203–216.

Küppers, G. & Lortz, D. 1969 Transition from laminar convection to thermal turbulence in a rotating fluid layer. *J. Fluid Mech.* **35**, 609–620.

Matthews, P.C., Hurlburt, N.E., Proctor, M.R.E. and Brownjohn, D.P. 1992 Compressible magnetoconvection in oblique fields: linearized theory and simple nonlinear models. *J. Fluid Mech.* **240**, 559–569.

Meyers, S.D., Sommeria, J. & Swinney, H.L. 1989 Laboratory study of the dynamics of Jovian-type vortices. *Physica D* **37**, 515–530.

Nagata, M. 1986 Bifurcations in Couette flow between almost corotating cylinders. *J. Fluid Mech.* **169**, 229–250.

Nagata, M. 1988 On wavy instabilities of the Taylor-vortex flow between corotating cylinders. *J. Fluid Mech.* **188**, 585–598.

Pedlosky, J. 1987 *Geophysical Fluid Dynamics.* Springer.

Proctor, M.R.E. & Jones, C.A. 1988 The interaction of two spatially resonant patterns in thermal convection. I. Exact 1:2 resonance. *J. Fluid Mech.* **188**, 301–335.

Rabaud, M. & Couder, Y. 1983 A shear-flow instability in a circular geometry. *J. Fluid Mech.* **136**, 291–319.

Rand, D. 1982 Dynamics and symmetry: Predictions for modulated waves in rotating fluids. *Arch. Rat. Mech. Anal.* **79**, 1–37.

Read, P.L., Bell, M.J., Johnson, D.W. & Small, R.M. 1992 Quasi-periodic and chaotic flow regimes in a thermally driven, rotating fluid annulus. *J. Fluid Mech.* **238**, 599–632.

Ruelle, D. 1973 Bifurcations in the presence of a symmetry group. *Arch. Ration. Mech. Anal.* **51**, 136–152.

Schmitt, D. & Schüssler, M. 1989 Nonlinear dynamos I. One-dimensional model of a thin layer dynamo. *Astron. Astrophys.* **223**, 341–351.

Steinberg, V., Ahlers, G. & Cannell, D.S. 1985 Pattern formation and wave-number selection by Rayleigh–Bénard convection in a cylindrical container. *Phys. Scr.* **32**, 534–547.

Zhang, K. 1992 Spiralling columnar convection in rapidly rotating spherical fluid shells. *J. Fluid Mech.* **236**, 535–556.

Zhang, K. & Busse, F.H. 1987 On the onset of convection in rotating spherical shells. *Geophys. Astrophys. Fluid Dynam.* **39**, 119–147.

Zhong, F., Ecke, R.E. & Steinberg, V. 1991 Asymmetric modes and the transition to vortex structures in rotating Rayleigh–Bénard convection. *Phys. Rev. Lett.* **67**, 2473–2476.

Index

Printed in the United States
By Bookmasters